Approaching Facial Difference

Facialities: Interdisciplinary Approaches to the Human Face

In this series, historians of all periods, experts in visual culture and literary scholars explore the many ways in which faces have been represented in the past and present, and in particular the issue of facial difference, disfigurement, 'beauty' and 'ugliness'. Faces are central to all human social interactions, yet have been neglected as a subject of study in themselves outside of the cognitive sciences and some work on aesthetics of the body. Titles in the series will range across themes such as approaching the difficult history of disfigurement, how facial difference and disability intersect, the changing norms of appearance relating to the face and other features such as the hair (facial and otherwise), violence targeted at the face, and the reception and representation of the face in art and literature.

Approaching Facial Difference

Past and Present

Edited by
Patricia Skinner and Emily Cock

BLOOMSBURY ACADEMIC
LONDON • NEW YORK • OXFORD • NEW DELHI • SYDNEY

BLOOMSBURY ACADEMIC
Bloomsbury Publishing Plc
50 Bedford Square, London, WC1B 3DP, UK

BLOOMSBURY, BLOOMSBURY ACADEMIC and the Diana logo are trademarks of Bloomsbury Publishing Plc

First published in Great Britain 2018

Cover design: Adriana Brioso
Cover image © Bones Of The Head. After a 19th Century Print. (© Universal History Archive/UIG via Getty Images)

A catalogue record for this book is available from the British Library.

A catalog record for this book is available from the Library of Congress.

ISBN: HB: 978-1-3500-2829-6
ePDF: 978-1-3500-2831-9
eBook: 978-1-3500-2830-2

Series: Facialities: Interdisciplinary Approaches to the Human Face

Typeset by Deanta Global Publishing Services, Chennai, India

To find out more about our authors and books visit www.bloomsbury.com and sign up for our newsletters.

Contents

List of Illustrations

List of Figures

Table

Notes on Contributors

Jason Bate is Lecturer in the School of Art and Design at Falmouth University. His PhD was on photography, disfiguration and reconstructive surgery in England during and after the First World War, and he has since published articles in *History and Technology* and *Visual Culture in Britain*.

Emily Cock is Leverhulme Early Career Fellow in the Department of History at Cardiff University, having previously worked as a research assistant on the *Effaced from History?* project. Her PhD investigated prostitution and plastic surgery in early modern England, and she has since published articles in *Social History of Medicine* and *BMJ Humanities*.

Jane Draycott is the Lord Kelvin Adam Smith Research Fellow in the Department of Classics at the University of Glasgow. She co-edited, with Emma-Jayne Graham, *Bodies of Evidence: Ancient Anatomical Votives, Past, Present and Future* (London: Routledge, 2017).

Jane Frances was Policy Advisor in Education for the *Changing Faces* charity. She is the author of *Educating Children with Facial Disfigurement* (London: Routledge/Falmer, 2004).

Morna Laing is Senior Lecturer at the Chelsea College of Arts, University of the Arts London. Her PhD research explored the representation of the 'woman-child' in fashion magazines, and she has published articles in *Critical Studies in Fashion and Beauty*.

Bonnie Millar is Research Assistant in the Faculty of Medicine and Health Sciences at the University of Nottingham. She specializes in late-medieval courtly literature, and has published numerous books and articles.

Patricia Neville is Lecturer in social sciences in the School of Oral and Dental Sciences at the University of Bristol. She has published numerous articles on the intersection between the social sciences and dental health and education.

Aidan Searle is Senior Research Associate in the School of Oral and Dental Sciences at the University of Bristol. His publications focus on facial disfigurement and on life course epidemiology and oral health.

Patricia Skinner holds a personal chair in history at Swansea University and is leader of the *Effaced from History?* project consortium. She is the author of *Living with*

Disfigurement in Early Medieval Europe (New York: Palgrave Macmillan, 2017) and numerous articles on acquired disfigurement in the Middle Ages.

Kathryn Smith is completing her PhD at Liverpool John Moores University. A member of FaceLab, her doctoral research project is a cross-cultural consideration of the ethics of professional and public attitudes to visual depictions of the dead in contemporary culture. The research is supported by the National Research Foundation (South Africa).

Theresa Tyers is an Honorary Research Associate in the Centre for Medieval and Early Modern Research at Swansea University, having completed her PhD at the University of Nottingham. She has published several articles on medieval vernacular healthcare manuals.

Andrea Waylen is Senior Lecturer in psychology at the University of Bristol. Her research focuses on dento-facial appearance and quality of life, particularly related to cleft lip and palate and head and neck cancer.

Michelle Webb completed her PhD in the Department of History at the University of Exeter. Her research focuses on facial disfigurement in England in the sixteenth and seventeenth centuries.

Caroline Wilkinson is the director of the School of Art and Design and of FaceLab at Liverpool John Moores University. She is the author of *Forensic Facial Reconstruction* (Cambridge: Cambridge University Press, 2004) and numerous articles.

Acknowledgements

The editors gratefully acknowledge the support of the Wellcome Trust, whose Seed Award 107780 facilitated the collaborative meeting from which this book emerged. We should also like to thank the anonymous reviewer of the book for their constructive suggestions, and the staff at Bloomsbury, particularly Beatriz Lopez, for their support in bringing these essays to publication.

1

Introduction: Situating the Different Face

Emily Cock and Patricia Skinner

What is a face and how does it relate to personhood? In this book, scholars in the humanities engage in dialogue with experts in modern medical and educational psychology to explore the many ways in which faces have been represented in the past and present, focusing on the issue of facial difference and disfigurement read in the light of shifting ideas of 'beauty' and 'ugliness'. Faces are central to all human social interactions, yet have been neglected as a subject of study in themselves, outside of the cognitive sciences and some work on aesthetics of the body.[1] Facial disfigurement is rarely discussed in disability studies unless accompanied by other impairment/s, and is overlooked by historians of medicine unless is the subject of surgical or other procedures. It has a central part to play, however, in exploring past and present human social interactions (individual and group), and the processes by which norms of appearance are determined and enforced, not by any legal frameworks, but by sociocultural practices. The latter include medical practices designed to screen out 'anomalies' at birth; the surgical urge to restore, replace or refine facial features; or 'scientific' or creative practices that use the disfigured human face to signal other qualities, inviting onlookers to judge character, or respond with horror, disgust or even amusement. Exploring the history and culture of the 'queer' face, one which may fall outside perceived 'norms', provokes questions about *why* such images have power, and ultimately provides insight into the lived experience of those with disfigurements. Their stories are not universally tales of disempowerment – the chapters in this book underline a need to move away from paradigms of 'sympathy' or 'pity' and instead invite the reader to consider the processes that have reinforced disfigurement as a disabling condition when in fact it has not always been identified as such.

That said, under the UK Equality Act (2010), people with 'severe disfigurement' are protected from discrimination as 'disabled', the disfigurement having a 'substantial adverse effect on the ability of the person concerned to carry out normal day-to-day activities'. A similar definition appears in the Americans with Disabilities Act (2010) that regards disfigurement as a disability if it 'substantially limits a major life activity'. But the relationship between 'disability' and 'disfigurement' is complex and problematic, and scholarship on this issue is patchy. Disfiguring conditions such as eye or ear loss may cause functional incapacities, but scars (e.g. from smallpox, as depicted in Lady Mary May's effigy in West Sussex) and skin conditions may be disabilities of appearance alone, and thus largely ignored by disability historians.[2] Disfigurement history overlaps with histories of disability in the sense that both deal with experiences of stigma resulting from physical difference, but people with facial deformations have often been regarded differently to those with other physical impairments in the past, and treated inconsistently.[3] To take just two British examples, in 1670/1 Parliament passed 'Lord Coventry's Act', which made it a capital offence to 'of malice aforethought, and by laying in wait, unlawfully cut or disable the tongue, put out an eye, slit the nose, cut off the nose or lip, or cut off or disable any limb, or member of any other person, with intent to maim or disfigure him'. But in 1897, British legislation, to compensate those disabled in the workplace, *excluded* facial disfigurement from its list of impairments qualifying a person for support, since it did not apply to 'any injury which does not disable the workman for a period of at least two weeks from earning full wages at the work at which he was employed'. Both examples, however, need context: one addresses injury to honour, while the other is driven by economic concerns reflecting the increasing connection (although with long-established roots) between impairment and ability to work.[4] Historiography of occupational health and disability has mirrored that tendency, paying less attention to appearance-related injuries such as the 'phossy jaw' experienced by those working in late Victorian match factories.[5] An important exception is the growing body of work on the experiences and representations of men who returned from the First World War with facial injuries: not only were they recognized as having experienced a unique form of injury in relation to selfhood and socialization, but commentators highlighted the difficulty that men with otherwise non-impedimentary facial injuries would encounter in obtaining public-facing employment in retail or other service industries.[6]

As Susan Schweik, Jenny Edkins and Heather Laine Talley have pointed out, facial appearance is also a political issue, inscribing key social differences and underpinning relations of power in contemporary society.[7] But facial difference needs to be disentangled from bodily impairment: commenting on the legacy of the London Paralympics, James Partridge, CEO of the charity *Changing Faces* remarked to the editors of this book that '2012 did nothing for us', and in 2016 *able-bodied* Olympians with disfiguring conditions (Joanna Rowsell Shand, *alopecia areata*) and facial accessories (Fiona Bigwood, eye-patch) still had to explain their appearance, regardless of their achievements in sport. Prosthetic limbs might now be a 'normal' sight, but facial difference remains a difficult topic of public discourse and acceptance. People with disfigurements continue to face barriers to full citizenship across the world: biometrics calibrated to a 'normal' face does not work when confronted with difference, and in an increasingly public, visual, online age, 'normal' includes airbrushing out the slightest facial imperfection in photographs, even of schoolchildren. In a recent report based on their large-scale survey of people living with a facial disfigurement in Britain today, *Changing Faces* provided startling evidence of the impact of disfiguring conditions. Some of their findings:

- 80 per cent of respondents reported receiving 'comments or unpleasantness from a stranger', and half from a shop assistant;
- half attested that reactions to their disfigurement affected their aspirations to stay in education past the age of sixteen;

- 80 per cent decided not to apply for a job because they believed that their appearance would affect their chance of employment, or through apprehension of being made uncomfortable by new colleagues;
- an Implicit Attitudes Test of the general public 'found that 67% of adults in the UK attach less positive attributes and characteristics to people who have a disfigurement'.[8]

People working in fields such as education and healthcare were also found to be in significant need of further awareness training about how to help and respond to individuals with disfigurements. Moreover, the report highlights the need for widespread awareness and understanding of the experiences and requirements of people living with disfigurements, and the effect of social reactions on these experiences. This book offers a useful contribution to this project, and to a wider understanding of the role of the non-normative face in identity-formation and social interactions in a range of historical contexts. While disfiguring injuries, punishments and diseases; the role of physiognomic characteristics in discourses of race, class and criminality; and the emergence of aesthetic surgery have received historical attention, there has been no sustained study of facial disfigurement and its medical, material, cultural and experiential aspects in a long-term historical perspective.[9]

But there was, and is, a counter-narrative to be considered. Disability bioethicists argue that the social benefits of a diverse society outweigh the perceived 'problems' of those born with conditions affecting their appearance and/or abilities, and strongly oppose modern screening procedures that test for such conditions.[10] From this perspective, past societies without such medical technologies enjoyed a richer and more variegated human experience, but in a world where more people may have looked facially 'different', was there more tolerance of that difference? In his important book on the physical, social and religious interpretations of 'face', Stephen Pattison identifies the strong communal bonds of pre-industrial society, and Christian religious practices that were not afraid to use intense visual images to stimulate faith. In such a world physical facial difference was seen as part of God's work, the 'image of God' residing in every human being, and did not require much attention.[11] Returning to the contemporary situation, there is a growing counterculture of artists and film-makers who reject conventional notions of 'beauty' and explore the aesthetics of difference through experiments with their own appearance and/or reuse and repurposing of images of disfigurement from the past. Siebers describes disability and disfigurement as acquiring 'aesthetic value' in modern art.[12] In feminist art, too, conventional beauty has often been repudiated.[13] Do such practices challenge prejudice, or is their purpose simply provocative? Gender theory exposes the power relationships inscribed in texts and images, and proposes queer readings starting from the perspective of the disempowered minority.[14] How did/does disfigurement de- or regender the person?

People with facial disfigurements remain one of the last minorities (perhaps 10 per cent of Western population, possibly higher in developing regions) *without* a history. This book represents an invitation to consider what has caused this effacement and presents a number of ways in which evidence might be gathered and read to link contemporary disfigurement to its rich history. A focus on the face is necessary to understand the fluid meanings of facial difference. Although negative stereotypes inscribing disgrace and social difference are most visible, non-normative faces can and have also signified honour (for example, in duelling scars), successful medical/surgical interventions (particularly, but not exclusively, in the case of military veterans) and celebrity status (from the 'ugly face' clubs of the eighteenth and nineteenth centuries to modern media figures such as Katie Piper).[15] That paradox lies at the heart of the present book. It is important to go beyond exceptionalism in discussing facial disfigurement: a person with an exceptional face might well be singled out for special attention (positive – 'triumph over adversity' – or negative – derided or feared), and may or may not give a first-person account of their life experiences.[16] But the ideologies that drive the responses of onlookers are more pervasive and may relate to facial appearance alone or be bound up in broader attitudes towards age, gender and/or skin colour. The intersection between these categories allows for a more nuanced understanding of both experience and perception. Note, however, that 'skin colour' here refers to anomalous colours, conditions or practices *within* ethnic or racial groups, rather than the long-established study of discrimination and stereotyping based on interracial prejudices.[17]

The chapters in this book use three approaches to examine facial disfigurement. First, *language* explores primarily written texts and descriptions. These often engage in what Skinner has recently defined as 'textual staring', where use of language mimics for the collaborating writer and reader the voyeuristic and gratifying act of staring with eye or camera.[18] Bonnie Millar mines medieval depictions of the Arthurian court and the ways in which facial deformity is used to generate revulsion and fear in the characters depicted, as well as in the minds of those reading and hearing the tales. Deformity here figures as a disruptive force, threatening courtly and chivalric society, but is safely contained within a fictional and supernatural world. Audiences can therefore engage in the *frisson* of imagining the spectacle without being confronted by it in real life. Michelle Webb continues this story of the fascinated but disgusted observer, moving into real life and focusing on descriptions of damaged faces in the sixteenth and early seventeenth centuries in England, drawing from a wide range of sources including diaries, medical texts, sermons and court records. She finds that the experience and recording of disfigurement in this period occurred on sharply gendered lines. Both chapters articulate the difficulty of getting beyond the views of onlookers and accessing the voice of the person who is disfigured. That this is not just a premodern phenomenon illustrated by Jane Frances, who finds that modern studies of disfigurement are just as likely to objectify those living with facial differences (the 'starees' to use Garland-Thomson's term, or 'inhabitants of faces' to use Pattison's), and fail to examine critically the ways in which starers are influenced by prejudices so embedded as to seem 'normal'. This tendency, she argues, permeates into academic and professional discourses that make assumptions about the lives of those they claim to study objectively.[19]

Second, *visibility* examines how gazing and staring have been at the heart of almost all the source materials relating to disfigurement. Numerous visual sources, for example, medical photography, funeral sculpture, portraiture, criminal catalogues and facial 'averaging' for psychological studies, depict facial difference for many different purposes. These offer rich territory for transhistorical analysis, from caricatures to freak shows to voyeuristic television programmes.[20] Jane Draycott examines representations of hair loss in ancient Rome, arguing that it represents as much a facial disfigurement as scars and blemishes on the face's surface. Drawing upon a rare, first-person narrative of living with hair loss, she explores the stigma and moral judgements surrounding the condition. The holistic approach to facial features also lies at the heart of Kathryn Smith's exploration of forensic art practices, reconstructing faces from memory or physical remains. Her chapter examines different approaches in psychology to obtaining more precise witness descriptions and the differing considerations of restoring facial features in forensic work, and usefully interrogates the ways in which distinctive elements might or might not influence recall. Normative visual representations of the face extend and complicate the ways in which stigma and exclusion are expressed in different media. Patricia Neville and her colleagues call for more qualitative research on the contemporary experience of disfigurement, in this case adults and children living with cleft lip and palate (CLP). Their study highlights and critiques the assumption that normalizing surgical interventions uniformly offer a better quality of life to CLP patients and their families, as their interviews reveal mixed feelings about the changes of face introduced by the surgery. Queering assumptions about 'correct' facial features also underpins Morna Laing's study of the work of experimental musicians CocoRosie. She explores how historic references suffuse their music videos, and how incorporating the non-normative face emphasizes the continuing dominance of certain forms of 'beauty'.

Finally, the *materiality* of facial difference is represented by chapters that examine the restoration, repair and representation of faces across the centuries. Caroline Wilkinson revisits the circumstances of reconstructing historic faces from archaeological remains, and the challenges of poor preservation and pervasive interests of those commissioning the reconstructions. Theresa Tyers explores the ambivalence displayed by medieval medical practitioners towards facial cosmetics and cures, and in particular the misogyny underlying male writers' attitudes towards women's use of such substances. Her texts were produced against a background of a period which Luke Demaitre has identified as one of increased self-awareness of facial appearance and colouring.[21] Finally, Jason Bate reflects on the photograph as a material object that does not simply capture the likeness of the First World War veterans in his study, but provides him (and us) with a direct, tangible relic of the experiences of both patients and practitioners in the hospitals where pioneering surgical reconstructions using prosthetics were carried out. The passage of time between then and now is bridged by handling albums of material and speaking to the families of those depicted in the images, as well as by the author's own emotional responses to the fractured faces on view.[22]

These discrete studies pose a series of questions which we hope will generate further research into the ways in which facial difference as a site of aversion (but also of fascination, beauty and self-awareness) has a long and complex history, and test the hypothesis that while the specific ways (and range of media) in which facial difference is represented have changed over time, the impact of disfigurement has remained remarkably constant. Knowing that constancy is a key driver for this research, only a long-term view can determine *why* different faces have maintained their power to shock.

By exposing expressions of prejudice (and alternative models of self-determination), the book aspires to create a greatly enriched understanding of the historically rooted but evolving societal and cultural contexts, shaping attitudes and responses to disfigurement that have resisted inclusion and equal treatment. The specific value of the study is that it will demonstrate to people living with disfigurement, and those who work beside and with them, that their experiences are not uniquely contemporary, and it will empower them to construct a historical identity (if they so choose) that can offer support and validation rather than invisibility and effacement. Attitudes to disfigurement are not fixed, but are historically and culturally contingent in subtle as well as obvious ways. Shining a light on these will go some way to dismantling their pervasive power.

Notes

1 Giorgio Agamben, 'The Face', in *id.*, *Means without End: Notes on Politics*, trans. V. Binetti and C. Casarino (Minneapolis: Minnesota University Press, 2000); Claudia Benthien, *Skin: On the Cultural Border Between Self and the World*, trans. T. Dunlap (New York: Columbia University Press, 2002); Daniel Black, 'What is a Face?' *Body and Society* 17 (2011): 1–25; Vicky Bruce and Andy Young, *Face Perception* (London and NY: Psychology Press, 2012).

2 Impairment: Jane Draycott, 'Reconstructing the Lived Experience of Disability in Antiquity: A Case Study from Roman Egypt', *Greece and Rome* 62 (2015): 189–205; Lady Mary May: https://effacedblog.wordpress.com/2016/06/16/effaced-fieldtrip-lady-mary-mays-statue-in-west-sussex/ (accessed 30 May 2017). See, however, the plethora of recent, more general work on skin: Claudia Benthien, *Skin: On the Cultural Border Between Self and the World*, trans. T. Dunlap (New York: Columbia University Press, 2002); Katie L. Walter, ed., *Reading Skin in Medieval Literature and Culture* (Basingstoke: Palgrave Macmillan, 2013); J. Reinarz and K. Siena, eds, *A Medical History of Skin: Scratching the Surface* (London: Pickering and Chatto, 2013).

3 Erving Goffman, *Stigma: Notes on the Management of Spoiled Identity* (London: Penguin, 1963); E. E. Jones, A. Farina, A. Hastorf, H. Markus, D. Miller and R. A. Scott, *Social Stigma: The Psychology of Marked Relationships* (New York: Freeman, 1984); Stephen C. Ainlay, Gaylene Becker and Lerita M. Coleman, eds, *The Dilemma of Difference: A Multidisciplinary View of Stigma* (New York and London: Plenum Press, 1986).

4 Charles II, 1670 and 1671: An Act to Prevent Malitious [*sic*] Maiming and Wounding: http://www.british-history.ac.uk/statutes-realm/vol5/pp691-692 (accessed 08 June 2017); M. Roberts-Jones, *Handbook on the Workmen's Compensation Act 1897* (Cardiff: Western Mail, 1898) https://archive.org/stream/handbookonworkme00robe#page/n5/mode/2up (accessed 08 June 2017); and A. T. Glegg, *Commentary on the Workmen's Compensation Act 1897* (Edinburgh: William Green and Sons, 1899), 71: 'The fact that a man is maimed, or it may be disfigured, for life, is not to be taken into account, except in so far as it affects his wage-earning capacity'; Irina Metzler, *A Social History of Disability in the Middle Ages* (New York and London: Routledge, 2013), 36–91; Roddy Slorach, *A Very Capitalist Condition: A History and Politics of Disability* (London: Bookmarks, 2015).

5 J. L. Bronstein, *Caught in the Machinery: Workplace Accidents and Injured Workers in Nineteenth-Century Britain* (Redwood City: Stanford University Press, 2008); J. Marx, 'Uncovering the Cause of "phossy jaw" 1858-1906', *Journal of Oral and Maxillofacial Surgery* 66.11 (2008): 2356–63.

6 In this book, see Jason Bate's chapter, and further Suzannah Biernoff, 'The Rhetoric of Disfigurement in First World War Britain', *Social History of Medicine* 24.3 (2011): 666–85, and *ead. Portraits of Violence: War and the Aesthetics of Disfigurement* (Ann Arbor: University of Michigan Press, 2017); Marjorie Gehrhardt, *The Men with Broken Faces:* Gueules Cassées *of the First World War* (Berlin: Peter Lang, 2015); Kerry Neale, '"Without the Faces of Men": The Return of Facially Disfigured Veterans from the Great War', in *When the Soldiers Return: November 2007 Conference Proceedings*, ed. Martin Crotty (St Lucia, QD: University of Queensland, 2008): 114–20.

7 Susan Schweik, *The Ugly Laws: Disability in Public* (New York: NYU Press, 2009); J. Edkins, 'Dismantling the Face: Landscape for Another Politics?', *Environment and Planning D: Society and Space* 31 (2013): 538–53; *ead.*, 'Facing and Defacing', in *International Politics and Performance: Critical Aesthetics and Creative Practice*, ed. J. Edkins and A. Kear (London: Taylor and Francis, 2013), 40–59; Heather Laine Talley, *Saving Face: Disfigurement and the Politics of Appearance* (New York: *New York University* Press, 2014).

8 Changing Faces, *Disfigurement in the UK* (London: May 2017), 5, 9, available at https://www.changingfaces.org.uk/wp-content/uploads/2017/05/DITUK.pdf (accessed 08 June 2017).

9 Guy Geltner, *Flogging Others: Corporal Punishment and Cultural Identity from Antiquity to the Present* (Amsterdam: Amsterdam University Press, 2014); Martin Porter, *Windows of the Soul: The Art of Physiognomy in European Culture, 1470-1780* (Oxford: Oxford University Press, 2005); Sander L. Gilman, *Creating Beauty to Cure the Soul: Race and Psychology in the Shaping of Aesthetic Surgery* (Durham, NC: Duke University Press, 1998); *id.*, *Making the Body Beautiful: A Cultural History of Aesthetic Surgery* (Princeton: Princeton University Press, 1999); Farhad B. Naini, *Facial Aesthetics: Concepts and Clinical Diagnosis* (Oxford: Wiley-Blackwell, 2011); David Shuttleton, *Smallpox and the Literary Imagination 1660–1820* (Cambridge: Cambridge University Press, 2007).

10 Rosemarie Garland-Thomson, 'The Case for Conserving Disability', *Journal of Bioethical Enquiry* 9 (2012): 339–55; Jackie Leach Scully, *Disability Bioethics: Moral Bodies, Moral Difference* (Lanham, MD: Rowman and Littlefield, 2008).

11 Stephen Pattison, *Saving Face: Enfacement, Shame, Theology* (Farnham: Ashgate, 2013), 119–25.

12 Tobin Siebers, *Disability Aesthetics* (Ann Arbor: The University of Michigan Press, 2010).

13 Wendy Steiner, *Venus in Exile: The Rejection of Beauty in Twentieth-Century Art* (Chicago: University of Chicago Press, 2002); and see Morna Laing's chapter in this book.

14 Judith Butler, *Gender Trouble: Feminism and the Subversion of Identity* (London and New York: Routledge, 1990).

15 On surgery, Emily Cock, '"Lead[ing] 'em by the nose into publick Shame and Derision": Gaspare Tagliacozzi, Alexander Read and the Lost History of Plastic Surgery', *Social History of Medicine* 28.1 (2015) and *ead.*, '"He would by no means risque his reputation": Patient and doctor shame in Daniel Turner's *De Morbis Cutaneis* (1714) and *Syphilis* (1717)', *BMJ Medical Humanities* online 29 November 2016, doi: 10.1136/medhum-2016-011057; on veterans Suzannah Biernoff, 'The Rhetoric of Disfigurement

in First World War Britain', *Social History of Medicine* 24.3 (2011), and *ead. Portraits of Violence: War and the Aesthetics of Disfigurement* (Ann Arbor: University of Michigan Press, 2017); Marjorie Gehrhardt, *The Men with Broken Faces:* Gueules Cassées *of the First World War* (Berlin: Peter Lang, 2015); on ugly face clubs Peter Clark, *British Clubs and Societies 1580-1800: The Origins of an Associational World* (Oxford: Oxford University Press, 2000), 71 and 194; on media figures Patricia Skinner, '"Better Off Dead than Disfigured?": The Challenges of Facial Injury in the Premodern Past', *Transactions of the Royal Historical Society*, 6th series 26 (2016): 26–41.

16 Skinner, '"Better off dead than disfigured?"'.

17 Thus Rana Hogarth, 'The Strange Case of Hannah West: Skin Colour and the Search for Racial Difference', *Social History of Medicine* 29 (2016): 557–72, might be considered as a contribution to disfigurement studies because it considers responses to a woman categorized as white who nevertheless had patches of black pigment on her skin and thus attracted the attention of medical practitioners.

18 Patricia Skinner, *Living with Disfigurement in the Early Middle Ages* (New York: Palgrave Macmillan, 2017), 161.

19 Rosemarie Garland-Thomson, *Staring: How we Look* (Oxford: Oxford University Press, 2009); Pattison, *Saving Face*, 52 and note 1.

20 Rosemarie Garland-Thomson, ed., *Freakery: Cultural Spectacles of the Extraordinary Body* (New York: New York University Press, 1996); Margrit Shildrick, *Embodying the Monster: Encounters with the Vulnerable Self* (London: Sage, 2002).

21 Luke Demaitre, 'Skin and the City: Cosmetic Medicine as an Urban Concern', in *Between Text and Patient: The Medical Enterprise in Medieval and Early Modern Europe*, ed. Florence Eliza Glaze and Brian K. Nance (Florence: SISMEL-Edizioni del Galluzzo, 2011), 97–120.

22 Marco Iacoboni, *Mirroring People: The Science of Empathy and How we Connect with Others* (London and New York: Picador, 2009).

Part One

Language

2

Dis/enabling Courtesy and Chivalry in the Middle English and Early Modern Gawain Romances and Ballads

Bonnie Millar

This chapter seeks to examine courtesy and chivalry in the middle English and early modern Gawain romances and ballads in the context of dis/enabling facial disfigurement. Disability and disfigurement offer a means to critique and illuminate the normative or the ableist. They instigate narratives or disrupt stories,[1] proffer opportunities for revisiting cultural, social and political behaviours, provide avenues to move beyond using 'aesthetic evidence',[2] and question people's embodiment as reliable indicators of merit. Indeed, we need to attend to the 'psychologisation of everyday life' and be aware of how the individual embraced as desirable by contemporary society has been constructed over a long period of 'oppression and domination, of medicalization, colonization, patriarchy and heteronormativity'.[3] Similar to disabilities, facial disfigurements and facial disfigurements framed by bodily deformity, which can cause social exclusion, provide impetus for actions, occasion storytelling and afford instances when one can 'pass through the [bodily] forms of a narrative in order to read knowledge'.[4]

Impairment and its associated disabling effects have become a popular topic of study for historians of the medieval period and for scholarly work on medieval literature.[5] While the lived experience of bodily impairment can sometimes be traced, the ontology of facial disfigurement (apart from deliberate blinding) remains rather more elusive. In literary texts, however, the impaired, grossly deformed or simply scarred face could often serve as plot device, a way of stigmatizing but also of enabling recognition, and a way to reinforce, among the readers and hearers of the tales, their own sense of what counted as 'normal' or 'different', often along gendered lines. This chapter explores an example of how this was done, using the 'loathly lady tales', referring to the mythical past of King Arthur and his court.

The tales were created against a background of political and social upheaval. Territorial ambition and sequestration were intrinsic to the political situation in fourteenth- and fifteenth-century England. The country had, after all, been intermittently involved in war with France since 1337, with varying degrees of success. Public opinion wavered between favouring the military campaigns with Edward III's initial successes, or Henry V's triumphs, and disapproving the tremendous demands made upon financial resources and upon the lives of knights, and Christian blood. Furthermore, there were expeditions against Ireland, difficulties with Wales and the ongoing problematic Anglo-Scottish relations with tensions in the border regions of both countries. Working through issues of difference, courtly behaviour and chivalry in literary form, set in a mythical past, allowed for boundary setting in the present.

Dis/enabling personal stories, tales of surmounting physical difference, recuperation and normalcy are used to incapacitate the Arthurian court in the loathly lady tales, *The Wedding of Sir Gawain and Dame Ragnelle* and *The Marriage of Sir Gawain* and to facilitate more pragmatic forms of courtliness and chivalry, while dis/enabling discourses of beauty and ugliness are key to the depiction of Guinevere's mother as a deformed wretch in *The Awntyrs off Arthur* when she warns of the dangers of materiality. In all cases, facial disfigurement is emphasized and generates revulsion and fear in onlookers and the court, with this troubling of the Arthurian world allowing disruptive voices to travel from beneath the surfaces and the edges and emerge, potentially rupturing and transforming the chivalric and courtly spaces of Arthurian society. The focus of the Arthurian court, the gold standard for chivalry, is King Arthur, who is surrounded by his noble knights, and has Queen Guinevere as companion. Many of the short middle English and early modern English romances and tales focus on Gawain and adventures or tests, rather than stories of love.

A deformed female protagonist furnishes the impulse for the narrative in two Gawain tales, the c.1500 tail-rhyme romance known as *The Wedding of Sir Gawain and Dame Ragnelle*, extant in a single manuscript copy dating from the sixteenth century, Oxford Bodleian 11951, and the later ballad entitled *The Marriage of Sir Gawain* from the regrettably damaged c.1650 Percy Folio volume, BL Additional 27879, with missing portions of text. The physical deformity of this figure mirrors the moral/ethical turpitude of Arthur's reign and forms a space where social relations between the male protagonists can be explored and resolved.[6] 'Representations of disability … allow an interrogation of static beliefs about the body while also erupting as the unseemly matter of narrative that cannot be textually contained. The coinage of the phrase "narrative prosthesis" argues that disability has been used throughout history as a crutch on which literary narratives lean for their representational power, disruptive potentiality, and social critique.'[7] This is certainly true here, as Arthur's encounter with Ragnelle sets in motion a train of events that end in the denouement as to what women most desire, that is mastery or supremacy. Once the conflicts have been negotiated, lessons learned from the moral turpitude, the brides are restored to their normative forms with their disfigurements having fulfilled their functions.

Turning to the physicality of Ragnelle, in *The Wedding of Sir Gawain and Dame Ragnelle* the emphasis is placed firmly on her face. Ragnelle's visage, framed as it is by wild ungroomed hair, strikes horror into the courtiers when she becomes subject to their gaze:

> Her face was red, her nose snotyd withalle,
> Her mowithe wyde, her tethe yalowe overe alle,
> With bleryd eyen gretter then a balle.
> Her mowithe was nott to lake:

Her tethe hyng overe her lyppes,
Her chekys syde as wemens hippes.
A lute she bare upon her bak;
Her nek long and therto greatt;
Her here cloteryd on an hepe;
In the sholders she was a yard brode.
Hangyng pappys to be an hors lode,
And lyke a barelle she was made.
And to reherse the fowlnesse of that lady,
Ther is no tung may telle, securly;
Of lothynesse inowghe she had.[8]

(Her face was red, her nose covered with snot as well, her mouth was wide, her yellow teeth over-hanging, with her bleary eyes greater in size than balls. Her mouth was not too small; her teeth hung over her lips, her cheeks were as broad as women's hips. She had a hump on her back; her neck was as broad as it was long; her hair was a tangled mess and her shoulders were a yard broad. She was built like a barrel with hanging breasts large enough to be those of a horse. No one can truly relay the foulness of that lady due to the extent of her ugliness.)

The ladies favour a secret wedding for her, to avoid the ignominy of having to admit that she is one of their fold. The detailed description of her physiognomy accentuates her large mouth, which incorporates huge lips, numerous hairs and tusk-like teeth protruding in different directions. With such a mouth Gawain could be devoured through embracing her. Indeed, the connection here between this foul mouth and kissing is indicative of lust, the destructive desires of the flesh and the difficulties inherent in surmounting them and maintaining societal decorum.

She was so fowlle and horyble.
She had two tethe on every syde
As borys tuskes, I wolle nott hyde,
Of lengthe a large handfulle.
The one tusk went up and the other doun.
A mowthe fulle wyde and fowlle igrown,
With grey herys many on.
Her lyppes laye lumpryd on her chyn;
Nek forsothe on her was none iseen –
She was a lothly on![9]

(She was so foul and horrible; she had two teeth on each side like boar's tusks, each the width of a hand, with one descending and one ascending. Her wide foully formed mouth had many grey hairs. Her lips were lumped on her chin preventing her neck from being seen – she was a truly loathly one!)

Although she is first introduced in the woods of Inglewood, a liminal space not encompassed by cultured social value systems, she appears to be of high estate, possessing splendid clothes and a valuable palfrey.[10] In the romance, Ragnelle nobly volunteers to aid the monarch in return for the gift of Gawain as husband, a deal to which Arthur consents, albeit uneasily. In the ballad Arthur is happy to accept assistance, though unwilling to publicly acknowledge his actions in awarding Gawain to a hideous hag. In truth, this incongruity in her uncouth physicality decked out in great finery signifies the gulf between art and nature, appearance and reality. Similarly, *The Marriage of Sir Gawain* provides a disturbing portrait of Ragnelle with her mouth and eye transposed, suggesting an all-consuming gaze:

> Then there as should have stood her mouth,
> Then there was sett her eye;
> The other was in her forhead fast,
> The way that she might see.
> Her nose was crooked and turnd outward,
> Her mouth stood foule awry;
> A worse formed lady than shee was,
> Never man saw with his eye.[11]

(One of her eyes was positioned where her mouth should have been; the other was safely in her forehead so that she might see. Her nose was crooked and protruded outwards, and her mouth was hideously twisted; no man has ever seen a more deformed lady.)

Gawain's comrades at arms fear that any man who should be so unfortunate as to become intimate with such a woman could be in mortal danger (lines 132–5). They appear to dread that he could be devoured, obliterated by this potent uncontrolled female, the *vagina dentata*,[12] a traditional figure whose origins stem back to the Greek Medusa and the Celtic Sheela-na-gigs.[13] Ragnelle's cultural and social power is evident at many levels; notably she is both the object of gazing by Arthur and the courtiers, and the subject as she gazes back upon the court, Arthur and Gawain. This linking of the senses, taste and sight, and by extension seeing and eating brings out the compulsive, detrimental nature of desires and interconnections between the senses, desires and people.

The association between seeing and consuming extends to the splendid wedding Ragnelle desires (lines 557–62, 590–4). She wishes to have her nuptials in full public view. Subject to the observations of the guests she devours copious amounts of food and drink, leading to comments as to how she eats as much as six people. Indeed, the prolonged accounts of her appearance and culinary adventures are innovations in this redaction.[14]

Following the wedding day festivities, Ragnelle and Gawain retire for the night. The bride requests Gawain's embraces and he acquiesces:

> I wolle do more
> Then for to kysse, and God before![15]

(I will do far more, than to merely kiss you, and in the presence of God!)

Once more there is an emphasis on mouths, with the references to kissing. Miraculously, Gawain turns to find his wife transformed into a beautiful young lady, with the choice of having her fair by night or fair by day. Graciously, he hands her the decision-making power, giving her mastery, thereby breaking the enchantment so she can remain fair always. She has been the victim of a spell which could only be broken when:

> the best of Englond
> Had weddyd me verament,
> And also he shold geve me the sovereynté
> Of alle his body and goodes, sycurly.
> Thus was I disformyd;
> And thou, Sir Knight, curteys Gawen,
> Has gevyn me the sovereynté sertyn,
> That woll nott wrothe the erly ne late.[16]

(the best knight of England had truly married me, and also that he should have given me mastery over all his body and possessions. In this manner I was deformed; and you, Sir Knight, courteous Gawain, have surely given mastery to one who will never hurt you.)

As Lennard Davis has highlighted, 'The disabled body must be explained, or at least tolerate the inquisitive gaze (or the averted glance) of the questioner,' and in Ragnelle's case the penetrating stares of lords and ladies fascinated and horrified in equal measure.[17] Her defacement, like disablement more widely, is associated with a 'story' and constitutes an element of a 'narrative',[18] the acquisition of familial and territorial lands. Five harmonious years together follow Ragnelle's recuperation, her surmounting of physical challenges, with Gawain the devoted husband, and the union is blessed with a son. Thereafter she departs this mortal life, and Gawain is freed of his matrimonial ties to adventure once more.

The Marriage of Sir Gawain recounts a slightly different version of Gawain's dilemma. In the ballad Gawain makes a decision – deciding to have his wife beautiful by night as the alternative would be difficult for him. Ragnelle objects on the grounds that it would be unfair on her as it would mean her being a court exile. Gawain relents immediately and allows her to make the ultimate choice and in an instant her disfigurement has been surmounted. The root cause of her socially debilitating defacement was the greed, ambition and jealousy of a young female with whom her father has replaced his previous wife, Ragnelle's mother. This act of replacement generates a dangerous familial structure which challenges the reader/listener to question the structures of power and authority beneath these intimate decisions, and by extension those prevalent in the court as a whole.

> My father was an old knight.
> And yett it chanced soe
> That he marryed a younge lady
> That brought me to this woe.
> Shee witched me, being a faire young lady,
> To the greene forrest to dwell,
> And there I must walke in womans liknesse,

Most like a feeind of hell.
She witched my brother to a carlish B[19]

(My father was an old knight and yet it happened that he married a young lady, thereby occasioning me this distress. She placed a spell on me, transforming me from a beautiful young lady to a woman most like a hellish fiend to live and walk in the green forest. She transformed my brother into loutish B)

Thus her true beauty is made visible through what Stephanie Hollis terms 'a co-operative decision negotiated between Equals'.[20]

These consuming visible desires are central to relations between the male characters in both texts. In *The Wedding of Sir Gawain and Dame Ragnelle* Arthur has given Gromer Somer Joure's lands to Gawain, publicly subjugating the knight; a situation Gromer seeks to ameliorate through combat. At Arthur's pleas to be spared the fighting, Gromer enigmatically offers to pardon him the battle in exchange for the answer as to what women most desire. The difference between the two narratives is that in the ballad Arthur is more explicit in his rejection of the combat, while in the romance he endeavours to excuse his actions or lack of actions.

To fight with him I saw noe cause,
Methought it was not meet,
For he was stiffe and strong withall,
His strokes were nothing sweete.[21]

(I did not see any need to fight with him, I thought it was not seemly, for he was fierce and strong as well, his blows would not be gentle.)

It is through female intercession, the restored Ragnelle, that political difference is checked. Once the loathly lady incident is resolved and the enchantment broken, disfigurement and aesthetic difference eliminated, the conflicts between the Arthurian world and the forest are overcome, at least on the surface.

Interestingly, although it is Gawain who has gained possession of Gromer's lands, it is the sovereign who is the focus of Gromer's discontent. Both versions illustrate that these destructive desires extend all the way to the king, the head of the community and the court. Fractures evident in the head destabilize the political/societal body as a whole and if the sovereign lacks competence and balance so will the totality. Significantly, the two redactions note the love and concern Gawain's comrades and fellow courtiers possess for him in contrast to the comparatively little thought spared for Arthur's welfare. However, both the ballad and romance focus less on the marriage as an instance of personal happiness, than on its ability to teach us about governance, how to obtain and maintain loyalty. Hence Arthur's uncouth behaviour, disregard for the rights of others and cowardice are denigrated, with Gawain's concern for appropriate behaviour and the rights of his wife upheld as a model of how to govern. Arthur's method of concealing what is shameful, preventing it from becoming public is shown to be ineffectual, in contrast to Gawain's behaving justly in private as well as public.

In part Ragnelle's physical impairment is socially constructed. Since she is found in the forest, a wild locale, the uncivilized, un-social potentially destructive forces of this space become mapped on her body. Her entry into court thus signifies how such forces can penetrate society, potentially disenabling the courtly realm. Her brother is also linked with this area and in the ballad his monstrosity is also highlighted, as he is depicted as a churlish brute bearing a club. Arthur fears the power of these aesthetically different bodies, he is unwilling to fight the brother, displaying little in the way of courage or conviction in either text. All the disruptive forces, the misuse of power are writ large on Ragnelle's impaired body. The ballad even explicates that her deformity is the result of her stepmother's attempt to take lands, property not rightfully hers, indicating that the source of Arthur's difficulties are probably the same as in the romance, the usurpation of territory. Thus Ragnelle's physicality mirrors the deformity of Arthur's and her stepmother's behaviour. By exploring these social constructs through troubling impaired figures the texts re-enforce the porous nature of such institutions and illustrate how the forces which can undermine them have a prior internal, external and continuing existence. In order for society to function, impairment and malformation must be controlled and channelled or it could disenable the community. Thus Arthur would do well to learn from Gawain's balanced behaviour.

It is not just in the marriage tales that Arthur would benefit from attending to Gawain's words and actions. In *The Awntyrs off Arthur* Gawain endeavours to guide and encourage the covetous king in alternate actions, behaviours and policies. The poem, dating from the late fourteenth century, survives in four fifteenth-century manuscripts and was composed in northern England. It is written in thirteen-line stanzas, with the first nine in alliterative long line, and last four much shorter and forming what is known as the 'wheel'.[22] Structurally, the poem divides evenly into two episodes, the first of which details the warnings and prophecies – concerning worldly vanities – of the bodily transformed and facially disfigured ghost at the Tarne Wathelene. The second incident is a combat between Galleroun and Gawain to decide the rightful lordship of territory.

Similar to the two loathly lady tales there are some disputed lands which have been bestowed upon Gawain by Arthur. The incident of territorial sequestration is resolved, as in the former texts, through female intercession, on this occasion provided by Galleroun's lady and Guinevere. The battle is abandoned and Galleroun regains his lands while Arthur gifts Gawain:

> with gerson and golde,
> Al the Glamergan londe with greves so grene;
> The worship of Wales at wil and at wolde,
> With Criffones Castelles curnelled ful clene;
> Eke Ulstur Halle to hafe and to holde,
> Wayford and Waterforde, wallede I wene;
> Two baronrees in Bretayne with burghes so bolde,
> That arn batailed abought and bigged ful bene.
> I shal doue the a duke and dubbe the with honde,

Withthi thou saghtil with the knighte,
That is so hardi and so wight,
And relese him his right,
And graunte him his londe.[23]

(with treasure and gold, all the lands of Glamorganshire with green groves, the honour of Wales at your command, including the beautifully crenellated Criffones castles; also Ulster hall, what I believe to be walled towns – Wayford and Waterford; two baronies in Brittany with fortified cities, with well-built battlements. I shall make you a duke and dub you with my own hand on the sole condition that you freely give back what is this brave and strong knight's right and grant him the return of his lands.)

This reconciliation takes place after a bitter-fought battle in which Galleroun and Gawain's 'blees wex blak' (line 658), their faces became black or darkened, bruised with all the fighting and the concomitant bleeding. Thus the effects of battle cause the knights to become facially disfigured; a transformation mirrored in the physically deformed female apparition in the poem's other episode. Chivalry and courtliness are measured through encounters with courtly outsiders and the critique they proffer. In this instance, Guinevere's mother with a blackened appearance and erased facial identity foretells the inglorious demise of the Arthurian Kingdom and represents a potential future manifestation of its queen, Guinevere:

Bare was the body and blak to the bone,
Al biclagged in clay uncomly cladde.
Hit waried, hit wayment as a woman,
But on hide ne on huwe no heling hit hadde.
Hit stemered, hit stonayde, hit stode as a stone,
Hit marred, hit memered, hit mused for madde.
Agayn the grisly goost Sir Gawayn is gone;
He rayked oute at a res, for he was never drad.
Drad was he never, ho so right redes.
On the chef of the cholle,
A pade pikes on the polle,
With eighen holked ful holle
That gloed as the gledes.[24]

(The body was bare and blackened to the bone, all caked in clay and foully covered. It cursed and wailed like a woman, but neither its skin nor complexion had any covering. It stammered, it was stunned, it stood like a stone. It grieved, it mumbled, it groaned as though it were mad. Sir Gawain hurriedly went to the hideous ghost, for he was never afraid. For those with good understanding are never fearful. On top of the ghost's neck was a toad biting the skull. The spirit's eyes were sunken and hollowed out and glowed like coals.)

Like Ragnelle the ghost is both the object of the gaze of others and a subject who looks outwards at those who gaze upon her. This detailed description draws attention to how her body has been consumed, 'blak to the bone'. Her eyes have become sunken bodily holes, and glow like burning coals. Again seeing and consuming are linked, this time with the devouring power of flame, and denigrating bestiality, with concomitant pain, suffering and exclusion.[25] She is encircled with serpents and toads, reminiscent of the Greek gorgons and Medusa. She wails and moans as though she were mad, as the toads masticate her skull. Focusing on her current torment, she situates the practice of material acquisition of furs, jewels and other luxuries as an act of futility and malprudence, endeavouring to inspire spirituality, seeking prayers to help her soul. It is the appearance as much as the rhetoric of the apparition, which brings out the theme of the ill consequences of fleshly desires, comparable to the functionality of Dame Ragnelle's depiction.

> Al glowed as a glede the goste there ho glides,
> Umbeclipped in a cloude of clethyng unclere,
> Serkeled with serpentes all aboute the sides –
> To tell the todes theron my tonge wer full tere.
> The burne braides oute the bronde, and the body bides;
> Therefor the chevalrous knight changed no chere.
> The hondes highen to the holtes, and her hede hides,
> For the grisly goost made a grym bere.
> The grete greundes wer agast of the grym bere.
> The birdes in the bowes,
> That on the goost glowes,
> Thei skryke in the skowes
> That hatheles may here.[26]

(The ghost glowed like a coal as she glided along encompassed in swathes of dark clothing, encircled with serpents and on all sides – and so many toads that it is impossible to count them. The chivalrous knight remained calm and drew his sword, and the corpse becomes still. The greyhounds rushed to the woods and hid their heads, because the gruesome ghost made such awful sounds. The great greyhounds were terrified of the corpse. The birds in the trees, staring down on the spirit, screeched so loudly in the woods that many noble men could hear them.)

The dogs take refuge and hide their heads when confronted with the corpse, while the screeches of the startled watching birds resound through the woods. The resultant disharmonious cacophony of shrieking birds, shocked animals and the groaning ghost intrudes into the elegant, decorous realms of the court, overwhelming the charming polyphony one would expect to find there. The toads and serpents both gaze upon and consume the spectre's body. They are observed by other animals and the courtiers, and in turn they can see right to the heart of the court as they cling to the exterior of the corpse, with eyes peering in all directions. The unhappy spectre is contrasted with Guinevere, Arthur, his court and also her former self. She was once a beautiful queen, of noble lineage revered by all who beheld her.

> I was of figure and face fairest of alle,
> Cristened and knowen with kinges in my kynne;

I have kinges in my kyn knowene for kene.
God has me gevene of his grace
To dre my paynes in this place[27]

(I was once in figure and face fairest of all, and I was baptized and honoured to have a lineage of kings, with many kings in my family renowned for their bravery. God has given me of his grace to endure torments in this place.)

She explains that once she had rose-red cheeks and a lily-white face, but that due to her licentious lifestyle she has become transformed through sin into a hideous creature with her face gruesomely eroded. The day of doom for Arthur's kingdom is not that far off; it will occur in his lifetime, when Mordred has matured:

In riche Arthures halle,
The barne playes at the balle
That outray shall you alle,
Delfully that day.[28]

(In Arthur's splendid hall, the child plays with a ball, who will one sad day destroy you all.)

The resplendent king is, to appropriate A. C. Spearing, 'enthroned at the exact centre of the poem, the twenty-eight stanza':[29]

The mon in his mantell sittis at his mete
In pal pured to pay, prodly pight,
Trofelyte and traverste with trewloves in trete;
The tasses were of topas that were thereto tight.
He gliffed up with his eighen that grey wer and grete,
With his beveren berde, on that burde bright.
He was the soveraynest of al sitting in sete
That ever segge had sen with eye sight[30]

(The king was seated at his meal, attired in a coat of exquisitely adorned fabrics, richly decorated with rows of interlacing love-knots, with tassels of topaz. He lifted up his great grey eyes, set off with his reddish beard, to look at a fair lady. He was the most resplendent of all, seated in his rightful place, that ever a man had seen.)

Once more we have a splendidly detailed portrait, this time of the magnificent king, a sovereign who looks up with his 'grey' and 'grete' eyes. By thus orchestrating his appearance Arthur is offering himself as object of the gaze of others. As Rob Imrie, following Paterson and Hughes, notes in relation to disabled people's mobility and movement, 'judgements' about people's ability are grounded in 'aesthetic evidence' as if this is 'an omnipotent guide to competence'.[31] Arthur is dressed in fantastically embroidered fabrics adorned with gems and gold.[32] He is endeavouring to control his image and reception. His finery mirrors Guinevere's, which in turn mirrors her mother's former glory, a veritable feast of riches. The earthly splendour of this ephemeral world is reiterated through the hunting narreme which reveals the passing rewards and 'delight in the transitory world and the joy in courtly pleasures'.[33] The hunt is an impetus to self-discovery and the deadly nature of desire.[34] Such a view is shared by Virginia Lowe who argues that Gawain's questioning of the ghost is indicative of how earthly glories and 'violent expenditure on the part of the king is morally undirected and will end in self-destruction and dishonour'.[35] The response Gawain gleans from the otherly spectre suggests as much:

> 'How shal we fare,' said the freke, 'that fonden to fight,
> And thus defoulene the folke on fele kinges londes,
> And riches ouer reymes withouten eny right,
> Wynnen worshipp in werre thorgh wightnesse of hondes?'
> 'Your kynge is to covetous, I warne the sir knighte.'[36]

('How shall we fare', said the knight, 'who undertakes to fight, and hence subdue the people in many kingly realms, entering countries without any right and winning honour through the force of their fighting?' 'Your king is too covetous, I warn you Sir Knight.')

Upon learning of the apparition's prophecies from Guinevere, Arthur attempts to ameliorate his deformed behaviour, by responding positively and alleviating political strife. These actions, as do the prophecies, suggest that change is always possible and that one can alter one's own behaviour, and have a beneficial effect on the actions of others as the intercessions of Gawain and the female protagonists do on Arthur. Chivalry and its representatives are neither fixed, nor beauteous, but they do offer a structured means for dealing with violence and power, moral and political foulness, fulfilling the need for reassurance and managing change.

> 'Have gode day, Gaynour, and Gawayn the gode;
> I have no lenger tome tidinges to telle.
> I mot walke on my wey thorgh this wilde wode
> In my wonyngstid in wo for to welle.
> Fore Him that rightwisly rose and rest on the Rode,
> Thenke on the danger and the dole that I yn dwell.
> Fede folke for my sake that failen the fode
> And menge me with matens and Masse in melle.
> Masses arn medecynes to us that bale bides;
> Us thenke a Masse as swete
> As eny spice that ever ye yete.'
> With a grisly grete
> The goste awey glides.[37]

('Good day Gaynour and Gawain the good; I have no more time to talk with you. I must walk through this wild wood to my dwelling place where I shall languish in woe. In the presence of Him who righteously rose and hung on the Cross, contemplate the danger and misery in which I live. Feed people, for my sake, who have not enough food and remember me with services and masses. Masses are the medicine for all us who dwell in pain; we consider a mass as sweet as any spice one could ever eat.' With a gruesome groan the ghost glides away.)

In the logic of *The Awntyrs off Arthur*, Arthur's gift of Wales to Gawain upon the return of Galleroun's lands is intended to redeem the political situation, paralleling the recuperation of Guinevere's mother and the alleviation of her pain by applying the medicine of masses.[38] Galleroun and the spirit's expeditions, their penetrations of the Arthurian court reveal the permeability of society, chivalry and their value systems. The Arthurian court may be repulsed by what it perceives to be the disabling physicality of others, but ultimately the court itself is immersed in impairment, dis/enabling perceived indicators of merit with desire.[39]

Conclusion

Tobin Siebers in his analysis of the disability theory discusses how disability augments 'representations of Otherness', lending added significance and how the non-normative nature of 'disabled bodies' traverses the 'rules of representation', generating sites of great meaning.[40] The intrusion of physically different, non-normative females into the elegant, decorous surrounds of the chivalric elite are instances of 'the radical individuality of disability', which can jeopardize 'the very possibility of a shared, political future among people, cutting an easy path to solipsism and political isolation'.[41] The actions of the noble ladies and knights in response to the visitors are both self-centred and isolationist. They endeavour to keep the loathly lady from public view and shield themselves from outside forces, while policies of imperialism and territorial sequestration remain unabated by the apparition's warnings. The encroachment of Guinevere's mother and Ragnelle in dis/abled feminine forms bestows heightened meaning to their actions and rhetoric, their prophecies and words of wisdom. Just as 'Oedipus's blindness becomes a metaphor and an enabling affliction that leads to his insightful recognition, first in Thebes and later at Colonus',[42] so do the loathly lady and the corpse's disfigurements lead to new insights into the nature of desires, desires evident in the social fabric as a whole and its individual constituents.

In a sense, discourses of disability are centred upon the concern to perpetuate 'the privileged status of the non-disabled individual', the perceived normative being, a 'need that … emerges from fears about the fragility and unpredictability of embodied identities'.[43] Goodley argues that 'non-disabled people's behaviour to the disabled … can be understood as a process of splitting (separating good (desired) and bad (not desired) areas of one's psyche), introjection (internalising those aspects of the good life (desired) that you want to keep dear to you) and projection (making sure you project out the bad (not desired) away from yourself on to others)'.[44] Here what is not deemed beneficial is projected upon Ragnelle and the ghost and that which is held dear upon Gawain, the epitome of chivalry. Ultimately, aesthetic appearance is no indication of merit, for the Arthurian court and its king with his reddish beard and grey eyes are morally and politically suspect, while Ragnelle and Guinevere's mother offer guidance, and Gawain occupies an anomalous position, with a foot in both camps, a member of the court yet mindful of the words of wisdom uttered by the disfigured. 'Disabled bodies exist as a stark reminder that the ableist Empire cannot conquer all,'[45] and through granting autonomy to the impaired ghost and lady Ragnelle, the precarious foundations of the ableist Arthurian court are laid bare.

By bringing the agency of gazing and eating, sight and taste, into play, the injurious nature of desire is exposed. The object of the gaze is empowered, levelling the dichotomies between beauty, deformity and dangerous appetites. Ultimately the ableist, norming project is destined to come up lacking as power and desire cannot truly be contained by human institutions such as chivalry and courtesy.

Notes

1 David T. Mitchell, 'Narrative Prosthesis and the Materiality of Metaphor', in *Disability Studies: Enabling the Humanities*, ed. Sharon L. Snyder, Brenda Jo Brueggemann and Rosemarie Garland-Thomson (New York: The Modern Language Association of America, 2002), 16–17.

2 Rob Imrie, 'Disability and Discourses of Mobility and Movement', *Environment and Planning* 32 (2000), 1650; Kevin Paterson and Bill Hughes, 'Disability Studies and Phenomenology: The Carnal Politics of Everyday Life', *Disability and Society* 14.5 (1999): 597–610.

3 Dan Goodley, *Disability Studies: An Interdisciplinary Introduction* (London: Sage, 2011), 82.

4 Mitchell, 'Narrative Prosthesis', 27.

5 For example, Irina Metzler, *Disability in Medieval Europe: Thinking about Physical Impairment during the High Middle Ages* (London: Routledge, 2010); Irina Metzler, *A Social History of Disability in the Middle Ages: Cultural Considerations of Physical Impairment* (London: Routledge, 2013); *Disability in the Middle Ages: Reconsiderations and Reverberations*, ed. Joshua Eyler (Aldershot: Ashgate, 2010); *The Treatment of Disabled Persons in Medieval Europe*, ed. Wendy Turner and Tory Pearman (Lampeter: Edwin Mellen, 2010); *Disability and Medieval Law: History, Literature, Society*, ed. Cory Rushton (Newcastle-upon-Tyne: Cambridge Scholars Publishing, 2013); Patricia Skinner, *Living with Disfigurement in Early Medieval Europe* (New York: Palgrave Macmillan, 2017).

6 The loathly lady commonly represents territory, the nation in Irish narratives, but medieval texts use this creature to explore not just political, but also personal, cultural and psychological issues, often as in the present cases, in a manner which permits multiple readings. Susan Carter, 'A Hymenation of Hags', in *The English 'Loathly Lady' Tales: Boundaries, Traditions, Motifs*, ed. S. Elizabeth Passmore and Susan Carter (Kalamazoo, MI: Medieval Institute Publications, 2007), 85; Russell A. Peck, 'Folklore and Powerful Women in Gower's *Tale of Florent*', in ibid., 101–2 and 125; Paul Gaffney, 'Controlling the Loathly Lady, or What Really Frees Dame Ragnelle', in ibid., 156–7. See, in addition, Jennifer R. Goodman, *Chivalry and Exploration 1298–1630* (Woodbridge: Boydell and Brewer, 1998), 186–7 on the trope of landscape as a woman's body.

7 Mitchell, 'Narrative Prosthesis', 17. For an overview of models of disability, see Jan Grue, 'Interdependent Discourses of Disability: A Critical Analysis of the Social/Medical Model Dichotomy' (PhD thesis, University of Oslo, 2011), 9–42.

8 *The Wedding of Sir Gawain and Dame Ragnelle* (lines 231–45). All quotations from this and *The Marriage of Sir Gawain* are from *Sir Gawain: Eleven Romances and Tales*, ed. Thomas Hahn (Kalamazoo, MI: Medieval Institute Publications, 1995).
9 *The Wedding of Sir Gawain and Dame Ragnelle* (lines 547–56).
10 Stephanie Hollis, '*The Marriage of Sir Gawain*: Piecing the Fragments Together', in *The English 'Loathly Lady' Tales*, ed. Passmore and Carter, 169; Mary Leech, 'Why Dame Ragnell had to Die: Feminine Usurpation of Masculine Authority in *The Wedding of Sir Gawain and Dame Ragnelle*', in ibid., 219.
11 *The Marriage of Sir Gawain* (lines 57–64).
12 Peck, 'Folklore and Powerful Women', 8.
13 Leech, 'Why Dame Ragnell had to Die', 215–6.
14 Ibid., 213.
15 *The Wedding of Sir Gawain and Dame Ragnelle* (lines 638–9).
16 Ibid. (lines 695–702).
17 Lennard J. Davis, *Enforcing Normalcy: Disability, Deafness and the Body* (London and New York: Verso, 1995), xvi.
18 Ibid., 3.
19 *The Marriage of Sir Gawain* (lines 175–83).
20 Hollis, '*The Marriage of Sir Gawain*', 173.
21 *The Marriage of Sir Gawain* (lines 40–4).
22 *Sir Gawain: Eleven Romances and Tales*, ed. Hahn, 169–226.
23 *The Awntyrs off Arthur* (lines 664–76).
24 Ibid. (lines 105–17). This incident is based on *The Trental of St Gregory*. While administrating mass his mother appears and confesses her incest and murder, asking him to say trentals for her soul: George Neilson, 'Cross-links between *Pearl* and *The Awntyrs off Arthure*', *Scottish Antiquary* 16 (1902): 67–78. Takami Matsuda, '*The Awntyrs off Arthure* and the Arthurian History', *Poetica* 19 (1984): 48–62 discusses the analogues of the work which are to be found among poems of a moralistic nature.
25 See further Rory Byrne, '"Muse on my mirror": Gawain's function in relation to the supernatural of Arthurian literature', *Innervate: Leading Undergraduate Work in English Studies* 3 (2010–11): 57–9; Corrine Saunders, *Magic and the Supernatural in Medieval English Romance* (Cambridge: D. S. Brewer, 2010), 222–3; Carl Gray Martin, '*The Awntyrs off Arthure*, an Economy of Pain', *Modern Philology* 108.2 (2010): 177–98; Susana G. Fein, 'The Ghoulish and the Ghastly: A Moral Aesthetic in Middle English Alliterative Verse', *Modern Language Quarterly* 48 (1987): 10; Leah Haught, 'Ghostly Mothers and Fated Fathers: Gender and Genre in the *Awntyrs off Arthure*', *Arthuriana* 20.1 (2010): 3–22; Richard James Moll, *Before Malory: Reading Arthur in Later Medieval England* (Toronto: University of Toronto Press, 2003), 125–40.
26 *The Awntyrs off Arthur* (lines 118–130).
27 Ibid. (lines 137–41).
28 Ibid. (lines 309–12). Krista Sue-Lo Two, '*The Awntyrs off Arthure at the Terne Wathelyne*: Reliquary for Romance', *Arthurian Literature* 20 (2003): 118, argues that this prophecy 'thwarts' any 'possibility' of redemption. In addition, see Randy P. Schiff, 'Borderland Subversions: Anti-Imperial Energies in *The Awntyrs off Arthure* and *Gologras and Gawane*', *Speculum* 84.3 (2009): 627–8.
29 A. C. Spearing, 'Central and Displaced Sovereignty in Three Medieval Poems', *Review of English Studies* n.s. 33 (1982): 251–2.

30 *The Awyntyrs off Arthur* (lines 352–9).

31 Imrie, 'Disability and Discourses of Mobility and Movement', 1650; Paterson and Hughes, 'Disability Studies and Phenomenology'.

32 Martin Connolly discusses the clothing motif in '*The Awntyrs off Arthure*: Beyond the Tapestry', *The Hiyoshi Review of English Studies* 60 (2012–3): 10–15.

33 Thorlac Turville-Petre, '*Summer Sonday, De Tribus Mortus*, and *The Awntyrs off Arthure*', *Review of English Studies* n.s 25 (1974): 10–11.

34 Virginia Lowe, 'Folklore as a Unifying Factor in *The Awntyrs off Arthure*', *Folklore Forum* 13 (1980): 209. Agatha Hansen uses Kristeva's and Žižek's theories to address the notion of desire in *The Awntyrs off Arthur* in 'The Characterization of Monstrous Femininity in the *Testament of Cresseid* and *The Awntyrs off Arthure*' (MA diss., Queen's University, Toronto, 2009), 29–54. See further Margaret Robson, 'From beyond the Grave: Darkness at Noon in *The Awntyrs off Arthure*', in *The Spirit of Medieval English Romance*, ed. Ad Putter and Jane Gilbert (New York: Pearson Educational Ltd., 2000), 219–36; Helen Phillips, 'The Ghost's Baptism in *The Awntyrs off Arthure*', *Medium Ævum* 58.1 (1989): 49–58; Michael Robinson, '*The Awntyrs Off Arthure*: Barren Does and a Motherly Ghost as Authorial Agent of "Arthurial" Parable and Reform', *Criterion* (Los Angeles: Loyola Marymount University, 2014): 32–40.

35 Lowe, 'Folklore', 215. See further D. N. Klausner, 'Exempla and *The Awntyrs off Arthure*', *Mediaeval Studies* 34 (1972): 307–25; Jeanne T. Mathewson, 'Displacement of the Feminine in *Golagras and Gawane* and *The Awntyrs off Arthure*', *Arthurian Interpretations* 1.2 (1987): 23–8; Robson, 'From beyond the Grave'; Helen Phillips, '*The Awntyrs off Arthure*: Structure and Meaning, a Reassessment', *Arthurian Literature* 14 (1993): 71–88; Turville-Petre, '*Summer Sonday*', 1–14 and Ralph Hanna, '*The Awntyrs off Arthure*: An Interpretation', *Modern Language Quarterly* 31 (1970): 275–97.

36 *The Awntyrs off Arthur* (lines 261–5).

37 Ibid. (lines 313–25).

38 See Schiff, 'Borderland Subversions', 618–21; Patricia Ingham, *Sovereign Fantasies: Arthurian Romance and the Making of Britain* (Philadelphia: University of Pennsylvania Press, 2001); Matsuda, '*The Awntyrs off Arthure*'; Stephen Shepherd, '"Heathenic" Catechesis and the Source of Awntyrs B', *Medium Ævum* 131.1 (2012): 1–17.

39 For an alternative reading of *The Awntyrs off Arthur* as the 'struggle for identity', see Jean E. Jost, 'Margins in Middle English Romance: Culture and Characterization in *The Awntyrs off Arthure at the Tarn Wathelyne* and *The Wedding of Sir Gawain and Dame Ragnelle*', in *Meeting the Foreign in the Middle Ages*, ed. Albrecht Classen (New York and London: Routledge, 2002), 133, 151.

40 Tobin Siebers, *Disability Theory* (Ann Arbor: The University of Michigan Press, 2008, reprint 2011).

41 Siebers, *Disability Theory*, 186.

42 Davis, *Enforcing Normalcy*, 100.

43 Goodley, *Disability Studies*, 78.

44 Ibid., 96.

45 Goodley, *Disability Studies*, 33.

3

'A Great Blemish to her Beauty': Female Facial Disfigurement in Early Modern England

Michelle Webb

In 1583 the courtier Sir Henry Sidney described, in a letter, the devastating after-effects of the smallpox which his wife Mary had caught from Elizabeth I some twenty years previously: 'I lefte her a full faire Ladye in myne eye at least the fayerest, and when I returned I found her as fowle a ladie as the smale pox could make her.'[1] The revulsion expressed in that statement remains raw and shocking, even more than four hundred years after it was written. This was a relatively private sentiment, expressed in a document which Sir Henry had no reason to suppose would ever be widely disseminated, and there is no way of knowing to what extent Lady Mary Sidney was aware of her husband's reaction to her disfigurement.[2] Like the vast majority of women who suffered damage to their faces in the sixteenth and seventeenth centuries, she left no record of her thoughts and feelings upon her altered appearance. However, while first-person accounts of acquired disfigurement are frustratingly largely absent from the writings of early modern women, it is still possible to gain insight into the experiences of, and reactions to, women with damaged faces. Comment upon female appearance was ubiquitous, and descriptions and judgements of pockmarked, scarred or otherwise marked faces were included in a vast range of texts.

In early modern England acquired facial disfigurement was not a rare phenomenon. Disease was responsible for much of the damage; smallpox routinely resulted in noticeable scarring and the pox, which was rife, could cause the ulceration and eventual destruction of the nose and palate. Little of this harm was either preventable or reversible, and facial injuries were similarly difficult to effectively treat. Disfigurement was an inescapable feature of many lives. This chapter will argue that, in common with almost every other facet of early modern experience, acquired facial difference was a sharply gendered experience. Women with disfigured faces were conceptualized as being both physically and socially damaged. Discussions about their faces frequently involved mournful comparisons with their previous appearance, and were often accompanied by dire prognostications about their marital prospects. Ideally, an early modern woman would compensate for her assumed spiritual and intellectual shortcomings with beauty and fecundity, but a damaged face was interpreted as beauty spoiled. Most observers clearly believed that an unmarred face was a prerequisite for a happy and successful female life. Disfigurement presents an opportunity for historians of the early modern period to explore reactions to individuals who could not conform to the expectations of society.

Yet despite the important role that facial difference played in the lives of many individuals, this is a subject which has been relatively little explored by scholars. The first significant mention of the subject came in 1986, when Margaret Pelling delineated the aspects of early modern society and belief which attributed exceptional importance to the face. Pelling convincingly argued that the prominence given to the face by clothing which concealed almost every other aspect of the body, combined with early modern medicine's preoccupation with complexion as a bellwether of internal disease, ensured that facial flaws were likely to be much commented upon.[3] Since the publication of Pelling's work on the implications of facial appearance, disfigurement has been researched primarily as an aspect of wider concerns, such as disability or beauty. David Turner has delineated the extent to which physical impairment was largely judged according to its visual impact, arguing that 'the concept of disability was subsumed under other categories, notably deformity and monstrosity'.[4] Naomi Baker, meanwhile, included disfigurement as one of the forms of unsightliness in her analysis of how ugliness was defined and interpreted in the sixteenth and seventeenth centuries; the anxiety, disgust or hilarity that a face or body could provoke; and the extent to which unattractiveness was considered to be potentially legible.[5] Other historians and literary scholars have investigated facial difference in the context of specific individuals, facial features or diseases. These studies include Anna Riehl's discussion of Elizabeth I's image and her brush with smallpox, and Marcus Nevitt's account of how the competitors of Sir William Davenant chose to attack his poetry and his politics by mocking his absent nose.[6] Sander Gilman has explored facial aesthetics in several studies, and Raymond Anselment and David Shuttleton have addressed both male and female experiences of smallpox scarring in the late seventeenth and eighteenth centuries.[7]

Research into the gendered implications of disfigurement has been somewhat sparse. Shuttleton has convincingly argued that a smallpox-scarred woman could experience a contraction of her social and marital horizons that was not generally matched by male experience.[8] Meanwhile, Emily Cock has explored the gendered aspects of the shame associated with noses lost to the pox.[9] However, further exploration of gender and disfigurement is warranted, and by considering facial difference of all types and origins, this chapter will contribute to our understanding of the position of stigmatized women within early modern society.

While scars and tumours were concrete, physical realities, disfigurement was a construct that owed much of its form to assumptions based upon gender. Appearance was conceived of as being central to female identity and to female prospects, and therefore anything which imperilled or damaged the appearance of a woman was treated as an extreme threat. Damage to female faces was framed as something which would profoundly alter a woman's future.

Women and beauty

In 1666 Samuel Pepys recorded in his diary that the household's cook-maid was lucky not to have broken her neck when she fell down the stairs late one night. He recounted how the 'poor girl' suffered a bruised head and 'did lie on the ground groaning and drawing her breath, like one a-dying', but he somewhat undermined this impression of sympathy by carelessly identifying her as 'our new ugly mayde, Luce'.[10] His contemporaries would have seen nothing unusual about this damning verdict on the girl whom he had just observed lying crumpled in the hallway, as there was apparently no occasion when it was inappropriate to comment upon the appearance of a woman. Family, rank and virtue were all integral to the ways in which women were identified and discussed, but arguably nothing was as central to a woman's perceived identity as her facial appearance. Pepys categorized almost every woman he met according to her physical characteristics. On just one day in August 1663 he mentioned 'Ballard's wife, a pretty and a very well-bred woman'; Mrs Lurkin 'a good, and merry poor woman, but a very tall woman'; 'a daughter of Mr. Brumfield's, black, but well-shaped and modest'; Kate Joyce 'being very handsome and well, that is, handsomely dressed to-day, and I grew mighty kind and familiar with her, and kissed her soundly, which she takes very well'; an unnamed 'pretty woman' who was the wife of his cousin's son; and the wife of the king's falconer 'an ugly pusse, but brought him money'.[11] None of the men whom he encountered that day warranted any physical description at all. Pepys recorded the appearance of the women whom he encountered regardless of their status or behaviour and frequently in scathing terms. He used the same pejorative words as either insults or blunt appraisals, dismissing his Aunt Wight as 'a silly, froward, ugly woman' but admitting that his acquaintance Mrs Parker was 'an ugly lass, but yet dances well, and speaks the best of them'.[12] Female appearance mattered to early modern men and remarking upon it was an entirely unremarkable act.

At the heart of most discussions of the face of a woman was a simple but vital question, was she beautiful? The characteristics most generally associated with beauty were closely allied to the flawlessness of youth: symmetry and a dewy complexion with a sharp contrast between the red of the cheeks and the white of the rest of the face.[13] Ideally, this was accompanied by a glow of goodness from within, as described in an English translation of the works of the Italian artist Gian Paolo Lomazzo: 'For it is not the red and white which giveth the gratious perfection of beauty, but certaine sparkling notes and touches of amiable cheerefulnesse accompanying the same.'[14]

However, beauty was not merely a pleasing arrangement of features or a by-product of youth. Instead, it was an active force and it bestowed upon those fortunate enough to possess it a considerable degree of power. In his funeral sermon for Lady Jane Cavendish, Adam Littleton characterized beauty as 'that which strikes the Eye and affects the Heart'.[15] Contemporaries would not have been in any doubt that it was the eyes of men that were being referred to here: the late-seventeenth-century reference work *The Ladies Dictionary* explained that 'the Harlot beautifies her Face to attract lascivious wandring Eyes, and the virtuous Lady to gain and keep the Love of a Chast Husband'.[16] Beauty was not a vague and ill-defined aesthetic quality; instead it was the ability to evoke male desire. Physical attractiveness was perceived as something which a young woman could, and undoubtedly would, direct and exploit to redress the power differential which existed between men and women. Contemporary theories relating to vision deemed that an observer played a predominantly passive role: a man was the helpless victim of a beautiful woman.[17] Beauty was also likely to be interpreted as being at least partially misleading, and concealing the true nature of a woman.[18] It was both an important form of cultural capital and the most effective resource in a woman's arsenal, and it was understood to be a valuable commodity.[19] An attractive face was often the only form of power allowed to young women, who typically lacked control over many aspects of their own lives. But the power bestowed by beauty was potentially destructive and transgressive, and it could entrap a man against his better judgement.[20] This is perhaps best expressed in the use of the term 'bait' to describe the attractive female face and form. According to the Welsh author Robert Parry, beauty was nothing but 'a pleasant bayte to intrappe the minde', and while the pamphleteer Joseph Swetnam's extreme misogyny was widely recognized by his contemporaries, large numbers of them nevertheless purchased his *Araignment of Lewd, Idle, Froward and Unconstant Women*, in which he bemoaned women's unwarranted and unearned power over men, warning that 'their faces are lures, their beauties are baytes, their looks are netts, and their wordes charmes, and all to bring men to ruine'.[21] Beauty was important because the faces of women mattered, they could enthral or repel men, and as such, they had the potential to determine the course which a young woman's life might take.

Descriptions of female facial disfigurement

Individuals with disfigured faces were frequently the objects of what Rosemarie Garland-Thomson has characterized as 'baroque staring': a gaze that was blatant and unapologetic.[22] Comments upon facial flaws, particularly those of women, were a recurrent feature of sixteenth- and seventeenth-century texts and tended to be constructed in similar ways. Female disfigurement was often described in detail, and in language which painted a vivid picture of the face in question. In certain cases this is attributable to the nature of the text concerned: legal records were particularly likely to include an almost forensic attention to the minute details of facial damage, as it could potentially be evidence. In 1677 three siblings, Robert, William and Elizabeth Dine were tried and convicted at the Old Bailey for mutilating the face of Jane King, a maid servant who had rejected Robert's advances. The official account recreated the crime in vivid detail:

> There was no body in Family but these two and an aged Master and Mistriss, who being both in bed, on the 20th of *February* last, as this innocent Virgin and her treacherous Sweet-heart were sitting together, his own Brother and Sister came and knock'd and call'd him by his name, who open'd the door and let them in; but no sooner were they entred, but this Sister, more like a Fiend than a Woman, fell upon this harmless Maid, and with a Knife, whilst the other Brother held her, and in a most inhumane manner cut out one of her Eyes, the sight whereof she has lost;

> cut her Nose towards the upper part so desperately, that two bones, produced in Court, were since taken out of it; her Lips they slash'd so, that with much difficulty, and not yet perfectly, she can use them in speech; two of her Teeth they beat out; gave her a lamentable wound in the Neck, and two stabs, besides a great wound in the Arm; all which was testified by the Chirurgeon: and so leaving her thus mangled as dead, went home.[23]

It is notable that this retelling of the attack combines a detailed list of injuries with the use of such emotive terms as 'lamentable', 'inhumane' and 'desperately'. A similarly vivid combination of description and reaction was also sometimes present in accounts of medical treatment. In her memoirs, Lady Anne Halkett, who had a keen interest in medical care, wrote about some of those who sought help from her while she was living in Scotland. For instance: 'A young woman who had beene very beautyfull, and her face became loathsomely deformed with a cancerous humour that had overspread itt, which deprived her of her nose and one of her eyes, and had eaten much of her forhead and cheecks away.'[24] While lacking the clinical detail present in legal and medical texts, pamphlet literature also often contained colourful or lurid pen portraits of female faces. These could be brief but memorable, such as the comment that a murderous female brothel owner had 'worne her nose thrid-bare', or they could involve a wealth of detail, such as the ballad description of the encounter between a 'North-Country Gentleman, and a Very Gaudy Gallant Lady of Pleasure', in which the horrified man relates that when he spied upon the lady undressing: 'Her haire being gone she seemed a Hagg, Her bald-pate did looke like an Estritches Egg'; in the next verse 'She put up her finger, and out dropt an eye'; in the next 'She fetches a yawn, and out fell her Teeth'; and finally 'She drew out her Handkercheif (as I suppose) To wipe her high forehead, and down dropps her Nose', so that 'she look'd just (if you will beleeve mee) Like a Lancashier Witch'.[25] Disfigured female faces evidently demanded detailed and affective descriptions.

Accounts of damage to women's faces were also notable for their tendency to frame female disfigurement in terms of beauty lost or imperilled.[26] The diary entry by Anne Halkett mentioned in the previous paragraph included the information that the cancer-stricken woman whom she treated 'had beene very beautyfull', and Jane King, the victim of the Dine siblings attack, was described as having been 'a very pretty young maid' before her face was mutilated.[27] When the face of a woman, or at least a young woman, had been damaged, her appearance tended to be discussed in very particular ways. Rather than a straightforward assessment of how they viewed her face in its current state, observers concentrated much of their attention upon the alteration that had taken place and upon how that alteration had obscured or nullified her previous state of beauty.[28] This was true of descriptions by both male and female commentators. When Lady Anne Clifford recorded in her diary that the Countess of Suffolk had contracted smallpox, she noted that it had 'spoiled that good Face of Hers which had brought to others much misery, and to Herself Greatness'.[29] Meanwhile, Fulke Greville attributed Lady Mary Sidney's decision to live in partial seclusion after she caught smallpox from Queen Elizabeth to 'the mischance of sicknesse having cast such a kind of veile over her excellent beauty'.[30] Kay Dennis has discussed how

Greville transformed the actual physical veil which Mary Sidney was later believed to have worn into a powerful metaphor and succeeded in implying that her innate beauty had survived her disfigurement.[31] The discourse of beauty spoiled, concealed or compromised was participated in by both those who intimately knew the women concerned and those who had an altogether more superficial relationship with them. As quoted at the very beginning of this chapter, Sir Henry Sidney prefaced his remark about the foulness of his wife's scarred face by remarking that she had previously been 'a full faire Ladye, in myne eye at least the fayerest'.[32] The beauty of his wife would have reflected well upon Sir Henry; however, the seventeenth-century surgeon Richard Wiseman had considerably less reason to comment upon the beauty of his patients. Yet when he wrote an account of how he had treated the inflamed facial tumour of an eleven-year-old girl, he recorded the primary reason why her family had sought his help: 'It was a great blemish to her beauty, upon which very account her parents had much endeavoured to have it cured, or at least to be disposed to such a calm temper, as it might bear a black patch.'[33] Evidently both parents and medical practitioners saw the preservation of beauty as a creditable aim.

Descriptions of female disfigurement depended very heavily upon comparisons.[34] Women with damaged faces were, however, not compared with their unmarred peers, but with their previous, undamaged selves. The discourse surrounding facially damaged women often seemed to assume that any woman, provided that she had not been irretrievably blighted by age, had an initial store of beauty to lose. But a face could be scarred, or otherwise disfigured, yet still be considered worthy of praise. The seventeenth-century clergyman Thomas Bayly described an encounter at court between the Marquess of Worcester and a young bride: 'A goodly proper woman, her face well feitured, an excellent eye she had, but was pitifully disfigured with the small Pox.'[35] Again, this type of classification was not confined to the elite; an anonymous humorous broadsheet listing women who were apparently available for purchase at the London Exchanges included this entry: 'One middle siz'd Gentlewoman about 23. Rough-cut, Pock-holes in her Face, not Ugly'.[36] Disfigurement and attractiveness could coexist in the same face; however, damage to the face was interpreted as a filter or impediment, which isolated a woman from her original or true physical self. It did not utterly destroy her beauty, but it did separate her from it, and most importantly, prevented her from exercising the power associated with the attractive female face. This disempowerment is perhaps most clearly expressed in the use of the term 'spoiled'. As mentioned earlier in this chapter, Anne Clifford described, with a certain amount of glee, how the Countess of Suffolk's face had been spoiled by smallpox, and when Elizabeth Pepys suffered from a temporarily swollen cheek, her husband Samuel worried that it 'may spoil her face, if not timely cured'.[37] This term, with all its implications of ruin and finality, was used in a diverse range of sources, including Bulwer's *Anthropometamorphosis,* where the consequences of using cosmetics were described: 'There being a venomous quality in the paint which wrinkleth the Face before its time, it dims the Eyes, and blacks the Teeth; with false colours they spoile their Face, and gaine nought but contempt and hatred of their Husbands.'[38]

The implication of the repeated use of the term 'spoiled' to describe female facial disfigurement was that a damaged face had lost its ability to beguile men.

The oft-acknowledged ability of the face to retain some attractive features despite disfigurement could not save a woman from the negative implications of having been marred by illness or injury, and a face did not have to be catastrophically damaged to be considered to have lost its power. Female faces, like female reputations, could be entirely ruined by one flaw. When writing about young maidens, the sixteenth-century author and playwright Robert Green stated that 'there is no face so faire, but the least Moale, the slenderest scarre, the smallest brunt of sicknesse, will quickly blemish'.[39] His contemporaries' descriptions of female facial difference largely support the belief that even a small facial flaw could render a woman's face undesirable. Thomas Jeamson's seventeenth-century guide to the use of cosmetics, *Artificiall embellishments*, rather theatrically declared that 'each wound is a grave where Loves dumb orator lyes inshrin'd', when discussing the damage left by smallpox or other disfiguring diseases.[40] And when Lady Anne Clifford recovered from the smallpox, she wrote in her diary that it 'did so marter my face, that it Confirmed more and more my mynd never to marrie againe'.[41]

Women whose faces had been somehow damaged were rarely dismissed by contemporaries as now falling into the category of ugliness. In 1668 Frances Stuart, Duchess of Lennox and Richmond, one of the most acclaimed beauties of the age and the object of Charles II's unrequited desires, contracted smallpox. Among those who took a keen interest in what would become of her was Samuel Pepys, who, despite being barely acquainted with Lady Frances, wrote repeatedly in his diary about her recovery. Significantly, he said very little about her health, instead concentrating his attention upon her face. On 26 March, 1668 Pepys bemoaned Lady Frances' presumed fate: 'She is pretty well, but mighty full of the smallpox, by which all do conclude she will be wholly spoiled, which is the greatest instance of the uncertainty of beauty that could be in this age.'[42] Four days later he mourned her impending disfigurement: 'I did see Mrs Stewart's picture as when a young maid, and now just done before her having the smallpox: and it would make a man weep to see what she was then, and what she is like to be, by people's discourse, now.'[43] By the end of August, Pepys confirmed that his fears had, at least to some extent, come to pass: 'So I to the Park, and there walk an hour or two; and in the King's garden, and saw the Queen and ladies walk; and I did steal some apples off the trees; and here did see my Lady Richmond, who is of a noble person as ever I saw, but her face worse than it was considerably by the smallpox.'[44] Yet less than a fortnight later he was disappointed at a missed opportunity to see 'the beautiful Mrs Stuart'.[45] Her scars had not rendered Frances Stuart ugly in the eyes of her admirer Pepys, a man who had no compunction in describing the faces of women with brutal honesty.

For Samuel Pepys, Frances Stuart's status as the aristocratic object of the king's desires undoubtedly contributed to her appeal. However, women from all ranks of society were described in ways which made clear that their disfigurement was not interpreted as ugliness. This may have been particularly true for those whose faces had been damaged by smallpox, since the frequency with which the disease caused facial scarring may have rendered those scars commonplace. It is also notable that their damaged faces were not considered to compromise their feminine status, and only in the very broadest of caricatures might a woman's damaged face lead to unflattering analogies with men. In the broadside ballad *Nevvs from Hide-Parke,* the 'Very Gaudy Gallant Lady of Pleasure', whose hair, eye, nose and teeth all transpired to be prosthetic, is a rare example of a disfigured woman whose appearance had apparently defeminized her. The ballad compares her to the infamous one-eyed regicide John Hewson: 'She look'd (with one eye) just like Hewson the Cobber.'[46]

Disfigured women were consigned to a third category: they were neither entirely beautiful nor irredeemably ugly, but instead damaged. The positioning of women with disfigured faces as being neither ugly nor monstrous but belonging to an entirely separate subset of those whose appearance was not wholly pleasing is confirmed by the way in which authors writing about the use of cosmetics discussed disfigurement. The cosmetic enhancement of the female face was a subject about which many sixteenth- and seventeenth-century men had very strong feelings indeed. Christian texts which persistently discussed women in terms of deception and hypocrisy perpetuated a discourse in which femininity and deceitfulness were so closely elided that they became indistinguishable from one another.[47] It seemed clear to many men that women who painted their faces did so in order to usurp the power of God to determine appearance, and the right of men to be able to easily categorize women according to that appearance.[48] The use of cosmetics was an area of female life which was apparently at risk of slipping entirely out of patriarchal control, and the opposition to this outrage was vehemently expressed and often harshly worded. The Portuguese author Francisco de Mello described cosmetics on an older face as 'no better than daubing a rotten post' and in Thomas Shadwell's *Bury-Fair,* a character proclaimed that 'your fine Women, are a Company of proud, vain, Fops and Jilts, abominably Daub'd and Painted; and I had rather kiss a Blackamoor, with a Natural Complexion, than any such'.[49] These were insults of genuine bile, and when authors railing against cosmetics suggested that disfigurement was a suitable punishment for this most rebellious of sins and that those wearing cosmetics 'invite the hand of God to strike them with deformity', it was an indication of what a disaster disfigurement was believed to be for a woman.[50] Yet despite all the horror which the use of cosmetics provoked, many of the authors who ranted against painted women were prepared to allow that those whose faces were disfigured, rather than merely ugly or old, should be permitted to attempt to improve their appearance.[51] A few pages after he had claimed that deformity was the just reward of those who painted their faces, John Bulwer seemingly reluctantly conceded that 'yet we cannot say, that it is absolutely unlawfull to use any Fucus, especially when any foule blemish doth disgrace the forme of modest Virgins or Matrons'.[52] Thomas Tuke, one of the most vitriolic opponents of cosmetics, also admitted that some few exceptions to their universal condemnation did exist, writing that theologians approved their use 'in Spaine, where the Sunne beame doth swart their women'.[53] Irina Metzler has noted a similar approach to disfigurement and cosmetics in medieval texts.[54] Any damage to the face, even if it was as mild as sunburn, placed a woman outside the usual strictures imposed upon female self-improvement.

While disfigurement was an independent category of appearance, separate from ugliness or monstrosity, it was sometimes closely connected to another form of female unattractiveness: old age.[55] When writing about smallpox Robert Gould lamented that 'thou dost at once what Age is doing long'; disfigurement rendered a woman undesirable, and there was nothing more undesirable in early modern Europe than an ageing woman.[56] As Naomi Baker has argued, a woman was very unlikely to ever be viewed as a Silenus figure, with her unprepossessing appearance concealing a noble mind or soul.[57] Rather, as Patrizia Bettella has noted, her most notable characteristic was likely to be grotesqueness.[58]

The connection which authors perceived between the effects of ageing and those of damage to the face draws attention to the extent to which life-stage was of overwhelming importance in determining the classification of female faces, especially if those faces had been disfigured in some way.[59] The interpretation of the impact that an atypical face was likely to have upon a woman's life was discussed almost entirely in terms of how it would affect her prospects of marriage, or how injurious it would be to the reputation or feelings of her husband. Sir Henry Sidney's reaction to his wife's disfigurement was one of outrage, and by stating that she had previously been 'in myne eye at least the fayerest', he privileged his own opinion of her face, although he did also state that her scars had not faded 'to her resolute discomforte', so that she now 'lyveth solitairilie', indicating that he was conscious of how devastated Mary Sidney was by her disfigurement.[60] In a discussion of the perils of maternal impression, the female author M. Marsin drily reminded her readers that 'there's ne'r a man in the world that cares to see his Daughter depriving her sweet-heart of his full kiss, by reason of the piece wanting in her harelip'.[61] Women themselves were aware that damage to a female's face was interpreted as affecting either the man who loved her or who might otherwise have done so.

While descriptions of disfigurement were often colourful and sometimes contemptuous, the language used to describe damaged female faces was often notably imprecise and lacking in nuance and subtlety. There was frequently little contrast drawn between different degrees or types of facial damage and, perhaps most importantly, different causes for facial disfigurement were often conflated, allowing the intense stigma associated with some diseases to become attached to women whose faces were atypical for other reasons.[62] It was not unusual for descriptions of women's disfigurement to display a slippage between different types of facial damage, with the damning language usually associated with the pox applied to other conditions. When Sir Henry Sidney wrote about the effects of his wife's smallpox, he described how 'I found her as fowle a ladie as the smale pox could make her'.[63] The word 'foul' was strongly associated with the pox and with the disgust which contemporaries felt for the sufferers of that disease, yet Sir Henry used it about scars which should theoretically have been subject to considerably less stigmatization.[64] David Shuttleton has suggested that the superficial visual resemblance between the damage caused by the pox and the smallpox could often result in some transference of stigma from one disease to the other.[65] The term 'foul' had also been used with reference to lepers, and Naomi Baker has noted that comparisons between ugly women and lepers were not uncommon in the early modern period.[66] When Gian Paolo Lomazzo described how women who used cosmetics containing mercury sublimate were liable to be left with 'an uncleane complexion', this was another example of the vocabulary associated with corruption and the pox being used to reference other types of damage to female faces.[67]

There was also a tendency to conflate different causes for the loss of the women's noses under the blanket heading of the 'whore's mark'.[68] The noseless whore was a recurrent feature in all types of written sources.[69] The term was most frequently used to describe the distinctive damage caused by the pox, such as in the case of the female brothel keeper described earlier in this chapter, who had 'worne her nose thrid-bare', but it was also used to denote the damage left by a deliberate injury.[70] In continental Europe the severing of the nose had long been a gendered punishment, intended to render the victim permanently separate from their peers.[71] However, in England this most damning of wounds was often associated with rivalry in love and with the punishment of a husband's mistress; the wronged wife in a popular seventeenth-century ballad vowed, 'For certain I will slit her nose since she hath wronged me, My husband hath full twenty pound upon her vainly spent, With feasting of her whoreish chops In mirth and merryment.'[72] Threatening to inflict a whore's mark upon a woman was not something that was confined to either fictional accounts of disputes or to the lower orders of society. In November 1668 Elizabeth Pepys discovered that her husband had spent the previous evening with her former maid Deb and 'she swore by all that was good that she would slit the nose of this girle.'[73] However, this type of injury could also be inflicted under a myriad of other circumstances and was not always directly connected to a woman's sexual conduct. In 1700 the Old Bailey heard how John Laing had taken drunken revenge upon his lover after he suspected her of theft: 'He seized upon her Head-Clothes, and gave her as he said he wou'd, an Whore's Mark for he almost Cut off her Nose and Gash'd her Cheek in that terrible manner as is not to be exprest.'[74] It is very likely that women such as Laing's victim would have been permanently stigmatized as immoral after their disfigurement, since any damage to the nose was likely to be interpreted as being due to the pox.[75] A damaged or missing nose proclaimed that a woman was guilty of sexual immorality, and in the absence of disease having ensured a women's legibility in this way, it was always possible that those who felt her alleged immorality should be made visible to all would take matters into their own hands. Unfortunately for those with disfigured faces, a woman's nose could easily designate her a whore.[76]

Female views on disfigurement

There were many similarities between the ways in which male and female commentators wrote about disfigured women and their faces, but there were also significant divergences. Although there are fewer female-authored texts, it is possible to gain some idea of how women viewed their own and other women's disfigurement, partly through female-authored texts and partly through the words and opinions of women as preserved, and mediated, by male authors. Women who wrote about disfigurement frequently did so in the context of discussions about illness. This ensured that they usually referred to the specific circumstances that had caused the facial damage in question, but, in addition to this, they had a marked tendency to speculate upon the ongoing implications of female disfigurement. There is evidence to suggest that women viewed facial damage predominantly as a causal event, which was important primarily because of the impact that it would continue to have upon a woman's life.

In his guide to cosmetics, Thomas Jeamson began his address to his readers by reminding them that 'deformity (fairest Ladies) is a single name, yet a complicated misery'.[77] It would appear that Jeamson knew his audience well, and had a sophisticated understanding of how to exploit their fears, because there is evidence to suggest that women were acutely aware of the potential ramifications of facial disfigurement.[78] The language in which they discussed damage to the face was frequently the language of catastrophe. As previously mentioned, Lady Anne Clifford described her face as having been martyred by smallpox.[79] Similarly, when writing about the face, Lady Grace Mildmay defined smallpox as 'the arch enemy of this masterpiece of nature'.[80] Other female diarists who wrote about the disease used emotive language to express their relief when it left no scars in its wake. After her two daughters had contracted smallpox, Lady Anne Fanshawe expressed her joy that 'both of them, God's name be praised! recovered perfectly well, without blemish'.[81] Anne Fanshawe's correspondence offers a further insight into the female dread of disfigurement, when she begs to be excused for failing to see Lord Arlington because she has 'a swelled face with the toothaich which I am ashambed to showe'.[82] Even temporary disfigurement had the potential to be shaming for a woman.[83] However, not every woman who wrote about disfiguring disease prioritized appearance above all else, although Raymond Anselment has emphasized the extent to which the damage caused to the face by smallpox often preoccupied those who wrote about the disease.[84] Alice Thornton recorded her daughter Katherine's 1666 bout of smallpox in some detail, and when she described the symptoms, it was her daughter's face on which she concentrated: 'She was all over her face in one scurfe, they running into each other.'[85] This is an excellent description of confluent smallpox and, as Alice Thornton recognized, Katherine's life and eyesight were both under threat. Her survival, with her eyes intact, caused her mother to write that 'by the goodnesse of God, for which I humbly blesse and praise His holy name, she passed the danger of death, beginning to heale', making clear that her daughter's health was her priority.[86] Later in the same diary entry she mentioned, almost as an afterthought, that Katherine 'only losst by her sickenesse her fair haire on her head, and that beautifull complection God had given her'.[87] If this had been a case of confluent smallpox, then the resulting scarring was potentially very severe.[88] We have no idea how Katherine herself felt about the loss of her complexion, but her mother clearly felt it to be secondary to the miracle of her survival.

For most women in early modern England, having their face disfigured was the stuff of nightmares, as it had the potential to cause a significant negative alteration in the way that they were viewed and categorized by men, upon whom so many features of their lives were dependent. But while there is considerable evidence to suggest that women mourned the disfigurement of their own faces, there are also reasons to suspect that they were not always entirely sympathetic when the faces of other women were damaged. The patriarchal nature of society encouraged women to compete for the attention and favour of men and this may explain why spite and envy were two of the many sins commonly attributed to women. Certainly Anne Clifford's comment about smallpox having 'spoilt that good face' of the Countess of Suffolk conveyed an unmistakable air of satisfaction at how the disease had stripped the countess of her visual appeal.[89] As we have seen, it was not unusual for women to threaten to disfigure their rivals in love by slitting their noses, although it is impossible to know how often, if ever, this threat was carried out. Certainly the apparent lack of prosecutions for this act of mutilation suggests that it was always more of a threat than a reality.

Male understandings of female disfigurement

While women recognized that the disfigurement of a female face was at least theoretically likely to have a catastrophic impact upon the ways in which that face could serve and benefit its owner, men were apparently more circumspect in their judgements upon disfigurement. This is not to suggest that they in any way disagreed with the idea that a damaged face usually rendered a woman fundamentally undesirable, but instead they recognized that this loss of visual capital had potential advantages for those men who wanted the security of mind which a tamed wife could provide. Joseph Swetnam, as ever, was prepared to say outright what many of his contemporaries merely hinted at: 'Hee that hath a faire wife and a whetstone, euery one will bee whetting thereon; and a Castle is hard to keepe when it is assaulted by many.'[90] General opinion was that a beautiful wife would be nothing but a source of worry to her husband; other men would constantly pursue her, and given that women were renowned for their inability to resist temptation and their lack of the masculine virtues of moderation and self-control she was almost guaranteed to eventually yield.[91] Women were weak, they were fickle, they were inherently unfaithful, and as Phyllis Rackin has described, they were 'appetitive creatures'.[92]

As historians including David Turner have noted, women were believed particularly likely to indulge in sexual sin if enabled by beauty.[93] For those men who feared being cuckolded, a disfigured wife was a potential solution, since scars would enforce domesticity upon a woman.[94] As Francisco de Mello remarked in *The Government of a Wife*, when choosing a spouse a man should remember that 'the security of his Thoughts is of much more value than the pleasure of his Eyes'.[95] De Mello was recommending that his readers consider the virtues of an ugly wife, but a character in the play *The Generous Enemies* was even more specific and lauded the advantages of a wife with an acquired disfigurement: 'I had rather marry a woman without a Nose, Then one that must be watch'd, to keep Her honest.'[96] The implication was that a wife with a damaged face would be unlikely to attract potential partners for adultery, due to both her disfigurement and the threat of disease implicit in her face. Therefore her passivity and her fidelity would be assured, as would her husband's peace of mind.[97] She would not require the constant surveillance that might otherwise be considered necessary to ensure a woman's fidelity.[98] There is no suggestion that a disfigured woman would be more intrinsically virtuous than one with an unmarred face, merely that she would lack the opportunity to stray and would passively avoid sin, conforming with the recognized pattern for female virtue that has been described by Laura Gowing.[99] Although the extent to which a woman with a damaged face was truly likely to be shunned by men is debatable, there is evidence to suggest that some women reacted to their disfigurement by retreating into seclusion. Sir Henry Sidney's disclosure that his wife Mary chose to 'lyveth solitairilie' after being scarred by the smallpox was elaborated upon by his son's biographer and childhood friend Fulke Greville, who poignantly explained that 'shee chose rather to hide her self from the curious eyes of a delicate time, than come up on the stage of the world with any manner of disparagement'.[100] The likely behaviour of a woman with a damaged face could further assuage her husband's fears of infidelity. A disfigured woman was conceptualized as someone who was likely to have been tamed by the damage to her face. Her visual capital had been diminished or destroyed, and as a result she had been stripped of a great deal of her power.

Conclusion

When Sir Henry Sidney commentated so damningly on the after-effects of his wife's bout of smallpox, his words corresponded with the sentiments expressed in many other early modern statements about female disfigurement. He structured his remark around an initial description of Lady Mary's former state of beauty, privileged his own viewpoint, framed her as a passive victim of disease, and then complained that she, rather than merely her face, had been spoiled by the damage to her complexion. According to her husband, Mary Sidney had been reduced to a state of all-encompassing foulness. Disfigured female faces attracted comment; this commentary was sometimes cruel but sometimes wistfully elegiac, mourning lost beauty, and spoiled faces and lives. For early modern women with disfigured faces, their appearance would impact upon their lives and their identities in a multitude of pervasive ways and a damaged face was likely to be accompanied by diminished prospects. A woman's face, identity and reputation could be all but indistinguishable from one another and it is therefore unsurprising that Sir Henry Sidney described a 'fowle' lady, rather than merely a 'fowle' face.

Notes

1 The National Archives, London: PRO, SP 12/159, fol. 38^{r-v}; Carole Levin and Catherine Medici, 'Lady Mary Dudley Sidney (c.1531–1586) and her Siblings', in *The Ashgate Research Companion to the Sidneys, 1500–1700,* ed. Margaret P. Hannay, Michael G. Brennan and Mary Ellen Lamb, 2 vols (Farnham: Ashgate, 2015), i, 31. All other sixteenth- and seventeenth-century primary sources, as well as nineteenth-century editions, are accessed from Early English Books Online: https://eebo.chadwyck.com/home (accessed 27 March 2017) unless otherwise stated.

2 *A Viceroy's Vindication? Sir Henry Sidney's Memoir of Service in Ireland, 1556–78*, ed. Ciaran Brady (Cork: Cork University Press, 2002), 1–4.

3 Margaret Pelling, 'Appearance and Reality: Barber-Surgeons, the Body and Disease', in *London 1500 –1700: the Making of the Metropolis,* ed. A. L. Beier and Roger Finlay (London: Longman, 1986), 82–122.

4 David M.Turner, 'Introduction: Approaching Anomalous Bodies', in *Social Histories of Disability and Deformity,* ed. David M. Turner and Kevin Stagg (London: Routledge, 2006), 4.

5 Naomi Baker, *Plain Ugly: the Unattractive Body in Early Modern England* (Manchester: Manchester University Press, 2010).

6 Anna Riehl, *The Face of Queenship: Early Modern Representations of Elizabeth I* (Basingstoke: Palgrave Macmillan, 2010), 53; Marcus Nevitt, 'The Insults of Defeat: Royalist Responses to Sir William Davenant's *Gondibert* (1651)', *The Seventeenth Century* 24.2 (2009): 287–304. On nose loss, see also Valentin Groebner, 'Losing Face, Saving Face: Noses and Honour in the Late Medieval Town', trans. Pamela Selwyn, *History Workshop Journal* 40 (1995): 1–15; Valentin Groebner, *Defaced: The Visual Culture of Violence in the Late Middle Ages,* trans. Pamela Selwyn (New York: MIT Press, 2004), 67–80.

7 Sander L. Gilman, *Disease and Representation: Images of Illness from Madness to AIDS* (Ithaca: Cornell University Press, 1988); Sander L. Gilman, *Health and Illness: Images of Difference* (London: Reaktion, 1995); Sander L. Gilman, *Making the Body Beautiful: A Cultural History of Aesthetic Surgery* (Princeton: Princeton University Press, 1999); Raymond Anselment, *The Realms of Apollo: Literature and Healing in Seventeenth-Century England,* (Newark, DE: University of Delaware Press, 1995); David Shuttleton, *Smallpox and the Literary Imagination* (Cambridge: Cambridge University Press, 2007).

8 Shuttleton, *Smallpox,* 115–36.

9 Emily Cock, '"Lead[ing] 'em by the Nose into Publick Shame and Derision": Gaspare Tagliacozzi, Alexander Read and the Lost History of Plastic Surgery, 1600–1800', *Social History of Medicine* 28.1 (2015): 1–21.

10 *The Diary of Samuel Pepys,* ed. Henry B. Wheatley, 8 vols. (London: George Bell and Sons, 1893, repr. 1946), v, 328.

11 *Diary of Samuel Pepys,* iii, 224–5.

12 Ibid., v, 230, and vii, 168.

13 See, for example, Adam Littleton, *A Sermon at the Funeral of the Right Honourable the Lady Jane Eldest Daughter to His Grace, William, Duke of Newcastle, and Wife to the Honourable Charles Cheyne, Esq* (London: John Macock, 1669), 11–12; Mary Rogers, 'Beauty and the Concepts of the Ideal', in *A Cultural History of the Human Body in the Renaissance,* ed. Linda Kalof and William Bynum (Oxford: Bloomsbury Academic, 2010), 127; David M. Turner, 'The Body Beautiful', in *A Cultural History of the Human Body in the Age of Enlightenment,* ed. Carole Reeves (Oxford: Bloomsbury Academic, 2010), 114; Farah Karim-Cooper, *Cosmetics in Shakespearean and Renaissance Drama* (Edinburgh: Edinburgh University Press, 2006), 7.

14 Gian Paolo Lomazzo, *A Tracte Containing the Artes of Curious Paintinge Caruinge Buildinge* (Oxford, 1598), 133.

15 Littleton, *Sermon,* 12; Emily Smith, 'The local popularity of *The Concealed Fansyes*', *Notes and Queries* 53.2 (2006): 190; Hannah Smith, 'Mary Astell, *A Serious Proposal to the Ladies* (1694), and the Anglican Reformation of Manners in Late-Seventeenth-Century England', in *Mary Astell: Reason, Gender, Faith,* ed. William Kolbrener and Michal Michelson (Aldershot: Routledge, 2007), 44.

16 N. H., *The Ladies Dictionary* (London: John Dunton 1694), 55.

17 A. Korhonen, 'To See and to be Seen: Beauty in the Early Modern London Street', in *Cultural History of Early Modern European Streets,* ed. Riitta Laitinen and Thomas V. Cohen (Leiden: Brill, 2009), 143.

18 Baker, *Plain Ugly,* 71.

19 Anselment, *Realms of Apollo,* 190; Rosemarie Garland-Thomson, *Staring: How We Look* (New York and Oxford: Oxford University Press, 2009), 37.

20 Joy Wittenburg, *Disorderly Women and Female Power in the Street Literature of Early Modern England and Germany* (Charlottesville: University of Virginia Press, 1992), 141–2.

21 Robert Parry, *Moderatus, the Most Delectable & Famous Historie of the Blacke Knight* (London: Richard Jhones, 1595), Sig. L1^{r}; Joseph Swetnam, *The Araignment of Lewd, Idle, Froward, and Vnconstant Women or the Vanitie of them, Choose you Whether: With a Commendation of Wise, Vertuous and Honest Women* (London: George Purslowe for Thomas Archer, 1615), 4; Katherine Usher Henderson and Barbara F. McManus, *Half Humankind: Contexts and Texts of the Controversy about Women in England 1540-1640* (Urbana: University of Illinois Press, 1985), 16; Diane Purkiss, 'Material Girls: The Seventeenth-Century Woman Debate', in *Women, Texts and Histories 1575 –1760*, ed. Clare Brant and Diane Purkiss (London: Routledge, 1992), 72.

22 Garland-Thomson, *Staring*, 50.

23 Anon., *A True Narrative of the Proceedings at the Sessions-house in the Old-Bayly, at a Sessions there held on April 25, and 26. 1677* (London, 1677), 7.

24 *The Autobiography of Anne Lady Halkett*, ed. John Gough Nichols (London: Camden Society, 1875), 67.

25 Anon., *A True and Sad Reration [sic] of the Great and Bloudy Murder Committed at Ratcliff in Stepney Parish neer the City of London, upon the body of John Hunter, a Sea Man, who was Stabbed to the Heart with a Long Knife, by One Mr. Smith and his Wife and a Young Maid* (London: B.A., 1647), sig A4r ; Anon., *Nevvs from Hide-Parke: Or A Very Merry Strange Passage which Happened Betwixt a North-Country Gentleman, and a Very Gaudy Gallant Lady of Pleasure* (London, 1640–3). On the appearance of witches, deployed for different purposes, see also Morna Laing's chapter in this book.

26 Shuttleton, *Smallpox*, 117.

27 *Autobiography of Lady Anne Halkett*, 67; Anon., *A True Narrative of the Proceedings*, 6.

28 For details of a similar phenomenon identified in a medieval literary source, see Julie Orlemanski, 'Desire and Defacement in the Testament of Cresseid', in *Reading Skin in Medieval Literature and Culture*, ed. Katie L. Walter (Basingstoke: Palgrave Macmillan, 2013), 170–1.

29 *Lady Anne Clifford, Countess of Dorset, Pembroke and Montgomery, 1590–1676. Her Life, Letters and Work*, ed. George C. Williamson (Kendal: Titus Wilson and Son, 1922), 130.

30 Fulke Greville, *The Life of the Renowned Sr Philip Sidney with the True Interest of England as it then Stood in Relation to all Forrain Princes* (London: Henry Seile, 1651), 6.

31 Kay Dennis, '"She was a Queen and therefore Beautiful": Sidney, his Mother and Queen Elizabeth', *The Review of English Studies* 169 (1992), 19.

32 The National Archives: PRO, SP 12/159, fol. 38^{r-v}.

33 Richard Wiseman, *Severall Chirurgicall Treatises* (London: E. Flesher and J. Macock, for R. Royston, 1676), 118.

34 Jo Eldridge Carney, *Fairy Tale Queens: Representations of Early Modern Queenship* (Basingstoke: Palgrave Macmillan, 2012), 102–4 and 110.

35 Thomas Bayly, *Witty Apophthegms Delivered at Several Times, and upon Several Occasions by King James, King Charls [sic] the Marquess of Worcester, Francis Lord Bacon, and Sir Thomas Moor* (London: W.R. for Matthew Smelt, 1669), 79.

36 Anon., *Charecters [sic] of Some Young Women, Belonging to the Changes, Who Are to be Dispos'd of by Way of Auction* (London: 1691).

37 Williamson, *Lady Anne Clifford*, 130; *Diary of Samuel Pepys*, vii, 228.

38 John Bulwer, *Anthropometamorphosis* (London, 1653), 265.

39 Robert Green, *A Disputation, betweene a Hee Conny-catcher, and a Shee Conny-catcher whether a Theefe or a Whoore, is most Hurtfull in Cousonage, to the Common-wealth* (London: A. Jeffes for T. Gubbin, 1592), sig. D2^{v}.

40 Thomas Jeamson, *Artificiall Embellishments, or Arts Best Directions How to Preserve Beauty or Procure It* (Oxford: William Hall, 1665), 54.
41 Williamson, *Lady Anne Clifford*, 156.
42 *Diary of Samuel Pepys*, vii, 353–4.
43 Ibid., vii, 357.
44 Ibid., viii, 88–9.
45 Ibid., viii, 94.
46 Anon., *Nevvs from Hide-Parke;* Christopher Durston, 'Hewson, John, appointed Lord Hewson under the protectorate (fl. 1630–1660)', in *Oxford Dictionary of National Biography*, see online ed. David Cannadine (Oxford: Oxford University Press, 2004), http://www.oxforddnb.com/view/article/13157 (accessed 12 February 2017).
47 Henderson and McManus, *Half Humankind*, 7. And see Theresa Tyers' chapter in this book for a medieval precursor of such commentary.
48 Frances E. Dolan, 'Taking the Pencil out of God's Hand: Art, Nature, and the Face-Painting Debate in Early Modern England', *PMLA* 108.2 (1993): 224–39; Farah Karim-Cooper, 'Fashioning the Face: Embodiment and Desire in Early Modern Poetry', in *Shakespeare and the Power of the Face*, ed. James A. Knapp (Farnham: Routledge, 2015), 29–42.
49 Francisco de Mello, *The Government of a Wife, or, Wholsom and Pleasant Advice for Married Men in a Letter to a Friend* (London: Jacob Tonson and R. Knaplock, 1697), 174–5; Thomas Shadwell, *Bury-fair a Comedy* (London: James Knapton, 1689), Act III, Scene i.
50 Bulwer, *Anthropometamorphosis*, 264.
51 Dolan, 'Taking the Pencil', 232.
52 Bulwer, *Anthropometamorphosis*, 270.
53 Thomas Tuke, *A Discourse against Painting and Tincturing of Women Wherein the Abominable Sinnes of Murther and Poysoning, Pride and Ambition, Adultery and Witchcraft are set Foorth & Discouered* (London: Thomas Creede and Bernard Alsop for Edward Marchant, 1616), 42.
54 Irina Metzler, *A Social History of Disability in the Middle Ages: Cultural Considerations of Physical Impairment* (New York: Routledge, 2013), 191.
55 Shuttleton, *Smallpox*, 124; Baker, *Plain Ugly*, 107.
56 Robert Gould, *A Poem Most Humbly Offered to the Memory of Her Late Sacred Majesty, Queen Mary* (London: Jacob Tonson and Francis Saunders, 1695), 13; Patrizia Bettella, *The Ugly Woman: Transgressive Aesthetic Models in Italian Poetry from the Middle Ages to the Baroque* (Toronto: University of Toronto Press, 2005), 14. The association had medieval precedents in poetry and narrative literature: see the hags described in Bonnie Millar's chapter in this book. An alternative viewpoint, put forward by female writers, was that an old woman acquired wisdom: Devoney Looser, *Women Writers and Old Age in Great Britain, 1750–1850* (Baltimore: Johns Hopkins University Press, 2008).
57 Baker, *Plain Ugly*, 69.
58 Patrizia Bettella, 'The Marked Body as Otherness in Renaissance Italian Culture', in *A Cultural History of the Human Body in the Renaissance*, ed. Linda Kalof and William Bynum (Oxford: Bloomsbury Academic, 2010), 152.
59 Regarding the need for scholars to pay greater attention to age as a category, see Sarah Toulalan and Kate Fisher, 'Introduction', in *The Routledge History of Sex and the Body 1500 to the Present*, ed. Sarah Toulalan and Kate Fisher (London: Routledge, 2013), 12.
60 The National Archives, London: PRO, SP 12/159, fol. 38^{r-v}.

61 M. Marsin, *The Womens Advocate, or, Fifteen Real Comforts of Matrimony being in Requital of the Late Fifteen Sham-Comforts: With Satyrical Reflections on Whoring, and the Debauchery of this Age* (London: Benjamin Alsop and Thomas Malthus, 1683), 22.
62 Riehl, *Face of Queenship*, 54.
63 The National Archives, London: PRO, SP 12/159, fol. 38[r–v].
64 Kevin P. Siena, '"The Foul Disease" and Privacy: The Effects of Venereal Disease and Patient Demand on the Medical Marketplace in Early Modern London', *Bulletin of the History of Medicine* 75.2 (2001): 200 and 206; Anselment, *Realms of Apollo*, 138.
65 Shuttleton, *Smallpox*, 9.
66 Baker, *Plain Ugly*, 105.
67 Lomazzo, *Tracte containing the Artes*, 130.
68 Garthine Walker, *Crime, Gender and Social Order in Early Modern England* (Cambridge: Cambridge University Press, 2009), 92–3.
69 See, for example, Thomas D'Urfey, *New Poems, Consisting of Satyrs, Elegies, and Odes Together with a Choice Collection of the Newest Court Songs set to Musick* (London: J. Bullord and A. Roper, 1690), 121.
70 Anon., *A true and sad reration*, sig A4r.
71 Patricia Skinner, 'The Gendered Nose and its Lack: "Medieval" Nose-Cutting and its Modern Manifestations', *Journal of Women's History* 26.1 (2014): 45–67; Valeria Finucci, *The Prince's Body: Vincenzo Gonzaga and Renaissance Medicine* (Cambridge, MA: Harvard University Press, 2015), 86–7; Groebner, 'Losing Face'; Groebner, *Defaced*, 67–80; Gilman, *Health and Illness*, 71–81.
72 Anon., *Five Merry Wives of Lambeth or, the Carpenter Cornuted* (London: R. Burton, 1680); Laura Gowing, 'Gender and the Language of Insult in Early Modern London', *History Workshop* 35 (1993), 10; Laura Gowing, 'Language, Power, and the Law: Women's Slander Litigation in Early Modern London', in *Women, Crime and the Courts in Early Modern England*, ed. Jenny Kermode and Garthine Walker (London: Routledge, 1994), 32.
73 *Diary of Samuel Pepys*, viii, 148; James Grantham Turner, *Libertines and Radicals in Early Modern London: Sexuality, Politics, and Literary Culture, 1630–1685* (Cambridge: Cambridge University Press, 2002), 216.
74 Ordinary's Account, 6 September, 1700, in *The Proceedings of the Old Bailey*, https://www.oldbaileyonline.org/browse.jsp?id=OA17000906n14-2&div=OA17000906#highlight (Last accessed 9 February 2017).
75 Laura Gowing, 'Marked Bodies and Social Meanings', in *A Cultural History of the Human Body in the Age of Enlightenment*, ed. Carole Reeves (Oxford: Bloomsbury Academic, 2010), 142.
76 Regarding the female body as a register of shame, see Laura Gowing, 'Women, Status, and the Popular Culture of Dishonour', *Transactions of the Royal Historical Society* 6th series 6 (1996), 232.
77 Jeamson, *Artificiall Embellishments*, sig. A3[r].
78 Dolan, 'Taking the Pencil', 235.
79 Williamson, *Lady Anne Clifford*, 156.
80 Quoted in Linda Pollock, *With Faith and Physic: The Life of a Tudor Gentlewoman, Lady Grace Mildmay 1552–1620* (London: Collins and Brown, 1993), 125.
81 Anne Fanshawe, *The Memoirs of Ann, Lady Fanshawe, Wife of the Right Honble. Sir Richard Fanshawe, Bart., 1600–72*, ed. Herbert Charles Fanshawe (London: John Lane The Bodley Head, 1907), 158.
82 Fanshawe, *Memoirs*, 584.

83 Susan E. Whyman, *Sociability and Power in Late Stuart England: The Cultural Worlds of the Verneys 1660–1720* (Oxford: Oxford University Press, 1999), 91.
84 Anselment, *Realms of Apollo*, 189.
85 Alice Thornton, *The Diary of Mrs Alice Thornton*, ed. Charles Jackson (Durham: Andrews and Co., 1875), 157.
86 Thornton, *The Diary of Mrs Alice Thornton*.
87 Ibid.
88 David Shuttleton, 'A Culture of Disfigurement: Imagining Smallpox in the Long Eighteenth Century', in *Framing and Imagining Disease in Cultural History*, ed. George Sebastian Rousseau, Miranda Gill, David Haycock and Malte Herwig (Basingstoke: Palgrave Macmillan, 2003), 69.
89 Williamson, *Lady Anne Clifford*, 130.
90 Swetnam, *Araignment* , 9.
91 Laura J. McGough, 'Quarantining Beauty: The French Disease in Early Modern Venice', in *Sins of the Flesh: Responding to Sexual Disease in Early Modern Europe*, ed. Kevin Siena (Toronto: Centre for Reformation and Renaissance Studies, 2005), 222.
92 Phyllis Rackin, 'Foreign Country: The Place of Women and Sexuality in Shakespeare's Historical World', in *Enclosure Acts: Sexuality, Property, and Culture in Early Modern England*, ed. Richard Burt and John Michael Archer (Ithaca: Cornell University Press, 1994), 76.
93 Robert B. Shoemaker, *Gender in English Society 1650–1850: The Emergence of Separate Spheres?* (Harlow: Routledge, 1998), 21; Henderson and McManus, *Half Humankind*, 56; Turner, 'Body Beautiful', 117; Ruth Kelso, *Doctrine for the Lady of the Renaissance* (Urbana: Illinois University Press, 1956), 25.
94 Shuttleton, *Smallpox*, 126.
95 De Mello, *Government of a Wife*, 23
96 John Corye, *The Generous Enemies, or, The Ridiculous Lovers a Comedy* (London: H. Lloyd for James Magnus, 1572), Act IV, Scene i.
97 Regarding the connection made between female honour and passivity, see Johanna Rickman, *Love, Lust, and License in Early Modern England* (Aldershot: Routledge, 2008), 175.
98 Peter Stallybrass, 'Patriarchal Territories: The Body Enclosed', in *Rewriting the Renaissance: The Discourses of Sexual Difference in Early Modern Europe*, ed. Margaret W. Ferguson, Maureen Quilligan and Nancy J. Vickers (Chicago: University of Chicago Press, 1986), 129.
99 Gowing, 'Women, Status, and the Popular Culture of Dishonour', 225–6.
100 The National Archives, London: PRO, SP 12/159, fol. 38[r–v]; Greville, *Life of the Renowned Sr Philip Sidney*, 5–6; Levin and Medici, 'Lady Mary Dudley Sidney', 36.

4

Does Researching Disfigurement Risk Perpetuating Stigma?

Jane Frances

It seems unnecessary to say that disfigurement affects how people look. *Obviously*, disfigurement describes the appearance, the face, for instance, of a person who has a condition, injury, illness or mark that changes their appearance, or who has undergone disfiguring medical treatment or punishment. But disfigurement also influences how people *look at* a person whose appearance is unusual. Out and about in public places, most of us routinely benefit from the urban social protocol that involves an anonymizing-not-noticing–civil inattention.[1] Accessed through having, or being able to create for oneself, an 'unremarkable'[2] outward appearance, civil inattention is interrupted by disfigurement: stares push the person being stared at away from the human community.[3] Along the high street, on the bus, in the café, when someone's face is disfigured, people *look*, and that look, seemingly automatic, unwitting, possibly fleeting and hastily modified, manifests as an objectifying stare.[4] Goffman finds that stigma is a process by which the reactions of others to a person's discredited attribute, such as facial disfigurement, spoil that person's identity.[5] Talley's exploration of the culture that generates our more or less medicalized modern Western approaches to repairing or cosmetically altering facial appearance shows how the 'disfigurement imaginary inspires "ways of seeing" that position the patient (whose face is deemed unattractive or disfigured) as already socially dead or with social death imminent'.[6] It is the way we look – our stare – however unwitting and momentary, before we gather ourselves and withdraw that look, which enacts the stigma that attaches to disfigurement.

This chapter concerns problems that arise when other people's reactions to someone's unusual facial appearance are unspecified or taken as given in disfigurement research. I argue that failure to be explicit about other people's reactions to a person's disfigurement perpetuates the stigma associated with facial disfigurement. First, I will set the scene with a consideration of staring from both sides, so to speak. This helps to illuminate examples of research writing about disfigurement in which other people's reactions to disfigurement are not made explicit, and which thereby become a tacit component of a stigmatizing compact between researcher/writer and reader. I will then draw upon other fields featuring stigma where contrasted perspectives and approaches may nourish discussion of the task of bringing clarity of purpose to research that is closely bound up with stigma.

Languaging 'the look'

Autobiographical writings of individuals who have a disfigured appearance always describe 'that look'. Fin Carling, whose gait was seriously affected by cerebral palsy, writes, 'I do not think it was fear so often as just uneasiness, or disgust, or insecurity. Anyway, I always had to be prepared for this look; prepared to meet the bewildered eyes of the shopgirl or the waiter or just the man in the street.'[7] Lucy Grealy, whose bone cancer in childhood was treated by removing part of her jaw, describes how 'the eyes of … perfectly formed children swiftly and deftly bored into the deepest part of me … (whereas their parents) … inexpertly pretend not to notice me … (their looks) … refracting away as quickly and predictably as light through a prism'.[8]

Such autobiographies not only reveal how little changes in reactions to visible difference, but also reflect upon the strategies they have had to develop to manage, somehow, this defining social response in order to achieve something resembling an ordinary day-to-day life. In his account of his own journey through serious burn injuries and their prolonged and complex medical treatment, James Partridge's main focus is the strange and difficult process of learning to live with his own radically changed and changing face. He notes that patients being treated in hospital for disfiguring injuries often experience the hospital as a safer, more comfortable environment than the outside world, due to the huge increase in staring reactions from others outside hospital.[9] He goes on to share a recollection of his brother's: 'I remember going with you into pubs and being aware of conversations stopping and eyes drilling into our backs. In one you completely disarmed a group of starers with the throwaway comment, "not looking my best today I'm afraid"'.[10]

In an unusual autobiographical reflection within a clinical research paper, Anne Marie Keary describes how, when she is out with her physically disabled daughter who uses a wheelchair, she and her daughter often find themselves on the receiving end of other people's 'furtive, confused looks'.[11] Keary is equally candid regarding her own response to disfigurement as an ordinary person who owns that she sometimes catches herself being the source of these looks: 'We don't know how to handle our desire to stare, or our fear of catching it by association, so we avert our eyes.'[12]

The language in all of these accounts lights up the incidence of staring as the common denominator of encounters with the social world for people whose appearance is disfigured. Rosemarie Garland-Thomson coins the term 'stareable sight' for those among us whose unusual face or body is apt to draw stares. Stareable sights, she finds, seduce the starer into 'an earlier developmental stage … befuddled, halted in mid-glance, mobility throttled, processing checked'.[13] The starer's focus is taken over by the stareable sight even if the starer tries to pay no attention: a stare can only be withdrawn once it has been established.[14] This is distressing for the staree whose heart rate increases and Electroencephalogram (EEG) registers distressed cortical activity.[15]

With her extensive experience of working with children and adults with a range of disabilities, Valerie Sinason finds in other people's reactions to visible disability a 'curiosity [that] is excited and voyeuristic, like the curiosity that can accompany a disaster. The sight of damage, of something gone wrong, induces an excited disturbance in such onlookers. Sometimes, there is a turning away, a fear and hostility, a sometimes spoken wish that such sights should be hidden from public view.'[16] We can begin to get a sense here of the proximity of social death which stalks anyone whose facial appearance is striking in such a way that civil inattention which, curiously, affirms mutual understanding of belonging is replaced by the look which others direct towards them.

Garland-Thomson's research into staring at unusual bodies – stareable sights – finds that staring, which can be described as uncivil attention, takes many forms, including the stare that manages to convert itself into information-gathering attention, and the blank stare where there is no processing towards finding meaning in what is being looked at.[17] The person being stared at may therefore 'read' the stares coming at them and may possibly make choices as to how they will respond.[18] In all my work since 1998 with people with disfigurements, I have not encountered anyone with an unusual facial appearance who found any kind of stare in any way OK. On the contrary, having an unusual appearance creates the task of learning to 'manage' unwanted attention, for example, during one's daily commute. This is achieved using a range of strategies from making eye contact and offering a disarming smile to pulling a face, to handing the starer an information card, to having a newspaper or book to read.

> In the past I have responded quite unfairly to staring people. One tiny child once asked me where my leg was, and I (sort of fed up that day) looked down and FREAKED OUT! 'Oh my God!' I exclaimed, 'I had it this morning!' Now I simply tell them I was sick and the doctor took it off and wait to see if they have any more questions. (Anonymous, 2006 interview)

This staree is quick to acknowledge her leadership in directing the staring encounter and the generous seriousness she brings to that responsibility. 'I know', she affirms, 'I haven't seen the end of that little devil inside me who now and then likes to play with people's fear and disgust factor.'[19]

Staring from children is often dealt with differently from adult staring by people for whom civil inattention is routinely replaced by uncivil attention when out and about. We see this, for instance, towards the end of the film *La Chambre des Officiers*. This dramatization of Marc Dugain's (French) novel of the same name follows Adrien, who has spent most of the war, and much of the film, in hospital undergoing experimental treatments for severe facial wounds. When he finally returns to civilian life and boards the Metro, he uses his hat to engage a staring child in a game of peekaboo.

The look, whatever form it takes, is a central experience for people who live with a facial disfigurement, and often replaces verbalization of the impact of their appearance on others. For them, finding ways to deal with uncivil attention is key to managing the task of going out and about in public. The social, cultural or psychological meaning or function of the look is a significant topic for research, but for another time. Here my purpose is to clarify the centrality of the look in the experience of anyone whose face is affected by a disfiguring condition, injury, mark or illness. This is the context in which we must consider any research into any aspect of disfigurement which misconstrues or underplays or simply does not describe or consider the look, or the loss of civil inattention that makes life so hard for people with facial disfigurement.

Obscuring 'the look' in psychoanalytic clinical research

As we have seen, people who have first-hand experience of facial disfigurement are unequivocal regarding the impact of the look in their lives. By contrast, my own psychoanalytic research into other people's reactions to disfigurement uncovered innumerable psychoanalytic clinical and research papers where clinicians and researchers whose work with people living with disfigurement never mentions staring. The focus is wholly upon the person whose appearance is disfigured, which has the effect of denying the reality or minimizing the negative impact upon the patient of staring and other forms of uncivil attention. The ways in which these psychoanalysts avoid their patients' or research subjects' experience of being stared at varies. But in all cases it has the effect of obscuring the nature or even the existence of social responses that enact the stigma associated with having a disfigured appearance. In the absence of a shared language for the look, patients who report being stared at, or who mask a facial birthmark whenever they go out, are understood by their analyst or researcher as having a problem with reality or with their own ego, or with both.

For example, a 'long study and observation' of patients (including young war veterans) attending a rehabilitation programme at a day centre in New York City discusses the impact of head injuries and disabilities on the ego:

> Any visible physical handicap or disfigurement adds to the feeling of social self-consciousness. Whenever the actual body-image diverges from the normal or ideal body-image, an additional source for anxiety is created and, to cope with it, the Ego has to use up a part of its resources.[20]

Some patients in this study have hemiplegia and, for others, their brain injuries have disfiguring sequelae such as facial paralysis. These have 'naturally a highly disturbing effect since they distort the social image of the patient'.[21] One patient 'feels that people whisper about and laugh at him'.[22] His frustrated, isolated reactions are likened to those of 'an individual surrounded by people speaking a foreign language whose mastery he has never fully achieved'.[23]

These patients' attempts to live with and articulate their unease or discomfort at other people's reactions to their changed appearance are understood in terms of disturbances affecting their own ego, their self-image and their failure to master a common language. Elsewhere in the psychoanalytic literature concerning disfigurement, other people's reactions such as staring and, at school, arguing and fighting, are attributed not to the patient's unusual appearance but to their unruliness or other faults of behaviour or character.[24] Patients' disfigurements are described as 'realistically trivial'[25] or they are troubled by 'minor physical anomalies'[26] for which 'the psychological reverberations go far beyond the actual physical disability'.[27]

In other psychoanalytic research papers, people's reactions to a patient's disfigured appearance are implied but not specified.[28] For example, Als and Brazelton's research into parents' responses and adaptations during the first year following the birth of babies with craniofacial conditions (which can also affect the hands and sometimes the heart as well) focuses on

- parents' mourning and adjustment, a process entered into and largely accomplished by all the parents apparently without any therapeutic support;
- parents' access to self-esteem, for example, through successful breast feeding or being a capable breadwinner;
- the importance for the infant's future development of the mother's success in separating herself and her infant (as compared with experiencing the infant as a projection of the bad parts of herself) in order to grieve and to develop a healthy mother–baby relationship;
- parents' magical thinking;
- parents' increasing access to reality.

Als and Brazelton's rendering of these topics for exploration reflects a clear but tacit understanding that the birth of a child with a craniofacial condition is to be not celebrated but mourned, even though baby Dan is a 'very calm and well-organized infant with an intense gaze, engaging smile, and vigorous suck'[29] and baby Al is 'a cuddly baby with large, compelling eyes'.[30] It is stated unequivocally that baby Al's 'facial disfigurement was obvious and painful to his parents' and it is tacitly understood that all the parents will experience the birth of their respective babies in this way, and that you, the reader, will see congenital craniofacial conditions in this way also.[31] This tacit collusion between writer and reader is further explored, below.

Als and Brazelton comply with the protocol prevalent in psychodynamic and psychological scholarship, that prohibits personal, subjective observations. (This seemingly self-effacing protocol is contrasted, below, with a very different anthropological approach.) Although the research methodology includes interviews and observation, there appears to be no interview question concerning other people's reactions, in public settings, upon seeing an infant whose appearance is strikingly unusual. Other people's reactions therefore emerge only obliquely:

> [Mr A] told his family and friends about Al, giving them the option to visit if they were comfortable.[32]
>
> For the lab visits, [Mrs A] dressed Al in pajamas rather than in street outfits. She avoided taking him out; when she had to, she would wrap his face or shield him from view to conceal his defect.[33]
>
> Once Al's [whole mouth] was repaired at 5½ months, both Mr. and Mrs. A reported experiencing more pleasure in being his parents. Although his mouth still looked disfigured to a stranger, his parents were desensitized. They had accomplished one of their goals: Al was 'all fixed up' by Christmas. Mrs. A began

to take him out, no longer covering his face to keep it from view, and dressed him in street clothes rather than pajamas.[34]

At times, all the parents looked to the future, hoping to experience a sense of relief. Often, however, this action stirred up new worries about how life would progress. How would their children feel about themselves as they grew up? *How would other people react to them?*[35] (My emphasis. This question is never explored in any way in this chapter.)

Here, as in so much psychoanalytic clinical and research material, it is taken for granted that the reader – a fellow clinician, scholar or researcher – will share the writer's tacit position and language regarding another's unusual appearance.

Anyone sharing this tacit view of disfigurement, using language that keeps the focus in the infant's unusual appearance and never mentions staring or 'the look', will proceed through the chapter as if everything is perfectly natural and reasonable. But teasing out our tacit assumptions is of the utmost importance because one of the ways that stigma works is through a universal tacit agreement not to think about it, let alone talk about it, but only to enact it through *normal* attitudes and behaviours. In this, as in so much psychoanalytic clinical and research writing and in most other contexts too, what is taken for granted concerning infants and children with disfigurement is that disfigurement is a bad thing:

- The birth of such a baby must be mourned.
- The infant's disfigurement/disability puts the mother at risk of imagining that some bad thing about herself has spoiled her baby.
- This child's life will surely be hard, especially at school because children can be so cruel.
- Medical science must do all that can be done to remove the appearance of the disfigurement/disability.

All these points may seem natural and true. If evidence is sought to back them up, it will surely be found and could be stated. Their unspecified, and therefore taken-for-granted position with regard to the stigmatizing effect of facial disfigurement, and their writers' unspoken assumption that the reader shares this position (indeed, no other position is conceivable) accords with a prevailing '"disfigurement imaginary" or a shared way of thinking about facial variance and interventions' that links facial disfigurement to social death but never states this using words.[36]

How might Als and Brazelton's rendering of the parents' concerns look if we re-language them in terms of stigma?

- When non-stigmatized parents have a baby who is born directly into a stigmatized category it is a huge personal, social and cultural shock. How is this profound change in circumstance to be understood and managed?
- What is a mother to think, to do, when her baby's birth leads to her being seen as belonging less to the category of ordinary people and more to the stigmatized category of 'the disfigured'?

- A child who looks like that cannot but be the butt of jokes and nicknames, bullied and ostracized. This is 'normal' behaviour.
- To the extent that medical science can re-form appearance, a child like this can be retrieved from the category of the stigmatized and can have a place in the vastly preferable category of 'normal'.

In this way we can begin to get a sense of the operation of language behind this tacit positioning that dogs disfigurement research.

Interestingly, another important research paper concerning the experiences of children undergoing not facial surgery but limb amputation ('mutilating surgery') for bone cancer, references research by Gilder et al. which describes

> the meaning of amputation to the patient in terms of threat to total body integration, phantom sensations, and the body image; they also discuss the effects of emotions on phantom behavior and the *emotional reactions of others on viewing amputees.*[37]

The term *viewing* does not signal staring, or even looking in the way it is experienced by people subject to the look. But it surely tilts in this direction. However, the psychoanalytic researcher herself does not pause where I have paused after 'the emotional reactions of others on viewing amputees', but goes straight on with her review of the Gilder's research:

> In two detailed case presentations, Gilder et al. demonstrate how important it is for patients to know whether the amputated limb has been cremated or buried. They also show that psychotherapy can help patients to get over the effects of painful phantom limbs.

Psychoanalytic research into disfigurement of face or body seems relentlessly drawn towards the experiences of the person whose appearance is affected (and, for newborns, their parents). The relevance of the reactions of others is somehow hidden in plain sight.[38]

I have traced Gilder et al.'s research to a US Navy research project.[39] An original copy is retained in a US Navy medical archive, but I have not been able find it published or archived elsewhere. However, an American training manual for staff working in rehabilitation gives a substantial summary of the 1954 research. This experimental research into other people's reactions to limblessness

> utilized full-sized mannequins as well as adult amputees as stimuli. One of their findings was that the perception of the amputee, and even the mutilated mannequin, was apt to provoke strong emotions of anxiety and even of anger. They interpret these reactions as a function of the threat that actual (or even simulated) amputation has on the integrity of body image of the normal person who identifies with the person viewed. The type of emotional response they encountered is seemingly dependent on the meaning the body image has for the

> subject, and on his particular way of dealing with such stress. If persons unrelated to amputees experience such threat and manifest such emotional reactions, then it would be much more likely that persons in direct contact with an amputee will experience threat and manifest their defensiveness by anger and rejection. It is this type of phenomenon that plays a major role in determining the social position and self-concept of the handicapped individual.[40]

Writing in 1979, Earle appears to have read Gilder's research without showing any interest, herself, in the impact upon the children she worked with of other people's reactions to their 'mutilated' appearance. This apparent lack of interest, I suggest, perpetuates the invisible, unheard misery of people whose disfigured appearance is made much harder to live with by the daily reality of other people's stares, disgust[41] and turnings away.

In a more recent (2002) clinical paper[42] concerning a patient with a congenital malformation, who is a singer, the clinician offers a substantial but partial account of Thomas Mann's *Confessions of Felix Krull, Confidence Man* (1911) to support his clinical assessment of his patient's mental pain. In the story, Krull, as a boy, had enjoyed a fabulous and moving performance by the great singer Müller-Rose. After the show he went back stage with his father, who knew Müller-Rose, to meet him. It was a terrible shock as the great singer, wig-less, grease-paint partly removed, was being massaged and wiped down after his huge exertion. Shirtless, his upper body was thickly covered in weeping red-rimmed pustules. 'Even today I cannot repress a shudder at the thought of them … . This grease-smeared and pimply individual is the charmer at whom the twilight crowd was just now gazing so soulfully!'[43] Without referencing Goffman, and without writing about stigma or spoiled identity as an aspect of his patient's social experience, the clinician's partial rendering of Mann's story nevertheless unwittingly conveys precisely the task faced by anyone with a disfigurement who must attempt to 'pass' if they can or experience the stigma with which society rewards their 'discredited attribute'. Like Müller-Rose, the patient whose treatment is the main subject of this chapter is able to 'pass'. Has she learnt that when someone finds out that you have been 'pretending' to be ok, they are even more disgusted with you than if your variation had been apparent from the outset? We cannot know because in 2002, as in psychoanalytic writing in previous decades, this is never discussed. This has the effect of absolving the disgusted onlooker: the source and resolution of the patient's psychic pain is located entirely within the patient's own psychic material and processes (which may include residues from early childhood experiences of parenting).

Finally, more recently still, here is an example of disfigurement research concerning the acceptability, in terms of medical ethics, of the life-limiting sequelae of immunosuppression necessitated by face transplantation (FT). Described in detail by Talley, this 2004 research is referred to as the 'RISK survey' by the FT development team exploring attitudes to transplantation of some three hundred people with 'real-life experiences' such as people who have lost a hand or who live with a transplanted kidney.[44] Talley's extensive research into the sociocultural meanings of FT includes substantial interviews with members of the FT research team to explore the meanings that the RISK survey findings have for them. She found that the RISK survey yielded the finding that FT is a valid intervention even with the risks and reduced life expectancy associated with immunosuppression. It appears that, unlike other organ transplants, which Talley describes as life-*saving*, FT addresses *quality* of life. Talley argues that finding this more limited gain to be worth the risk 'positions disfigurement as a uniquely threatening bodily impairment'.[45]

Talley notes that this research

> tacitly … relied on and actively constructed meanings attributed to facial disfigurement. While constructing 'disfigurement' was not an explicit objective of the team, their research implicitly relied on notions of how facial disfigurement devastates human life. This case vividly demonstrates the subtle yet powerful processes through which the disfigurement imaginary is socially constituted and facial difference emerges as a deadly condition.

From my own limited work linked to renal failure and renal transplant at Addenbrookes Hospital in Cambridgeshire, it seems likely that kidney transplantation does in practice address quality of life for many patients whose lives would otherwise be sustained by thrice-weekly kidney dialysis. This onerous ongoing treatment does not call for immunosuppression, and so does not carry with it the increased risk of serious illnesses such as cancers which immunosuppression always entails. I conclude that Talley makes the contrast between kidney transplant and face transplant more clear-cut than it is in practice (perhaps following the FT research team's thinking). Also, it is not clear whether the RISK survey includes patients whose renal failure is being treated with dialysis. However, the FT team's apparent underplaying of these points may in fact increase the strength of Talley's argument concerning the seemingly unequivocal RISK research findings that 'everyone [if facially disfigured] would risk absolutely the most to get a face transplant than any other procedure'.[46] Thus the (seemingly democratically derived) ethical case for FT positions facial disfigurement as a uniquely terrible bodily impairment which thereby justifies a uniquely novel and dangerous medical procedure. The team never interrogate the link they make between facial disfigurement and social death.

At the very least, then, Talley's exploration further highlights a crucial task for researchers looking at any aspect of facial appearance and facial difference: at every stage, from outline proposal, through methodology and results analysis, to write-up and dissemination, researchers must give meticulous attention to teasing out tacit assumptions, taken-for-granted perspectives, and seemingly *natural* or common-sense meanings.

A refreshingly nuanced exploration of parental responses to a newborn with a cleft is offered by Neville, Waylen and Searle in this book. In this contemporary exploration, the parents of newborns, and the children and young people with CLP express complex, negotiated and ambivalent positioning with regard to stigma that a range of other research, as we have seen, above, does not capture.

Learning to be specific

From the 1940s to the 1990s, with a view to improving nursing care, sociologist Frances Cooke Macgregor researched the social experiences of people with facial disfigurements.

In 1953, concerning the impact on a patient's lifestyle of the amputation of part of her nose, Macgregor writes:

> Before her disfigurement Mrs Dover, who lived with one of her two married daughters, had been an independent, warm and friendly woman who enjoyed travelling, shopping and visiting her many relatives. The disfigurement of her face, however, resulted in a definite alteration in her way of living. The first two or three years she seldom left her daughter's home, preferring to remain in her room or to sit in the backyard. 'I was heartsick,' she said, 'the door had been shut on my life.'[47]

Writing nearly forty years later in 1990, the same researcher describes how people with disfigurements, when out and about, face

> visual and verbal assaults and a level of familiarity from strangers not otherwise dared: naked stares, startled reactions, 'double takes', whispering, remarks, furtive looks, curiosity, personal questions, advice, manifestations of pity or aversion, laughter, ridicule and outright avoidance. Whatever form the behaviours may take, they generate feelings of shame, impotence, anger and humiliation in their victims.[48]

In the first example, the huge change in lifestyle is wordlessly attributed, perhaps by Mrs Dover herself as well as by the researcher, simply to the acquisition of a facial disfigurement. We cannot know what might have passed through the minds of the readers (clinicians) upon reading this in 1953, but as a general rule the reduced quality of life of people with disfigurement has continuously led plastic surgeons to redouble their efforts towards better aesthetic outcomes. The second example was similarly addressed to clinicians (with a view, again, to their doing more) but there is a huge change of perspective regarding what exactly goes on that is so distressing for their patients.

Without considerable research into Macgregor's work, influences and clinical writing, it is not possible to say to what extent this change – a kind of clarification – across time accompanied the growing voice of disability rights in the United States, or how much it derived from Macgregor's own professional involvement over many decades with a great many people affected by disfigurement. Did her research take her out walking along busy streets and into cafés with these people so that she herself witnessed the stares, the double-takes and the manifestations of pity and aversion? Or did she ask specific questions about other people's reactions, and record their answers?

When other people's reactions to disfigurement go unrecorded or unremarked, what is going on? Does the way 'the look' is rendered, or omitted, reflect *simply* differing professional or academic perspectives, or the particular focus of the research? Or is it a peculiarity of the stigma of disfigurement that manifestations of the enactment of stigma are obscured?

Staring happens

Even when considering the 'social value of unremarkability', Heather Laine Talley is not explicit about the role of other people's reactions such as staring.[49] Her exploration of the ideological function of 'disfigurement' as a cultural idea – the disfigurement imaginary – focuses on discourse rather than social interaction. Talley might argue that the disfigurement imaginary naturalizes or legitimizes a particular way of looking – with pity, with horror, and so forth. But in the absence of any explicit indication of the way that pity and horror are enacted in social encounters between those whose appearance is unremarkable, and anyone whose face is disfigured, means that she continually but silently relies on her readers' understanding that these enactments do occur and require no elaboration – they are naturalized and legitimized.

Let us consider this from a reader's perspective. With other people's reactions implied but not described, Reader A may bring an appropriate understanding to her engagement with Talley's text, without even noticing that this understanding is implicit. Yes, of course staring happens. Reader B, however, may notice that the nature of the *unmentioned responses* which are the main driver of the facially disfigured person's pressing need for facial work, is here being *taken for granted* without being stated explicitly. But, 'what is the world doing that turns a person's lack of unremarkableness into social death?' remains a curiously unasked question. Is this question unasked because everyone already knows the answer (staring, avoidance – all the things listed by Macgregor) and also knows that these are behaviours we have been socialized not to do?

What might lie behind Talley's decision (or unwitting oversight) in never referring explicitly to the social responses to disfigurement that Macgregor lists? Talley is clear that her book's focus is *facial work* and states that this can be understood as the 'material practice' of what Goffman termed 'face-work', being 'a technique of interaction through which individuals gain recognition as social beings'.[50] She examines in exceptional depth and detail a range of medical and cosmetic interventions in their various reality TV, transgender and charity mission contexts. Talley does not explicitly state that she is not concerned with onlookers' responses to the person who cannot access unremarkableness of appearance, but her subject matter is deeply enmeshed in the variously experienced social unacceptability of certain kinds of facial appearance.

We can similarly imagine Reader A and Reader B turning to Als and Brazelton's research, outlined above. Presenting the issues with only a tacit acknowledgement of other people's reactions to unusual remarkable appearance leaves Reader A undisturbed regarding the way people's widespread responses to certain kinds of facial difference are not *owned*. It is as if the staring that goes with not being unremarkable, has no subject but only an object – *staring happens* – so that *unusual-looking people are stared at*, but never *I stare at unusual-looking people*. This could turn Reader B into a kind of nuisance or even a kind of delinquent, liable to call out 'Hey, you're staring!' to someone – everyone – who knows they are staring, doesn't want to stare and feels uncomfortable thinking about it. Reader B may qualify as what Erich Fromm terms the 'Revolutionary Character' – someone who is 'identified with humanity and therefore transcends the narrow limits of his [*sic*] own society and who is therefore able to criticize his or any other society. He is not caught in the parochial culture in which he happens to be born.'[51]

Keeping the focus very much on people whose distinctive appearance makes them vulnerable to being stared at (and thereby off the people doing the staring) may not actively foster or perpetuate the stigmatizing of certain kinds of facial appearance or disfigurement. However, this focus surely fails to explore the stigmatization which *that look* enacts. What would be the effect of clear curiosity regarding whatever may lie behind this unowned, *subjectless* approach to *that look* ('staring happens')? I suggest that, at the very least, it would tend to open up new possibilities for thinking about and exploring our understanding of enactments that perpetuate stigma and our and society's resistance to and capacity for change.

Disfigurement researchers in other fields will surely encounter other kinds of screening around the enactment of stigma. I do not know if, or how, the psychoanalytic example outlined above might help with probing beyond such veils. For historians, for example, no or little record of the enacting of stigma (as in the psychoanalytic examples above) is unlikely to mean that stigma was not enacted. But it will surely be easier to uncover evidence of stigmatizing ideas and behaviours than to reconstruct the experiences of stigmatized persons or non-persons. In the final part of this chapter I broaden the field of enquiry to examine other fields in which this screening occurs.

Writing from the outside

In the context of enhancing inclusion by equipping schools with books attractive to the whole range of children, well-meaning and seemingly inclusive novels came under scrutiny.[52] Research found that their white authors were

> committed to creating … black characters who would elicit their readers' sympathy and approval. But 'the trap' that all fell into was perceiving their readers also to be all white: the ways in which they justified the worth of their characters were ways that would be wholly acceptable to white values and standards.[53]

The novelists' surely unwitting assumption concerning their readers would be likely to go unnoticed by white readers. We may imagine Reader A, again, reading the novel in the way the writer intended and appreciating the opportunity to see black characters in a better light. Reader B meanwhile is again at risk of calling out – 'Hey, you're a white person writing about black people!' to the novelist who knows this, and who may feel unfairly challenged regarding this seemingly unavoidable aspect of not just their work – imagining other people's lives is surely the core work for writers of fiction – but of their very being. Cultural appropriation is now a point of contention among writers from different cultural and ethnic backgrounds, and the debate includes an essay which clarifies the crucial part played by the audience – the customers – of fiction who are overwhelmingly white Western women.[54]

All this may suggest that anyone researching disfigurement from the outside, so to speak, will do well to bear in mind that their work will be read by people for whom disfigurement is a personal experience as well as a something to be researched, and therefore they should avoid the trap of assuming that their audience approaches disfigurement from the same outside as they do – as if this is all there is. (Perhaps we can see E. M. Forster writing from both the outside and the inside in his novel *A Passage to India.* To the colonial English, Indians are other, and Forster writes that the English 'pose of "seeing India" ... was only a form of ruling India; *no sympathy lay behind it*'.[55] Othered by his sexuality, I imagine E. M. Forster knew all about a look that was without sympathy.)

Comparative social status of researcher and researched

When research into the 'Super Rich', described as 'the 1% of the 1%', was aired on BBC radio the introduction reminded us of our habituation to research into the life experiences and subjectivities of people and communities whose educational, financial or social status is lower than that of the researchers.[56] Is there an unspoken belief, or perhaps hope, that research findings concerning any aspect of the lives of these lower-status research subjects will in due course lead to interventions that improve their lives? Elsewhere I have contrasted relational seeing and feeling, with objectifying staring that manifests a kind of power.[57] In this context, might we reflect upon the relative social statuses of the researcher and the person or persons being researched when the latter is or are affected by – and in fact stigmatized by – facial disfigurement?

Anthropological studies often involve well-resourced and highly educated westerners spending a significant amount of time living among and studying people whose educational experiences and resources are very different from their own. They wrestle with the discovery that

> people who find themselves being researched are rarely content with academic studies of their communities. They want information that can improve their lives rather than furthering someone's career.[58]

Within anthropology these disparities fuel a vigorous debate encompassing activism as well as advocacy. Is reciprocity a legitimate expectation for the subjects of anthropological study? What are we to make of an anthropologist's anguish at having not done more to protect her young informants from HIV?[59]

The challenge social anthropologists face concerning their status relative to that of the people or communities they research resonates with Edward Said's theorization of the other. In *Orientalism* he describes a pervasive Western tradition, both academic and artistic, of prejudiced outsider interpretations of the East, shaped by the attitudes of European imperialism in the eighteenth and nineteenth centuries.[60] The 'Oriental' cultures, presented by scholars, writers, designers and artists from the West, include Middle Eastern, South Asian, Oceanian, African and East Asian cultures. Said argues that the prevalent – and therefore invisible to itself – Western position takes for granted that the West is thinking, growing and evolving while everything Eastern is static, objectified, exoticized and othered. Orientalism represents a critical response to a huge range of mainstream Western activity, including scholarly activity.

Researching disfigurement

The perceived inequality of researcher and researched, hotly debated by anthropologists and theorized in Orientalism, and the apparent objectifying of those who are researched, is not discussed in disfigurement research where the experience of disfigurement resides almost exclusively in the person whose appearance is unusual. Yet terms such as evidence-based practice imply the practical usefulness of at least some research findings. Every clinical service development, educational intervention and campaign to improve the personal well-being and social experiences of people whose appearance is disfigured draws (even if selectively) on research findings that may, as I have endeavoured to show, be imbued with a kind of unconscious face-ism that positions people who look unusual as unfortunate people who need help.[61]

How might the people who are Changing Faces regard a purely scholarly research project into disfigurement, its psychosocial sequelae, and so forth? What if the researcher also has a facial disfigurement? In the case of disability rather than disfigurement, we can see how the social model of disability (as compared with the medical model) and the development of disability studies as an academic and research discipline emerged through disabled people themselves articulating their anger and frustration at being positioned as individuals affected by individual tragedy.[62] Concerning disability specifically as a stareable sight, Kevin Connolly, who was born without legs, has made an extensive photographic study of the starers he encounters everywhere he goes, alongside the conversations that develop whenever they are willing to be engaged beyond their stare. His finds that starers want a story to enable them make sense of the sight of a legless man, and he also reports discovering a kinship with othered people, such as blacks and women, that he is able to access through the othering of being stared at.[63]

Concerning facial disfigurement from the inside, so to speak, James Partridge, in his groundbreaking work for people with facially disfiguring injuries, maintains a strong focus on the need to develop a positive stance alongside practical strategies to manage other people's reactions: 'There are a number of general responses … you are likely to come across. I've called the group of them the SCARED syndrome – Staring, Curiosity, Anguish, Recoil, Embarrassment and Dread.'[64] This is very different from research into the experience of disfigurement that focuses from the outside and is liable to incorporate tacit assumptions that obscure the specific behaviours – other people's behaviours – that constitute a major everyday challenge facing people living with disfigurement.

The many ways in which people with facial disfigurements are disadvantaged have been extensively researched by psychologists and social psychologists.[65] Research reveals that these people are less dateable, less employable, less socially able and so forth. They are seen as morally inferior because of the enduring association between outward appearance and inner character, and are responded to with disgust presumably because of the association of disfigurement with infection. But these researchers rarely conclude with reflections upon how mainstream people might be enabled to reconfigure their responses to people who look unusual. Rather, they recommend extra support, counselling, Cognitive Behavioral Therapy, desensitization, social skills training and self-esteem building for the people who look unusual. Do these psychology and social psychology findings contribute to the case for more and better surgery to reduce, if possible, the appearance of the disfigurement? By contrast, having arrived inside the challenge of disfigurement, James Partridge writes:

> The common immediate response to disfigurement tends to be a desire to recapture your past visage … . For some this is the start of a lifelong quest for perfection: they are willing to undergo any number of surgical procedures to 'fine tune' their new face. Unfortunately they may become so obsessed with this that they fail to make any meaningful effort to rebuild their lives. Their surgical repair supersedes their social rehabilitation and makes that transition impossibly difficult. In effect these people get stuck in the belief that until they have cleansed their face, they stand no social chance whatsoever.[66]

The 'mammoth time commitment' required for plastic surgery procedures can be used to 'justify self-isolation'. Partridge counsels against retreating from social contact – a retreat that can be reinforced by the assumption that until one is facially restored, one cannot access social acceptability.[67] What emerges is a series of choices that each person who has acquired a facial disfigurement will need to make for themselves. To do this they will need, as well as information about what plastic surgery can and cannot do, clear information about other people's reactions and effective social strategies for dealing with these.

James Partridge's book *Changing Faces* can be thought of as a handbook for living with acquired facial disfigurement. As well as carrying forward this work, the organization Changing Faces seeks to change the way everyone faces disfigurement. This is what much disfigurement research tends to overlook. Can we imagine psychologists and social psychologists researching interventions that would enable the person doing the looking – the staring – to see and recognize the ordinary personhood of anyone whose unusual appearance simply broadens the diversity of how we look? What might historians discover through attending to the stigmatizing effacements enacted by others upon those who look different? When Dr Treves arranged for Joseph Merrick to live at the London Hospital, what were the responses and the thoughts of the wealthy individuals who visited him there? Did they make financial donations? What did they write in the privacy of their own diaries?

Mainstream people tend not to reflect much upon their responses to unusual-looking people. When I started to research 'reactions to disfigurement', everyone outside Changing Faces, and even my eminent supervisor, thought I was researching the reactions of the people whose appearance was unusual. For a while we spoke at cross-purposes. I did not mention this anywhere in my write-up which reads as if it never happened.

Conclusions and further questions

The stigma associated with disfigurement is ripe for research. How does it come about that having a disfigurement leads to 'social death' such that a quality-of-life intervention such as FT which reduces length of life is to be preferred to living with disfigurement? The difficulties faced by people with unusual faces, and the many medical, cosmetic and psychological interventions to ameliorate these, have been much researched, but our socialized and acculturated responses to disfigurement both currently and across the millennia much less so.

Transformation from being an ordinary, unremarkable-looking person to being a stareable sight happens all the time. Car accidents, house fires, skin cancer, head and neck cancer are not uncommon human experiences. Movement in the other direction is rarer but, with ongoing medical advances and cosmetic procedures, is an increasingly compelling goal – characterized as evading or being rescued from the 'profound social exclusion' and 'social death' associated with disfigurement.[68]

It seems helpful, wherever possible, to enable people to make the move away from the stigma of disfigurement. With an unremarkable appearance a person can manifest all kinds of traits and qualities. But if your appearance is disfigured your personal qualities, your individuality, your achievements will be at best distorted and patronized as heroic, your identity lost beneath a particularity of outward appearance that homogenizes you into 'the disfigured'. Recent research reveals that this is not new but was also a feature of the lives of people in medieval times who survived facial injury.[69] Research into this process is long overdue.

The categories of 'normal'- or 'unremarkable'-looking, to use Talley's term, and 'disfigured' may seem quite natural – a simple matter of experience of the range of human appearance that we can take for granted. Can we envisage that they are constructs? The 'taken for granted always pushes in certain directions, always towards certain ends, *but always surreptitiously*'.[70] If normal or unremarkable-looking and disfigured are deemed 'natural' or 'natural-seeming' categories, then we must ask, were they constructed in prehistory, and why? (In *Purity and Danger* Mary Douglas links the understanding of clear categories to the development of human civilization and culture, but her work is not specific to facial appearance.[71]) How and why have they been perpetuated ever since through myriad psychological, social and cultural mechanisms?

This chapter has explored the look, and thus might equally well have featured in the section on 'visibility' in this book. However, the ways in which the look is described – or not – in research papers on impairment and disfigurement is a matter of language and communication. By omitting to acknowledge staring and the possible discomfiture of the *researcher*, approaches to disfigurement research obscure the very mechanisms that maintain the stigmatization of disfigurement and which are ultimately enacted by individuals, including researchers. People stare, not just the 'normals'. Kevin Connolly, a staree (see above), is a starer too and 'would stare at someone like himself if he were to see him on the street'.[72] What could facilitate constructive engagement between starer and staree? Would this offer a way out of this perpetuation of stigma? Should we include our own response to disfigurement in our research? Its omission by psychoanalytic researchers formed a plank in the false floor which pathologized starees' discomfort as social anxiety. Frances Macgregor and Heather Laine Talley do not do this. Here, I have not done this. The stigmatizing responses that they and I have written about are thereby rendered as other people's or no-one's – 'staring happens' – never our own.

Notes

1 F. Macgregor, 'Facial Disfigurement: Problems and Management of Social Interaction and Implications for Mental Health', *Aesthetic Plastic Surgery* 14.4 (1990): 249–57.
2 H. L. Talley, *Saving Face: Disfigurement and the Politics of Appearance* (New York: New York University Press, 2014), 74.
3 R. Garland-Thomson, *Staring: How We Look* (New York and Oxford: Oxford University Press, 2009), 93.
4 J. Frances, 'Damaged or Unusual Bodies: Staring, or Seeing and Feeling', *Body, Movement and Dance in Psychotherapy: An International Journal for Theory, Research and Practice* 9.4 (2014): 198–210. doi: 10.1080/17432979.2014.931887.
5 E. Goffman, *Stigma: Notes on the Management of Spoiled Identity* (Harmondsworth: Penguin, 1963).
6 Talley, *Saving Face*, 22.
7 F. Carling, *And Yet We Are Human* (London: Chatto and Windus, 1962), 54–5.
8 L. Grealy, *Autobiography of a Face* (New York: Harper Perennial, 1994), 10.
9 J. Partridge, *Changing Faces: The Challenge of Facial Disfigurement* (London: Penguin, 1990), 65.
10 Partridge, *Changing Faces*.
11 A. M. Keary, 'Do We? Can We Look at the Disabled Body?' in *Contemporary Body Psychotherapy: The Chiron Approach*, ed. L. Hartley (London: Routledge, 2009), 170.
12 Keary, 'Do We?', 165.
13 Garland-Thomson, *Staring*, 21–2.
14 Ibid., 17 and 21, citing Steven Yantis's research on control of visual attention.
15 Ibid., 17, citing Simon Baron-Cohen, *Mindblindness: An Essay on Autism and Theory of Mind* (Cambridge, MA: MIT Press, 1997), 98.
16 V. Sinason, *Mental Handicap and the Human Condition: New Approaches from the Tavistock* (London: Free Association, 1992), 27.
17 Garland-Thomson, *Staring*, 21–2.
18 Ibid., 79–94.
19 Ibid., 89.
20 G. Bychowski, 'The Ego of the Brain Wounded', *Psychoanalytic Review* 36 (1949): 333–43.
21 Ibid., 334.
22 Ibid., 342.
23 Ibid., 336.
24 S. L. Dubovsky and S. E. Groban, 'Congenital Absence of Sensation', *Psychoanalytic Study of the Child* 30 (1975): 49–73; E. M. Earle, 'The Psychological Effects of Mutilating Surgery in Children and Adolescents', ibid., 34 (1979): 527–46; W. G. Niederland, 'Narcissistic Ego Impairment in Patients with Early Physical Malformations', ibid., 20 (1965): 518–53; *id.*, 'Scarred: A Contribution to the Study of Facial Disfigurement', *Psychoanalytic Quarterly* 44 (1975): 450–9.
25 S. J. Coen, 'The Sense of Defect', *Journal of the American Psychoanalytical Association* 34 (1986): 47–67.
26 Niederland, 'Narcissistic Ego Impairments', 519.
27 Ibid., 519–20 citing Kaplan 'The Role of Birth Injury in a Patient's Character Development and his Neurosis', *Bulletin of the Philadelphia Association of Psychoanalysis* 9 (1959): 1–18.

28 H. Als and T. B. Brazelton, 'Parenting an Infant with a Birth Defect: The Regulation of Self-Esteem', *Psychoanalytic Study of the Child* 39 (1984): 561–89; L. A. Nabors, H. D. Lehmkuhl and J. S. Warm, 'Children's Acceptance Ratings of a Child with a Facial Scar: The Impact of Positive Scripts', *Early Education and Development* 15.1 (2004): 79–92; S. M. Mojon-Azzil et al., 'Strabismus and Discrimination in Children: Are Children with Strabismus invited to Fewer Birthday Parties?' *British Journal of Ophthalmology* 95.4 (2010). doi: http://dx.doi.org/10.1136/bjo.2010.185793.
29 Als and Brazelton, 'Parenting an Infant with a Birth Defect', 576.
30 Ibid., 563.
31 Ibid.
32 Ibid., 564.
33 Ibid.
34 Ibid., 567.
35 Ibid., 584.
36 Talley, *Saving Face*, 18.
37 Earle, 'Psychological Effects', 530 (my emphasis).
38 Compared with Als and Brazelton's qualitative research, that of Neville et al. in this book, gains much from the researchers' open curiosity concerning children and parents' ideas about whether or not, and if so why, stigma attaches to having a facial appearance affected by cleft.
39 R. Gilder, S. V. Thompson, C. W. Slack and K. B. Radcliffe, 'Amputation, Body Image and Perceptual Distortion', US Naval Medical Research Institute Project N. M. 004 008 04 03 (1954).
40 J. F. McGowan and T. L. Porter, *An Introduction to the Vocational Rehabilitation Process. A Training Manual. Rehabilitation Series No.68–32* (Washington, DC: Rehabilitation Services Administration, 1967) at http://files.eric.ed.gov/fulltext/ED042011.pdf (accessed 30 November 2013), 73.
41 Ryan, S., M. Oaten, R. J. Stevenson and T. I. Case, 'Facial Disfigurement is Treated Like an Infectious Disease', *Evolution and Human Behaviour* 33.6 (2012): 639–46. doi: http://dx.doi.org/10.1016/j.evolhumbehav.2012.04.001.
42 Robinson, K. 'The Character of the Exception as a Defense against Deformity', *Psychoanalytic Study of the Child* 57 (2002): 305–26.
43 Ibid., 306 quoting Mann page not given.
44 Talley, *Saving Face*, 154.
45 Ibid.
46 Ibid., 153, quoting the head of the FT team.
47 F. Macgregor et al., *Facial Deformities and Plastic Surgery* (Springfield, IL: Thomas, 1953), 91–2, quoted in Goffman, *Stigma*, 23–4.
48 Macgregor, 'Facial Disfigurement', 254.
49 Talley, *Saving Face*, 76.
50 Ibid., 17.
51 Fromm cited by E. Hopper, 'The Social Unconscious in Clinical Work' [1996], in E. Hopper, *The Social Unconscious: Selected Papers* (London: Jessica Kingsley, 2003), 150. Hopper worries that the term 'Revolutionary Character' is problematic and proposes the term 'Mature Person', which may seem equally problematic.
52 G. Klein, *Reading into Racism: Bias in Children's Literature and Learning Materials* (London: Routledge and Kegan Paul, 1985). For example, see https://www.theguardian.com/books/2016/sep/15/we-need-to-talk-about-cultural-appropriation-why-lionel-shrivers-speech-touched-a-nerve (accessed 28 February 2017).

53 Klein, *Reading into Racism*, 2.
54 Clare Vaye Watkins, 'On Pandering', (2015), at http://www.tinhouse.com/blog/41314/on-pandering.html (accessed 3 February 2017).
55 E. M. Forster, *A Passage to India* (London: Penguin, 1924 repr. 1961), 301–2, my emphasis.
56 'Thinking Allowed', BBC Radio 4, 9 January 2017: Laurie Taylor in discussion with Rowland Atkinson, research chair in Inclusive Society at the University of Sheffield, Roger Burrows, professor of Cities at Newcastle University and Emma Spence, PhD researcher at Cardiff University: http://www.bbc.co.uk/programmes/b086s8yv (accessed 7 February 2017).
57 Frances, 'Damaged or Unusual Bodies'.
58 Alexander M. Ervin, *Applied Anthropology* (Boston and London: Allyn and Bacon, 2000), 129, cited by P. Kellett, 'Advocacy in Anthropology: Active Engagement or Passive Scholarship?' *Durham Anthropology Journal* 16.1 (2009): 22.
59 P. Kellett, 'Advocacy in Anthropology' 28, citing Rachel Burr, 'Shaming of the Anthropologist: Ethical Dilemmas during and in the Aftermath of the Fieldwork Process', *Anthropology Matters Journal* (2002): 6.
60 Edward Said, *Orientalism* (London: Pantheon, 1978).
61 As previously mentioned, Neville et al.'s chapter in this book relays research that is a valuable exception.
62 See, for example, the work of Michael Oliver, both disability rights activist and university-based researcher-scholar.
63 Garland-Thomson, *Staring*, 89–94.
64 Partridge, *Changing Faces*, 65.
65 R. Bull and N. Rumsey, *The Social Psychology of Facial Appearance* (New York: Springer-Verlag, 1988).
66 Partridge, *Changing Faces*, 49.
67 Ibid.
68 Talley, *Saving Face*, 130, 150.
69 P. Skinner, '"Better off Dead than Disfigured"? The Challenges of Facial Injury in the Premodern Past', *Transactions of the Royal Historical Society* 26 (2016): 25–41. doi:10.1017/S0080440116000037; P. Skinner, *Living with Disfigurement in Early Medieval Europe* (New York: Palgrave Macmillan, 2017).
70 F. Dalal, 'The Social Unconscious and Ideology in Clinical Theory and Practice', in *The Social Unconscious in Persons, Groups, and Societies Vol.1: Mainly Theory*, ed. E. Hopper and H. Weinberg (London: Karnac, 2011), 243–64, 246.
71 M. Douglas, *Purity and Danger* (London: Routledge Kegan Paul, 1966).
72 Garland-Thomson, *Staring*, 89.

Part Two

Visibility

5

Hair Loss as Facial Disfigurement in Ancient Rome?

Jane Draycott

Hair loss, whatever the cause, significantly affects an individual's appearance and, depending upon the individual (for example, their sex, gender, age, health, livelihood), the ramifications of it can be more or less severe. The National Health Service's statistics of the UK indicate that hair loss is an issue that affects both men and women in contemporary Britain. Around half of men experience male pattern baldness by the age of fifty, although it usually begins in the late twenties or early thirties, while women can experience temporary hair loss after childbirth and female pattern baldness after menopause, usually in the early fifties, and this is known as androgenetic alopecia.[1] Another common type of premature hair loss, affecting one or two people in every thousand, is alopecia areata (loss of small patches of hair from the scalp or other parts of the body), which for the most part affects people between the ages of fifteen and twenty-nine and can lead to the more severe hair loss of alopecia totalis (total loss of scalp hair) and alopecia universalis (total loss of scalp and body hair).[2]

According to numerous works of ancient literature, hair loss was an issue that affected both men and women in the Roman republic and empire just as it affects both men and women today. While a certain degree of hair loss was recognized as being a normal part of the natural ageing process, and thus something that the majority of men and women could anticipate experiencing at some point in their lives, during the Roman period the bald man or woman was the butt of numerous jokes at best and subject to outright hostility at worst.[3] An individual so afflicted could expect to be subjected to public mockery and humiliation on a regular basis, no matter what their social status. It has been suggested that such behaviour was a means of bolstering group cohesion at the expense of the outsider, expiating fear and embarrassment and ensuring an outlet for sublimated aggression.[4] But why should bald people be seen as outsiders, objects of fear and embarrassment, and a trigger for sublimated aggression? This chapter will focus on premature hair loss, whatever the perceived cause might have been, arguing that to the Romans this was a type of facial disfigurement, and this disfigurement was viewed not only as a physical failing but also as a moral one, and conclude by examining a rare first-person narrative account of living with such a condition in order to explore the unique perspective on living with facial disfigurement in antiquity that this work provides.[5]

Physical appearance

The ancient Roman conception of physical appearance was a conscious construction predicated upon the desire to fuse natural law with cultural practice.[6] The body was viewed as the place where nature and culture met, which made it a text to be deciphered and read.[7] Consequently, an individual's body was constantly subjected to scrutiny, and judgements were constantly being made regarding it.[8] A naturally beautiful exterior was considered to reflect a morally sound interior. This was particularly important with regard to the practice of oratory; it was imperative that an orator's body was seen to represent the orator's virtue if his words were to be taken seriously.[9] The face was of particular importance: according to Cicero, nature

> has so formed [a man's] features as to portray therein the character that lies hidden deep within him; for not only do the eyes declare with exceeding clearness the innermost feelings of our hearts, but also the countenance, as we Romans call it, which can be found in no living thing save man, reveals the character.[10]

Ancient oratorical treatises taught individuals how to best present themselves, providing step-by-step instructions on how to correct any potentially problematic features of their appearance and manner, tending to start at the head and working downwards.[11] Deviations from what was considered the norm marked an individual out and rendered them 'Other'. The severity of the deviation was less important than the fact of the deviation, and consequently it is difficult to establish a hierarchy of severity. Such deviations – often referred to pejoratively as deformities – were considered serious enough to potentially bar an individual from holding public office, and were certainly weaponized as a means of blackening someone's character.[12] According to oratorical treatises, it was considered an acceptable rhetorical strategy to draw attention to an opponent's appearance and use it against them, and this seems to have been common practice in the late republic and into the imperial period (i.e. the first century BCE to the first century CE).

According to Cicero, 'In ugliness (*deformitatis*) too and in physical blemishes (*vitiorum*) there is good enough (*bella*) matter for jesting, but here as elsewhere the limits of licence are the main question.'[13] This is a deliberate juxtaposition of the ugly body and the pretty jokes that could be made at its expense, and a demonstration of Ciceronian wit and wordplay. There is a lot of supporting evidence for this practice in surviving works of late republican literature, such as the poetry of Catullus.[14] Physical disparities were not considered to be accidents of nature, but rather the direct results of behaviour that was deviant in some way, whether one's own behaviour or the behaviour of one's ancestors.[15] The latter in particular are evident in the Roman institution of the *cognomen*, an extra personal name or nickname, as in the republican period it was frequently a pejorative reference to an aspect of an individual's physique, which in turn could be read as a pejorative reference to their character.[16] Consequently, physical differences were fair game for ancient orators, and a mockery of them was a common feature of ancient oratory; the orator wanted to prove the moral deviance of his opponent, win the argument and thus win the case. According to Cicero, '[One should speak of] his personal endowments, among which it is easiest to praise a handsome appearance, as providing a very great indication of virtue.'[17] The reverse was also true: a lack of physical beauty reflects a deficiency in virtue.

Cicero's defence speech *For Quintus Roscius*, one of his earliest extant orations, puts the strategies and techniques set out in his rhetorical treatise *On Rhetorical Invention* into practice.[18] He focuses on the plaintiff, Gaius Fannius, and invites his audience to compare the physical appearance of Quintus Roscius with that of Gaius Fannius, to Roscius' advantage.[19] Gaius Fannius seems to have voluntarily rendered himself bald through depilation, taking Roman grooming practices to extremes.[20] Of course, it is possible that Gaius Fannius had gone bald for other reasons and that Cicero claimed it was the result of depilation to blacken his character even further.[21] Cicero's descriptions of Roscius and Fannius must have been accurate, since both men would have been present when the speech was delivered, and thus visible to Cicero's audience.

As we have seen, physical appearance is due to a combination of nature and culture, and hair is a useful means of mediating between nature and culture, with the hair itself representing nature and the ways in which it is styled or treated representing culture.[22] Many Roman discussions of physical appearance paid particular attention to hair, and an individual's hair was frequently utilized as a means of drawing conclusions about that individual's character and morals. But what if an individual had no hair? Could a lack of hair and the way in which this was treated likewise mediate between nature and culture?

Hair loss

In Graeco-Roman antiquity, while it was understood that some peoples, such as the Myconians, were naturally bald, it was recognized as early as the fourth century BCE that a certain degree of hair loss occurred as a part of the natural ageing process.[23] The Greek philosopher Aristotle, drawing on the humoral theory of balanced heat and cold, moisture and dryness in a healthy body, attributed this to the body's cooling and drying out.[24] Consequently different parts of the head were considered to be more or less dry than others, based on the progression of the hair loss.[25] The same substance that was responsible for the growth of hair was thought to be responsible for the production of semen.[26] This led to numerous assumptions regarding who could suffer from hair loss: neither children, nor eunuchs, nor women should suffer from hair loss.[27] Hair was expected to be at its best during the prime of a man's life when he was at his physical peak.[28] Thus we see a connection being made between hair, masculinity and virility, a connection that would continue to be made throughout classical antiquity and into the fifth century CE.[29] Myths offered cautionary tales of men who lost their hair prematurely and subsequently lost their power and authority too.[30]

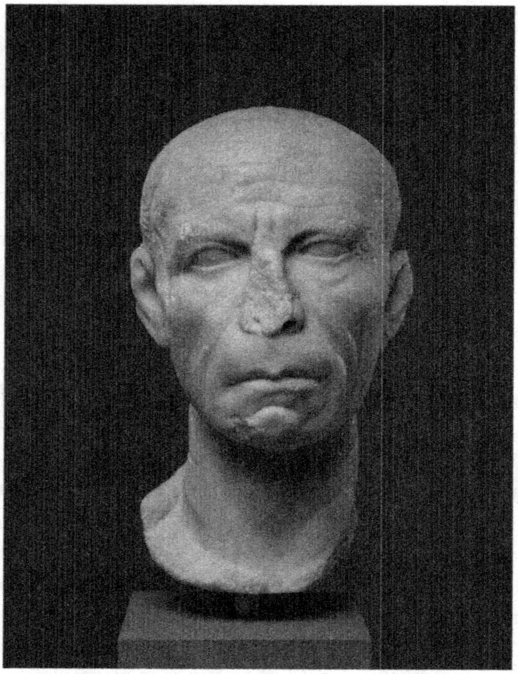

Figure 5.1 Marble bust of an old man, late first century BCE, Metropolitan Museum of Art inv. 21.88.14. Image courtesy of Metropolitan Museum of Art.

It was also recognized that once hair loss occurred as part of the natural ageing process, hair was unlikely to grow back.[31] The Roman medical writer Celsus differentiated between hair loss as a result of illness and hair loss as a result of ageing, suggesting remedies for the former but considering the latter incurable.[32] Hair loss as a part of the natural ageing process was not necessarily viewed in a negative light, particularly not during the Roman republic and early empire, as age was valued for the wisdom and experience that it brought.[33] Thus we see honorific portrait busts of elderly men dating from the first century BCE that seem to emphasize the features of old age (see Figure 5.1).[34] This is referred to as the veristic style of Roman portraiture. A bald head is combined with wrinkles and a severe expression, all of which indicate *dignitas*, *gravitas* and *auctoritas*.

What we do not see, however, are honorific portrait busts of younger men with bald heads. A young man who experienced premature hair loss, be it the result of male pattern baldness or illness, was not viewed in the same way as an old man who experienced hair loss as part of the ageing process. A woman of any age who experienced hair loss, whatever the cause, was also treated differently. This type of hair loss was in no way desirable. Perhaps this explains the context in which we do see portraits of younger men with bald heads. Baldness is the most common physical deviation depicted in portrait graffiti, and the frequency with which names accompany these drawings indicates that they were intended to represent real people (see Figure 5.2, the text says '*Rufus est*', 'this is Rufus') and so be recognizable to the viewer.[35] It has been

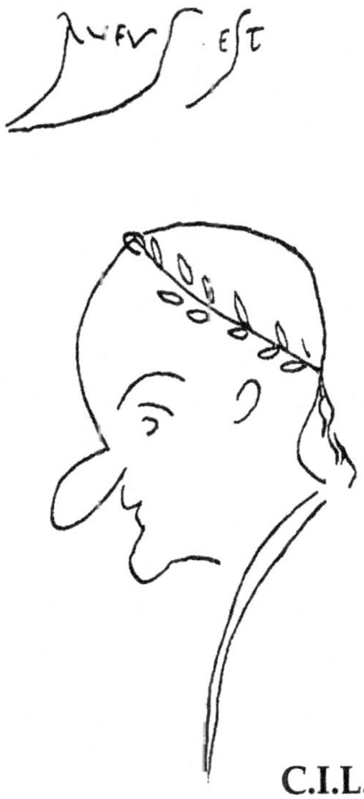

C.I.L. IV 9226

Figure 5.2 Graffito, late first century CE, Villa of the Mysteries at Pompeii, *CIL* IV 9226.

suggested that these drawings are an attempt at simplistic humour, a humour that is at its sharpest when it is directed at a specific person.[36] So we see another way in which bald people could be publicly mocked and potentially humiliated: a graffito was more durable and so could last far longer than a joke made at someone's expense.

There were many reasons why someone might lose their hair prematurely. However, frequently the loss of hair was viewed as something which the victim was directly responsible for, as certain types of behaviour were thought to cause hair loss. Some of this involved the hair directly, and the loss of hair is an obvious and understandable consequence: plucking out grey hairs could lead to baldness, while dyeing grey or white hair in an attempt to disguise its colour could lead to all hair falling out.[37] In other cases, the connection between the behaviour and the hair loss is less clear with regard to cause and effect, but it seems to have involved excess, such as an excess of sexual activity or an excess of wine consumption.[38] Consequently, those thought to be engaging in these types of behaviour and showing evidence of it were subjected to moralizing and judgement. According to Seneca:

> What wonder, then, that we can trip up the statement of the greatest and most skilled physician, when so many women are gouty and bald! Because of their vices, women have ceased to deserve the privileges of their sex; they have put off their womanly nature and are therefore condemned to suffer the diseases of men.[39]

This practice of drawing conclusions about an individual's behaviour and moral calibre from their physical appearance extended beyond casual inference, and was in fact a highly technical scientific discipline known as physiognomy.

Physiognomy

It has been suggested that the practice of divination was a ubiquitous reflex in response to uncertainty.[40] Ancient Mediterranean society was a 'face-to-face' society, and in such a society scrutiny of one's fellow man was an essential survival skill.[41] The discipline of physiognomy, a set of techniques which attempted to assess a person's personality and the nature of their character through the observation of their physical characteristics and behaviour, drawing on beliefs regarding the interdependence of the soul and the body, was one way in which this scrutiny could be exercised. According to pseudo-Aristotle:

> Among things that come about by nature one might especially see that the body and soul are connate with each other to the extent that in the case of most affections they are causally active on each other. No animal has ever been born that has the appearance of one animal but the mind of another, but soul and body are always of the same animal – so that, necessarily, such-and-such a mind follows such-and-such a body.[42]

The specifics of ancient Greek and Roman physiognomic techniques are attested by several surviving treatises: pseudo-Aristotle's, once thought to be authored by Aristotle but now dated to the third century BCE; Polemon's, written between 133 and 136 CE and later adapted by Adamantius; a Latin work once attributed to Apuleius now dated to the second half of the fourth century CE; and a Byzantine adaptation of Adamantius. It is likely that many more circulated and simply did not survive. It is also likely that there were considerably more practitioners than theorists, operating at all levels of ancient society.

It is clear from the content of the surviving treatises that the head was particularly important in physiognomy. This could be attributed to the fact that it was the most consistently visible part of the human body and was crucial in personal interaction. As a result, an individual's hair was worthy of attention. For Polemon, lacking hair was indicative of treachery and wickedness of action.[43] For Adamantius, baldness was a sign of malice and deceit.[44] It has been observed that elderly individuals were at a distinct disadvantage when subjected to physiognomic analysis, and certainly their tendency towards hair loss would be part of this.[45] At even more of a disadvantage were young men who had suffered premature hair loss as they did not have the excuse of old age to explain the loss away, which would lead observers to seek an alternative explanation or draw their own conclusions regarding it.

The Roman historians and biographers of the late first and early second century CE utilized certain aspects of physiognomic theory in their writings as a means of elucidating the true natures of influential individuals, and this is particularly apparent in relation to the Julio-Claudian and Flavian emperors.[46] The emperors are the Romans about whose physical appearances we have the most comprehensive information, both written descriptions and (presumably) somewhat accurate portraits. Suetonius' *Twelve Caesars*, a collection of biographies of the first twelve Roman emperors, contains detailed descriptions of each of them. Suetonius' decision to include these descriptions has been linked to his knowledge of the ancient discipline of physiognomy, with which his sophisticated readers would also have been familiar.[47] Hair – or lack of hair – is a prominent feature of each description. It is, however, the emperors who lost their hair prematurely that concern us here, as the comments made regarding this give us an insight into how premature hair loss was viewed, which seems to have been as a type of facial disfigurement: Julius Caesar, Gaius (Caligula), Galba and Domitian are all described as suffering from hair loss.[48] According to Suetonius, Caesar's 'baldness was a disfigurement (*deformitatem*) which troubled him greatly, since he found that it was often the subject of the gibes of his detractors'.[49] While Suetonius does not specify how old Caesar was when the hair loss first occurred, he was apparently mocked for it by his own soldiers during the triumph of his conquests in Gaul, which took place when he was fifty-four.[50] The fact that Caesar himself apparently considered the loss of his hair as detrimental to his looks, and went to some trouble to attempt to disguise this loss, first through a comb over and then through wearing a laurel wreath, despite the positive connotations of hair loss for an older man, demonstrates how negatively premature hair loss was viewed. The emperor Gaius, assassinated at the age of twenty-nine, seems to have suffered from almost total hair loss.[51] According to Suetonius, Gaius' 'hair [was] thin and entirely gone on the top of his head, though his body was hairy. Because of this to look upon him from a higher place as he passed by … was treated as a capital offence'.[52] This description is confirmed by Seneca, a contemporary of Gaius, who observed that

> amid the multitude of his other vices [he] had a bent for insult [and] was moved by the strange desire to brand every one with some stigma, while he himself was a most fruitful source of ridicule; such was the ugliness of his pale face bespeaking his madness, such the wildness of his eyes lurking beneath the brow of an old hag, such the hideousness (*deformitas*) of his bald head with its sprinkling of beggarly hairs.[53]

However, it is the description given by Suetonius of the final emperor in the sequence, Domitian, that is particularly interesting in relation to his premature hair loss:

> He was handsome and graceful too, especially when a young man, and indeed in his whole body with the exception of his feet, the toes of which were somewhat cramped. In later life he had the further disfigurement (*deformis*) of baldness, a protruding belly, and spindling legs, though the latter had become thin from a long illness.[54]

How accurate was Suetonius' description? Just how bald was Domitian? And to what extent was this baldness seen as disfiguring?

The emperors of the Flavian dynasty favoured a more realistic style of portraiture than their Julio-Claudian predecessors, harking back to veristic republican portraiture. Consequently, it is possible to observe a tendency towards male pattern baldness in the Flavian family.[55] While this seems to have been less of an issue with regard to the founder of the dynasty Vespasian, who was sixty years old at the time of his acclamation and therefore could hardly be said to been suffering from premature hair loss, it is something that seems to have bothered his sons. The portraits of Titus, although difficult to date precisely because he died two years into his reign at the age of forty-one, show his hair receding at the temples and brushed forward to cover a bald spot on the crown of the head.[56] The portraits of Domitian that date from his father's reign show him as a young man with a full head of hair, while his mature portraits, like his brother's, show his hairline gradually receding, his hair thinning, and longer strands of hair at the back of his head combed forward and arranged in an attempt to disguise this.[57] Thus we have evidence not only of Domitian's baldness, but also his attempts to hide it. The Latin word used by Suetonius to describe Domitian's baldness is *deformis*, which can be translated literally as 'misshapen', 'deformed', 'unsightly', 'ugly', 'odious', 'disgusting' or 'disgraceful', and figuratively as 'departing either physically or morally from the right shape or quality'. Behind this stands the idea of a norm from which some sort of deviation has occurred.[58] Concurrently, words denoting moral character often refer to an implied norm of perfection and balance.[59] For Domitian, as for the other 'bad' emperors, it was all downhill. He started off beautiful but losing his hair meant not only that, like Julius Caesar, he lost his beauty, but potentially also his virtue. In point of fact, Suetonius devotes a significant amount of space to Domitian's virtues and vices, and the way that the balance of these shifted over the course of his reign.[60]

According to Suetonius, and perhaps unsurprisingly considering the emphasis placed upon physical appearance by the Romans, Domitian 'was so sensitive about his baldness, that he regarded it as a personal insult if anyone else was twitted with that defect in jest or in earnest'.[61] The extent to which this was true has been debated.[62] An inability to tolerate jokes made at one's expense is a key indicator of a 'bad' emperor, and so it is not surprising that Suetonius would use it here.[63] As we have seen, bald people were the butt of numerous jokes both in relation to their baldness and their attempts to hide it; so it is feasible that Domitian was exposed to some of this even if he was not the one on the receiving end of it.[64] Yet the poet Martial mocked men and women suffering from premature hair loss in numerous epigrams, and did so during Domitian's reign and in a book of poetry dedicated to him, which it is likely that Domitian was at least aware of, even if he did not read the poems himself.[65] The fact that Martial does not seem to have suffered any kind of punishment or penalty for his mockery indicates that Domitian was not overly sensitive about his baldness. That is not to say that he was not distressed by it. In a book *On the Care of the Hair*, which he published and dedicated to a friend, he wrote the following by way of consolation to the man and himself: 'Do you not see that I am too tall and comely to look on? And yet the same fate awaits my hair, and I bear with resignation the ageing of my locks in youth. Be assured that nothing is more pleasing than beauty, but nothing shorter-lived.'[66]

This last line is a reference to the Greek hero Achilles, an individual famous for his glorious hair and frequently placed in opposition to another character from the *Iliad*,

Thersites, due to the extreme differences in their physical appearances.[67] This anecdote gives us an insight into the response of a Roman man to premature hair loss. It is clear that Domitian, despite his high status and the unlikelihood that anyone would have dared to comment on his hair loss to his face, was bothered by it and his unnamed friend was likewise bothered about his hair loss. Hence Domitian's treatise, an attempt to console the pair of them through rationalizing that hair loss was part of the ageing process, was inevitable, but ironically resulted in Domitian being remembered specifically for it.[68] It is clear from surviving medical and cosmetic treatises that people went to considerable lengths to maintain their hair.[69] If they were unsuccessful in these endeavours, they could utilize paint, hair pieces, wigs and headdresses in an attempt to disguise their loss.[70] They could even offer prayers and *ex-votos* in an attempt to engage the help of the gods in reversing it.[71] Such a purpose has been suggested for an unusual *ex-voto* recovered from a ritual deposit on the Esquiline Hill, perhaps originally from the temple of Minerva Medica in Rome; it has been proposed that it depicts an individual suffering from alopecia areata (see Figure 5.3).[72]

But just why was the presence of hair so important? And could the loss of it be overcome? In attempt to understand why premature hair loss was considered to be a

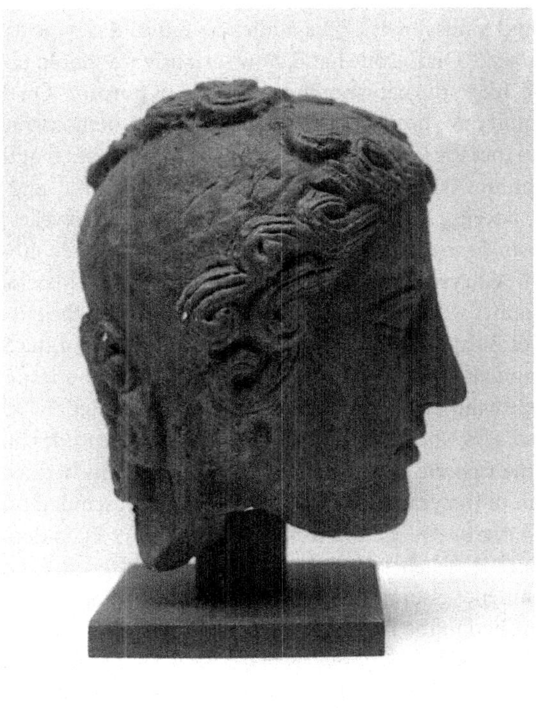

Figure 5.3 *Ex-voto*, late first century BCE, Antiquario Communale inv. 2652. Courtesy of the Wellcome Library, London.

form of facial disfigurement by the ancient Romans, the remainder of this chapter will examine two ancient treatises, one devoted to hair, and the other, written in reply to the first, devoted to baldness.

Dio Chrysostom's *In Praise of Hair* and Synesius' *In Praise of Baldness*

The two treatises in question are examples of a literary genre known as the encomium. Encomia were originally written for the purpose of praising the virtuous actions or other good qualities of a particular individual. However, the genre evolved over time and versions known as paradoxical encomia came to be written for the purpose of praising objects and inanimate beings, things not normally considered worthy of praise, and the two treatises to be examined fall into this latter category as neither hair nor baldness would, in all seriousness, have been considered worthy subjects for such a treatment.[73] Thus, their authors had the unenviable job of needing to justify their chosen subjects and essentially make them worthy of praise through the process of praising them. Successfully doing so would have been seen as proof of the writers' abilities and so would have enhanced their reputations. Texts like these were 'designed for the pleasure and amusement of the audience, full of flashy, witty arguments and clever turns of phrase'.[74] On the one hand, how seriously we should take either of these works is debatable due to the nature of the genre and its purpose. On the other, the fact that the works would have been performed in public as a demonstration of oratorical expertise indicates that there is at least some truth in the points that they make.

The first encomium is Dio Chrysostom's *In Praise of Hair*, and it survives in a fragmentary form because it was replied to some three hundred years later by Synesius in his encomium *In Praise of Baldness*. Dion Cocceianus (*circa* 40 – after 111 CE), also known as Dio Chrysostom ('Golden-mouth' because of his eloquence), was an orator and philosopher who came from a wealthy family from the city of Prussia in the Roman province of Bithynia. He wrote 'display pieces' typical of the Second Sophistic on a variety of popular themes which he would have delivered in person as a means of informing, improving and entertaining a variety of audiences while he travelled around the empire. His work is considered to reflect the attitudes and culture of the upper classes of the eastern half of the empire. He was banished from Rome by the emperor Domitian in the year 82 CE, a banishment not rescinded until the accession of the emperor Nerva in 96 CE, and it is tempting to link Dio's decision to write on this subject to Domitian's hair loss and a desire to wreak revenge, however small, on him for the banishment (assuming, of course, that Dio writing the treatise was not the reason he was banished in the first place).

Dio's work is described by Synesius as being 'so brilliant that the bald man, in the face of its arguments, must be covered with shame'.[75] Unfortunately, the treatise does not survive in its entirety, but the sections that do survive are useful to contemporary readers attempting to understand Roman attitudes to hair. In the surviving fragments, Dio focuses on male hair specifically, and uses the examples of the three hundred Spartans and the Homeric heroes to argue that hair makes a man both beautiful and terrifying, and is indicative of masculinity and virility.[76] The examples would have been extremely familiar to Dio's audience, the highly educated Greek elite of the eastern Roman Empire.[77] Dio's contemporary Plutarch wrote a treatise about the Spartans and discussed their grooming practices, particularly in relation to how they differed to Athenian ones.[78] The Spartans wore their hair long while the Athenians, except for a small number of mostly young men who deliberately aped Spartan styles and were consequently known as 'Laconizers', wore theirs short.[79] For the Spartans, hair was emblematic of maturity and all of the responsibilities that came with it; boys' heads would be shaved around the age of twelve, prior to puberty, and then, around the age of twenty, once they were considered to have reached manhood, they were permitted to grow their hair long. This long hair was linked directly with virility and other qualities desirous in a Spartan male, and was carefully groomed and styled before battle.[80]

Synesius (370–413 CE) came from a wealthy family from the city of Cyrene in the Roman province of Cyrenaica, and was supposedly descended from the Heraclidae, the legendary founders of the city.[81] He travelled in Greece and pursued his education in Alexandria, where he was a pupil of the philosopher Hypatia. He married a Christian and was appointed bishop of Ptolemais before converting to Christianity. *In Praise of Baldness*, written in reply to Dio's *In Praise of Hair* between 400 and 409 CE, offers a unique testimony of the experience of suffering from the disfigurement of premature hair loss under the later Roman Empire. Although both Dio's and Synesius' treatises can be seen as rhetorical exercises or even showcases, ways for the orators to display their eloquence, if we visualize Synesius delivering this work as a speech, it would be significantly less powerful were he to be delivering it with a full head of hair. If the point of the speech is to argue a case that the audience are inclined to take against from the start, and argue it so persuasively that the audience becomes convinced of the case's merit, then the entire argument would be undermined if the speaker were hirsute. Synesius himself comments on the difficulty of arguing such a case effectively, and even uses this as a means of denigrating Dio's rhetorical skill in comparison to his own.[82]

Synesius' attempts to refute the arguments put forward by Dio provide an insight into the sorts of ways in which bald men were perceived specifically because of their hair loss, and it is clear that they were perceived to be disfigured. For baldness was not considered to be a natural state or even a state in its own right, rather it was always considered to be the absence of hair. He mentions the shame that a bald man feels on several occasions, and the reproaches that he receives from others because of his baldness.[83] The treatise also suggests how a bald man might respond to jokes made at his expense, and there is a clear element of verbal sparring and competition. He does confirm that the quality of an individual's hair contributes significantly to that individual's physical attractiveness: 'By nature we all desire to be beautiful – to which end, in large measure, the tresses with which Nature has rendered us familiar since childhood, contribute.'[84] This is particularly important when it comes to the opposite sex, and this is an interesting – and unusual – expression of the realities of female sexual desire.[85] He states that he is bald due to illness, but based on the selection of examples and arguments he uses, his treatise was not intended to be restricted to the temporarily bald. He argues that baldness is a sign of intelligence.[86] He uses the example of the philosophers Diogenes and Socrates, sculptures of whom he has observed.[87] He argues that, since priests frequently removed their body hair in pursuit of purity, baldness is a sign of purity.[88] He also discusses the shape of the bald head, likening it to a sphere, a shape long considered to have divine qualities.[89]

Perhaps most interestingly, and potentially controversially, Synesius argues against judging someone according to their appearance. This flies in the face of the discipline of physiognomy that had been so popular in earlier centuries. He feels that hair is praised by the ignorant and superficial, and this preference results in those with full heads of hair being represented in works of art where bald heads would be more appropriate.[90] However, elsewhere in the treatise he does engage with divination; he argues that hair is negative because the comet, with its hairy tail, is a negative portent.[91] This might be a reference to an anecdote regarding the emperor Vespasian found in Suetonius' *Twelve Caesars*. Apparently, when on his deathbed, Vespasian was informed of the appearance of a comet in the sky, a portent usually associated with the death of a ruler; he claimed that it referred not to him, as he was bald, but to the king of the Parthians, who was long haired.[92] He also associates baldness with the moon.[93] This seems to have been a common association; Varro recommends not cutting one's hair when the moon is waxing for fear of going bald, while Ausonius states that Mars loved the beardless while the moon loved the bald.[94] He also observes that baldness has a role to play in the maintenance and restoration of health, as it is frequently recommended as a starting point for therapy.[95]

Synesius engages directly with the examples of famous hirsute individuals that Dio cites and refutes each one. When Dio refers to the Spartan practice of maintaining their hair on the eve of battle, and the fact that this adds to their fearsome reputation, Synesius counters that all three hundred Spartans died.[96] He follows that with a series of examples in which hair and military activity were a bad mix.[97] When Dio uses Achilles, who was, as we have seen, particularly famed for his hair in antiquity, Synesius counters that with the suggestion that, based on the description that Homer gives, Achilles was likely balding.[98] He concludes his treatise by proposing that it is the nefarious characters of ancient society who pay the most attention to their hair: the adulterous, the luxurious, the effeminate, the perverted, the self-mutilated.[99] This is a standard listing of individuals who were considered deplorable by the elite members of ancient Roman society, and certainly their hair was used as a means of identifying them.

Conclusion

The Romans considered premature hair loss, whether experienced by men or by women, disfiguring. However, an individual suffering from premature hair loss was not only literally but also figuratively disfigured. Premature hair loss was a significant deviation from the natural order of things, and the way that both men and women were supposed to look prior to old age. In a society that utilized physical appearance as an indicator of any number of things, an individual's hair and the way that it was styled told someone looking at them everything they needed to know. Someone suffering from premature hair loss was, quite literally, not conforming to Roman society's unofficial rules regarding how they were supposed to look for the purpose of fulfilling the social contract. Additionally, the belief in a direct connection between an individual's physical appearance and their character and morality meant that those whose physical appearance deviated from the desirable norm were thought to possess a character and morality that likewise deviated from the desirable norm.

Finally, let us return to Domitian. The Flavian dynasty had previously been associated with physiognomy in a positive way. Domitian's elder brother Titus, who had been educated alongside Emperor Claudius' son Britannicus, had had his face examined by a physiognomist hired to examine Britannicus', and the physiognomist declared that while Britannicus would never rule, Titus would.[100] Additionally, Vespasian himself gave considerable credence to divination through astrology and dream interpretation.[101] Was it feasible to endorse divination through physiognomy in the case of one Flavian emperor and reject it in the case of another? Domitian's hair loss was certainly, as far as the Romans were concerned, a facial disfigurement. However, it was also indicative of a much more severe moral disfigurement, so it is not surprising that it was this physical feature that the 'bald Nero' was remembered for.

Notes

1 http://www.nhs.uk/conditions/hair-loss/Pages/Introduction.aspx (accessed December 2016).
2 Ibid.
3 On the bald man as a figure of fun, see Mary Beard, *Laughter in Ancient Rome: On Joking, Tickling, and Cracking Up* (Berkeley and Los Angeles and London: University of California Press, 2014), 51, 132–3, 146.
4 On public mockery and humiliation more generally, see R. Garland, 'The Mockery of the Deformed and the Disabled in Graeco-Roman Culture', in *Laughter Down the Centuries*, ed. S. Jäkel and A. Timonen (Turku: Turun Yliopisto, 1994), 74.
5 Ancient accounts of impairment and/or disability written from the first-person perspective are exceedingly rare. For discussion, see J. Draycott, 'The Lived Experience of Disability in Antiquity: A Case Study from Roman Egypt', *Greece & Rome* 62.2 (2015): 189–205.
6 A. Corbeill, *Controlling Laughter: Political Humour in the Late Roman Republic* (Princeton: Princeton University Press, 1996), 15.
7 C. Edwards, *The Politics of Immorality in Ancient Rome* (Cambridge: Cambridge University Press, 1993), 90.
8 E. Gunderson, *Staging Masculinity: The Rhetoric of Performance in the Roman World* (Ann Arbor: University of Michigan Press, 2000), 70.
9 Ibid., 61.
10 Cicero, *On the Laws* 1.27, trans. C. W. Keyes, Loeb Classical Library 213 (Cambridge, MA: Harvard University Press, 1928).
11 See, for example, Quintilian, *The Orator's Education* 11.3.70, trans. D. A. Russell, Loeb Classical Library 494 (Cambridge, MA: Harvard University Press, 2002).
12 See, for example, the case of Marcus Sergius Silus who lost his right hand in the Second Punic War and was subsequently victimized by contemporaries who tried to exclude him from religious rites on the grounds of his impairment. His speech refuting their right to do so is discussed by Pliny the Elder, *Natural History*

7.28.104–5, trans. H. Rackham, Loeb Classical Library 352 (Cambridge, MA: Harvard University Press, 1942).

13 Cicero, *On the Orator* 2.239, trans. E. W. Sutton and H. Rackham, Loeb Classical Library 348 (Cambridge, MA: Harvard University Press, 1942).

14 J. P. Cèbe, 'Catulle et la physiognomie', *Annales de la Faculté des Lettres d'Aix* 43 (1967): 174–8.

15 See, for example, Aristotle, *Generation of Animals* 1.721b29–35, trans. A. L. Peck, Loeb Classical Library 366 (Cambridge, MA: Harvard University Press, 1942); Pliny the Elder, *Natural History* 7.11.50 for the belief that impairments and disfigurements could be inherited. See also N. Vlahogiannis, 'Curing Disability', in *Health in Antiquity*, ed. H. King (London: Routledge, 2005), 180–91, for the suggestion that impairments were interpreted as divine punishment for transgression.

16 Cicero, *On Invention* 2.28, trans. H. M. Hubbell, Loeb Classical Library 386 (Cambridge, MA: Harvard University Press, 1949); Quintilian, *The Orator's Education* 1.4.25.

17 Cicero, *Divisions of Oratory* 74, trans. H. Rackham, Loeb Classical Library 349 (Cambridge, MA: Harvard University Press, 1942).

18 Cicero, *On Invention* 1.34–6.

19 Cicero, *Speech in Defence of Quintus Roscius the Comedian* 20, trans. J. H. Freese, Loeb Classical Library 240 (Cambridge, MA: Harvard University Press, 1930).

20 For a similar accusation of extreme depilation, see Scipio Aemilianus' invective against Publius Sulpicius Gallus at Aulus Gellius, *Attic Nights* 6.12.5, trans. J. C. Rolfe, Loeb Classical Library 200 (Cambridge, MA: Harvard University Press, 1927). For the Roman practice of depilation more generally, see Seneca, *Natural Questions* 1.16–17, trans. T. H. Corcoran, Loeb Classical Library 450 (Cambridge, MA: Harvard University Press, 1971); Martial, *Epigrams* 3.74 and 10.65, trans. D. R. Shackleton Bailey, Loeb Classical Library 94 and 95 (Cambridge, MA: Harvard University Press, 1993); Athenaeus, *Dinner Sophists* 13.564f–5f, trans. S. D. Olson, Loeb Classical Library 327 (Cambridge, MA: Harvard University Press, 2010).

21 On accusations of effeminacy as an attempt to blacken someone's character, see C. Edwards, *The Politics of Immorality in Ancient Rome* (Cambridge: Cambridge University Press, 1993), 65; M. Gleason, *Making Men: Sophists and Self-presentation in Ancient Rome* (Princeton: Princeton University Press, 1995), 67–70; C. Williams, *Roman Homosexuality* (Oxford: Oxford University Press, 1999), 129–31. The line between effeminacy and urbanity was extremely thin, see Edwards, *Politics*, 67–8; this is explored in more depth in K. Olson, 'Masculinity, Appearance, and Sexuality: Dandies in Roman Antiquity', *Journal of the History of Sexuality* 23.2 (2014): 182–205.

22 M. M. Levine, 'The Gendered Grammar of Ancient Mediterranean Hair', in *Off with Her Head! The Denial of Women's Identity in Myth, Religion, and Culture*, ed. H. Eilberg-Schwartz and W. Doniger (Berkeley: University of California Press, 1995), 87–8.

23 On the Myconians, see Strabo, *Geography* 10.9, trans. H. L. Jones, Loeb Classical Library 211 (Cambridge, MA: Harvard University Press, 1928); Pliny the Elder, *Natural History* 11.47, trans. H. Rackham, Loeb Classical Library 353 (Cambridge, MA: Harvard University Press, 1940).

24 Aristotle, *Generation of Animals* 5.783b8–9, 5.784a–85a; Aristotle, *History of Animals* 3.11.518a 1–23, trans. A. L. Peck, Loeb Classical Library 437 (Cambridge, MA: Harvard University Press, 1965); Aristotle, *Problems* 17.26, trans. R. Mayhew, Loeb Classical Library 38 (Cambridge, MA: Harvard University Press, 2011).

25 Aristotle, *History of Animals* 3.11.518a; Macrobius, *Saturnalia* 6.4–5, trans. R. A. Kaster, Loeb Classical Library 512 (Cambridge, MA: Harvard University Press, 2011).
26 Aristotle, *Generation of Animals* 4.765b, 4.783b; Aristotle, *Problems* 10.24; *Greek Anthology* 11.190 and 11.368, trans. W. R. Paton, Loeb Classical Library 85 (Cambridge, MA: Harvard University Press, 1918).
27 Puberty: Aristotle, *Generation of Animals* 5.783b8; Aristotle, *Problems* 4.4; Pliny the Elder, *Natural History* 11.47. Eunuchs: Hippocrates, *Aphorisms* 28, trans. W. H. S. Jones, Loeb Classical Library 150 (Cambridge, MA: Harvard University Press, 1931); Hippocrates, *Nature of the Child* 510, trans. P. Potter, Loeb Classical Library 520 (Cambridge, MA: Harvard University Press, 2012); Aristotle, *Problems* 10.57; Aristotle, *History of Animals* 8.632a; Pliny the Elder, *Natural History* 11.47. Women: Aristotle, *Generation of Animals* 5.783b; Aristotle, *Problems* 10.57; Aristotle, *History of Animals* 3.11.518a; Pliny the Elder, *Natural History* 11.47.
28 Galen, *Mixtures* 2.621, trans. P. N. Singer (Oxford: Oxford University Press, 1997). While male hair loss and baldness in Graeco-Roman antiquity has not been extensively discussed, male hair loss and baldness in the early modern period has been explored in A. Korhonen, 'Strange Things Out of Hair: Baldness and Masculinity in Early Modern England', *The Sixteenth Century Journal* 41.2 (2010): 371–91, with a focus on what this means for constructions of masculinity.
29 Gleason, *Making Men*.
30 See, for example, Pterelaus and Nisus and Apollodorus, *The Library* 2.4.7 and 3.15.8 respectively, trans. J. G. Frazer, Loeb Classical Library 121 and 122 (Cambridge, MA: Harvard University Press, 1921); Ovid, *Metamorphoses* 8.1–151, trans. F. J. Miller, Loeb Classical Library 42 (Cambridge, MA: Harvard University Press, 1916); Aeschylus, *Libation Bearers* 612–22, trans. A. H. Sommerstein, Loeb Classical Library 146 (Cambridge, MA: Harvard University Press, 2009); Hyginus, *Fables* 198, 242, trans. M. Grant (Lawrence: University of Kansas Press, 1960); Pausanias, *Description of Greece* 1.19.4, 2.34.7, trans. W. H. S. Jones, Loeb Classical Library 93 (Cambridge, MA: Harvard University Press, 1918); Propertius, *Elegies* 3.19.21–8, trans. G. P. Goold, Loeb Classical Library 18 (Cambridge, MA: Harvard University Press, 1990); Vergil, *Georgics* 1.404–9, trans. H. R. Fairclough, Loeb Classical Library 63 (Cambridge, MA: Harvard University Press, 1999).
31 Aristotle, *Categories* 10.13a, trans. H. P. Cooke and H. Tredennick, Loeb Classical Library 325 (Cambridge, MA: Harvard University Press, 1938).
32 Celsus, *On Medicine* 6.1, trans. W. G. Spencer, Loeb Classical Library 304 (Cambridge, MA: Harvard University Press, 1938).
33 See T. Parkin, *Old Age in the Roman World* (Baltimore: Johns Hopkins University Press, 2003), 82–3 on baldness in old age viewed negatively; see K. Cokayne, *Experiencing Old Age in Ancient Rome* (London and New York: Routledge, 2003), 11–23 on baldness in old age viewed first negatively, and then positively.
34 Cokayne, *Experiencing Old Age*, 18–23.
35 J. Clark, *Looking at Laughter: Humor, Power and Transgression in Roman Visual Culture, 100 B.C. -A.D. 250* (Berkeley: University of California Press, 2007), 45, states that a quarter of the examples have bald heads.
36 Ibid., 47.
37 Plucking: Diodorus Siculus, *Library* 33.7.6–7, trans. F. R. Walton, Loeb Classical Library 423 (Cambridge, MA: Harvard University Press, 1967); Phaedrus, *Fables* 2.2, trans. B. E. Perry, Loeb Classical Library 436 (Cambridge, MA: Harvard University Press, 1965); Macrobius, *Saturnalia* 2.7, trans. R. A. Kaster, Loeb Classical Library 510

(Cambridge, MA: Harvard University Press, 2011). Dying: Ovid, *Amores* 1.14, trans. G. Showerman, Loeb Classical Library 41 (Cambridge, MA: Harvard University Press, 1914); Martial, *Epigrams* 14.27.

38 Sex: Aristotle, *Problems* 4.18. Wine consumption: Plutarch, *Table Talk* 652f, trans. P. A. Clement and H. B. Hoffleit, Loeb Classical Library 424 (Cambridge, MA: Harvard University Press, 1969); Macrobius, *Saturnalia* 6.11.

39 Seneca, *Moral Epistles* 95.21, trans. R. M. Gummere, Loeb Classical Library 77 (Cambridge, MA: Harvard University Press, 1925).

40 M. Gleason, 'The Semiotics of Gender: Physiognomy and Self-Fashioning in the Second Century AD', in *Before Sexuality*, ed. D. Halperin, J. Winkler and F. Zeitlin (Princeton: Princeton University Press, 1991), 389.

41 Gleason, 'The Semiotics of Gender'.

42 Pseudo-Aristotle, *Physiognomics* 1 (trans. S. Swain), in S. Swain, 'Appendix: The Physiognomy Attributed to Aristotle', in *Seeing the Face, Seeing the Soul: Polemon's Physiognomy from Classical Antiquity to Medieval Islam*, ed. S. Swain (Oxford: Oxford University Press, 2007), 637–61.

43 Polemon, *Physiognomy* 40, trans. R. Hoyland, in *Seeing the Face, Seeing the Soul: Polemon's Physiognomy from Classical Antiquity to Medieval Islam*, ed. S. Swain (Oxford: Oxford University Press, 2007), 329–463.

44 Adamantius, *Physiognomy* B37, trans. I. Repath, in *Seeing the Face, Seeing the Soul: Polemon's Physiognomy from Classical Antiquity to Medieval Islam*, ed. S. Swain (Oxford: Oxford University Press, 2007), 487–547.

45 Cokayne, *Experiencing Old Age*, 12–15.

46 For a survey of the physical appearances of the Roman emperors from the first century BCE to the fourth century CE, see H. V. Canter, 'Personal Appearance in the Biography of the Roman Emperors', *Studies in Philology* 25.3 (1928): 385–99; E. Evans, 'Roman Descriptions of Personal Appearance in History and Biography', *Harvard Studies in Classical Philology* 46 (1935): 43–84; A. E. Wardman, 'Description of Personal Appearance in Plutarch and Suetonius: The Use of Statues as Evidence', *Classical Quarterly* 17 (1967): 414–20.

47 B. Gladhill, 'The Emperor's No Clothes: Suetonius and the Dynamics of Corporeal Ecphrasis', *Classical Antiquity* 31.2 (2012): 315–48; D. Rohrbacher, 'Physiognomics in Imperial Latin Biography', *Classical Antiquity* 29.1 (2010): 92–116; A. Wallace-Hadrill, *Suetonius: The Scholar and his Caesars* (London: Duckworth, 1983), 196. Suetonius is also known to have written a treatise, now lost, called *On Physical Defects*.

48 Suetonius, *Julius Caesar* 45, and *Gaius* 50, trans. J. C. Rolfe, Loeb Classical Library 31 (Cambridge, MA: Harvard University Press, 1914); Suetonius, *Galba* 21 and *Domitian* 18, trans. J. C. Rolfe, Loeb Classical Library 38 (Cambridge, MA: Harvard University Press, 1914). The emperor Otho is described as wearing a wig that was of such high quality that no one knew he suffered from thinning hair until after his death, Suetonius, *Otho* 12, trans. J. C. Rolfe, Loeb Classical Library 38 (Cambridge, MA: Harvard University Press, 1914).

49 Suetonius, *Julius Caesar* 45.

50 Ibid., 51.

51 Suetonius, *Gaius* 59.

52 Ibid., 50.

53 Seneca, *On Firmness* 18.1, trans. J. W. Basore, Loeb Classical Library 214 (Cambridge, MA: Harvard University Press, 1928).

54 Suetonius, *Domitian* 18.1.

55 S. Wood, 'Public Images of the Flavian Dynasty: Sculpture and Coinage', in *A Companion to the Flavian Age of Imperial Rome*, ed. A. Zissos (Chichester and Malden: John Wiley and Sons Ltd, 2016), 130.
56 Wood, 'Public Images of the Flavian Dynasty', 132.
57 Ibid., 139.
58 Corbeill, *Controlling Laughter*, 34.
59 Ibid.
60 Suetonius, *Domitian* 3.2; see also B. W. Jones, *The Emperor Domitian* (London: Routledge, 1992); P. Southern, *Domitian: Tragic Tyrant* (London: Routledge, 1997).
61 Suetonius, *Domitian* 18.2.
62 Jones, *Emperor Domitian*, 31 believes it is true. Southern, *Domitian*, 119 does not.
63 Beard, *Laughter in Ancient Rome*, 132; on Suetonius' 'concealed invective', see T. Barton, *Power and Knowledge: Astrology, Physiognomics, and Medicine under the Roman Empire* (Ann Arbor: University of Michigan Press, 1994).
64 See, for example, *Philogelos* 56, trans. B. Baldwin (Amsterdam: Gieben, 1983).
65 Southern, *Domitian*, 119.
66 Suetonius, *Domitian* 18.2. See also L. Morgan, 'Hair and Heroism According to Domitian', *Classical Quarterly* 47.1 (1997): 209–14; K. M. Coleman, 'The Emperor Domitian and Literature', *ANRW* 2nd series 32.5 (1986): 3087–115.
67 Plutarch, *On Common Conceptions* 1065C, trans. H. Cherniss, Loeb Classical Library 470 (Cambridge, MA: Harvard University Press, 1976). On Thersites' physical appearance, see Homer, *Iliad* 2.211–77, trans. A. T. Murray, Loeb Classical Library 170 (Cambridge, MA: Harvard University Press, 1924); Epictetus, *Discourses* 4.2.10, trans. W. C. Oldfather, Loeb Classical Library 218 (Cambridge, MA: Harvard University Press, 1928); Plutarch, *How the Young Man Should Study Poetry* 35c, trans. F. C. Babbitt, Loeb Classical Library 197 (Cambridge, MA: Harvard University Press, 1927). See also Martial, *Epigrams* 12.82 for an account of the sycophant Menogenes complimenting a virtually bald man on his luxurious head of hair and resultant resemblance to Achilles.
68 See Juvenal, *Satires* 4.37–8, trans. S. M. Braund, Loeb Classical Library 91 (Cambridge, MA: Harvard University Press, 2004) and Ausonius, *Twelve Caesars* 2.10–12, trans. H. G. Evelyn-White, Loeb Classical Library 96 (Cambridge, MA: Harvard University Press, 1919) for Domitian being referred to by his subjects as 'bald Nero'.
69 On hair loss: Celsus, *On Medicine* 6.4; Pliny the Elder, *Natural History* 28.163–6, trans. W. H. S. Jones, Loeb Classical Library 418 (Cambridge, MA: Harvard University Press, 1963); Galen 12.403–5 K, in *Claudii Galeni Opera Omnia* Volume 12, trans. K. G. Kühn (Cambridge: Cambridge University Press, 2011); Aëtius, *Sixteen Books on Medicine* 6.65, trans. A. Olivieri (Berlin: Corpus Medicorum Graecorum, 1935); Paul of Aegina, *Medical Compendium in Seven Books* 3.1.1–2, trans. F. Adams (London: The Sydenham Society, 1844); on encouraging hair growth: Galen 13.432–4 K, in *Claudii Galeni Opera Omnia* Volume 13, trans. K. G. Kühn (Cambridge: Cambridge University Press, 2011).
70 J. Draycott, 'Prosthetic Hair in Ancient Rome?' in *Prostheses in Antiquity*, ed. J. Draycott (London: Routledge, forthcoming).
71 J. Draycott, 'Hair Today, Gone Tomorrow: The Use of Real, False and Artificial Hair as Votive Offerings', in *Bodies of Evidence: The Anatomical Votive Past, Present and Future*, ed. J. Draycott and E.-J. Graham (London: Routledge, 2017), 77–94.
72 M. D. Grmek and D. Gourevitch, *Les maladies dans l'art antique* (Paris: Fayard, 1998), 345–6.

73 Such treatises were also published in early modern England, Korhonen, 'Strange Things', 388–90.
74 S. Goldhill, 'Rhetoric and the Second Sophistic', in *The Cambridge Companion to Ancient Rhetoric*, ed. E. Gunderson (Cambridge: Cambridge University Press, 2009), 232.
75 Synesius, *Oration* 40.1.1, trans. A. Fitzgerald (London: Oxford University Press, 1930).
76 Synesius, *Oration* 40.3.3–9; on the three hundred Spartans, see Herodotus, *Histories* 7.208–9, trans. A. D. Godley, Loeb Classical Library 119 (Cambridge, MA: Harvard University Press, 1922); on the Homeric heroes, see, for example, Homer, *Iliad* 3.43 specifically on the Achaeans.
77 On Greek education during the Roman period, see R. Cribiore, *Gymnastics of the Mind. Greek Education in Hellenistic and Roman Egypt* (Princeton: Princeton University Press, 2001).
78 Plutarch, *Lycurgus* 22.1, trans. B. Perrin, Loeb Classical Library 46 (Cambridge, MA: Harvard University Press, 1914).
79 D. Leitao, 'Adolescent Hair-growing and Hair-cutting Rituals in Ancient Greece: A Sociological Approach', in *Initiation in Ancient Greek Rituals and Narratives: New Critical Perspectives*, ed. D. Dodd and C. A. Faraone (London: Routledge, 2003), 124–6.
80 Leitao, 'Adolescent Hair-growing', 119.
81 On the life of Synesius, see J. Bregman, *Synesius of Cyrene: Philosopher-Bishop* (Berkeley, Los Angeles and London: University of California Press, 1982).
82 Synesius, *Oration* 40.4.1–2; 4.1, 19.1.
83 Synesius, *Oration* 40.1.1, 5.1, 24.2.
84 Synesius, *Oration* 40.1.1.
85 For the explicit linking of female sex appeal and hair, see Ovid, *Art of Love* 3.242–50, trans. J. H. Mozley, Loeb Classical Library 232 (Cambridge, MA: Harvard University Press, 1929); Apuleius, *Metamorphosis* 2.8, trans. J. A. Hanson, Loeb Classical Library 44 (Cambridge, MA: Harvard University Press, 1996). For discussion, see L. Sensi, 'Ornatus e status sociale delle donne romane', *Annali della Faculta di Lettere e Filosofia Perugia-Sezione Studi Classici* n.s. 4 (1980–1): 55–102.
86 Synesius, *Oration* 40.5.1–5, 6.1–6.
87 P. Zanker, *The Mask of Socrates: The Image of the Intellectual in Antiquity* (Berkeley: University of California Press, 1995), 224.
88 Synesius, *Oration* 40.5.6, 7.1–6, 8.1. Levine, 'Gendered Grammar', 86 suggests that 'ritual baldness' mimics the liminal states of infancy and old age.
89 Synesius, *Oration* 40.8.1, 8.5.
90 Synesius, *Oration* 40.9.2–3; 9.1, 10.1–3.
91 Synesius, *Oration* 40.10.5.
92 Suetonius, *Vespasian* 23, trans. J. C. Rolfe, Loeb Classical Library 38 (Cambridge, MA: Harvard University Press, 1914); Aurelius Victor, *On the Caesars* 9, trans. H. W. Bird (Liverpool: Liverpool University Press, 1994); the Latin word for comet, *comata*, literally means 'long-haired star'. This comet is also recorded in Cassius Dio, *Roman History* 66.17.2, trans. E. Cary and H. B. Foster, Loeb Classical Library 176 (Cambridge, MA: Harvard University Press, 1925).
93 Synesius, *Oration* 40.10.6.
94 Varro, *On Agriculture* 1.37.2, trans. W. D. Hooper and H. B. Ash, Loeb Classical Library 283 (Cambridge, MA: Harvard University Press, 1934); Ausonius, *Eclogues* 26.5–6, trans. H. G. Evelyn-White, Loeb Classical Library 96 (Cambridge, MA: Harvard University Press, 1919).

95 Synesius, *Oration* 40.12.1–3; see Celsus, *On Medicine* 1.4.1, 3.18.8, 3.20.4, 3.23.7, 4.2.6, 4.2.9, 4.11.8, 6.2.2, 6.6.8 E, 6.6.15, 6.7.1 D, 6.7.4, 6.7.8 B, 7.7.15 D for recommendations of therapeutic head shaving.

96 For the Spartan practice of maintaining their hair on the eve of battle, and the famed three hundred doing so on the eve of the Battle of Thermopylae, see Plutarch, *Lycurgus* 22.1; Herodotus, *Histories* 7.208–9. For Dio's reference, see Synesius, *Oration* 40.3.1; for Synesius' refutation, see *Oration* 40.15.1.

97 Synesius, *Oration* 40.15.3–16.4.

98 For Dio's reference, see Synesius, *Oration* 40.3.4; for Synesius' refutation, see *Oration* 40.17.2–5, 6; for Achilles' supposed baldness, see *Oration* 40.18.3–4. The state of Achilles' head is discussed in Morgan, 'Hair and Heroism', 210–13.

99 Synesius, *Oration* 40.21.1–3.

100 Suetonius, *Titus* 2.1, trans. J. C. Rolfe, Loeb Classical Library 38 (Cambridge, MA: Harvard University Press, 1914).

101 Suetonius, *Vespasian* 25.

6

Portraits, Likenesses, Composites? Facial Difference in Forensic Art

Kathryn Smith

Introduction

Reflecting on recognition, Ludwig Wittgenstein commented:

> It is easy to have a false picture of the process called 'recognizing'; as if recognizing always consisted in comparing two impressions with one another. It is as if I carried a picture of an object with me and used it to perform an identification of an object as the one represented by the picture … . And it is not so much as if I were comparing the object with the picture beside it, but as if the object coincided with the picture. I see only one thing not two.[1]

A crime is witnessed, a suspect is sought. A witness is interviewed, ideally within forty-eight hours but usually much longer after the event. They may be traumatized. The structured conversation between witness and forensic artist produces an image, usually rendered as a greyscale image of limited tonal range that has a loose association with photo-realism. Sometimes these images are very detailed, demonstrating skilled draughtsmanship and embodying a distinct affect. More often they resemble a barely believable character from an early-generation video game, their facial parts floating on a basic head shape, sometimes without a neck but perhaps wearing a hat or glasses; whatever the witness regarded as salient and specific details that distinguished the individual. These are the images that you might see flashed up on the evening news, or pinned to the wall of your neighbourhood convenience store or police station.

The police composite sketch is arguably the most fundamental example of forensic art, with its origins in the most primary method of visual representation: drawing. Even though contemporary composites are digital images, produced using computer software, the 'sketch' descriptor often persists. These images possess considerable cultural cachet, an iconic visual shorthand for 'criminal investigation'. Based on verbal descriptions drawn from memory deriving from highly contingent events, composites are by definition unique and precarious forensic objects, representing an epistemological paradox in their definition as simultaneous 'artistic impression' and 'pictorial statement'. As artefacts of an eyewitness account, they lend support to an investigation yet should not be considered hard evidence, although the eyewitness account itself carries significant evidentiary weight. Contingent, contested and controversial, despite decades of operational use only in recent years has the field of cognitive psychology begun to fully understand and address the conditions that affect recognition rates both positively and negatively.[2]

This chapter advances an interpretation of 'facial difference' by transposing this idea onto the theory and practice of forensic art, in order to explore variation in facial depictions produced by and for the forensic context, with specific reference to the police composite. Forensic art describes a range of practices which produce facial images to aid processes of human identification. The quantification and reproducibility of facial likeness has indisputable forensic value. A core practice within human identification – police have long employed artists or officers with artistic ability to assist with aspects of criminal investigations. This forensic value is produced via its claims to scientific truth, created via scientific experimentation and validation in which repeatability is a core principle. Intended to produce a positive identification of a specific individual, forensic facial depictions are a form of visual intelligence and in theory, should not be considered evidence that 'proves' identification. This makes sense, given the often apparent difference between the published facial image and the person eventually identified. Others exhibit uncanny similarities to their intended target.

As an interdisciplinary practitioner working across both contemporary art and forensic contexts, I have an enduring interest in how contemporary art (and culture at large) is increasingly informed by forensic practices. This chapter represents some first steps in the context of a broader project that develops a theory of forensic art written from an interdisciplinary, practitioner perspective, in which I attend to ways in which knowledge transfers from one area of praxis to the other, and how these respective concerns are shaped in turn. In this transdisciplinary space, methods and ideals conventionally associated with laboratory research (empiricism, quantitative assessment, objectivity) are always in conversation with the subjective contingencies and desires, as well as critical rigour (formal analysis, aesthetic evaluation and interpretation) of the visual arts studio.

The human face is a highly complex organ, the anatomy, function and appearance of which are the domain of a host of individual disciplinary or specialist fields, from human identification and biometrics; maxillofacial, reconstructive and aesthetic surgery; neuroscience and cognitive psychology; computer vision, machine learning and artificial intelligence engineers; animators, game designers and of course, visual artists. Representations of the face exist as some of oldest examples of visual material culture (consider the Jericho skulls or Fayyum portraits), and the portrait remains an enduring genre in figurative art forms, especially photography. Yet manual and mechanical techniques still embody equivocal attitudes to mimetic truth, with painting and sculpture considered inherently subjective forms of visual representation, and mechanical processes continuing to enjoy a more intimate relationship to scientific objectivity and of course, reproductive efficiency. As a 'visual record', photography became the dominant mode of truthful and accurate visual representation, from bureaucratic structures to creative endeavours, clinical imaging and the domestic realm, theorized with reference to its 'indexicality' in the semiotic sense. That which it depicts once existed in the world, in front of a camera's lens.[3] It is widely accepted that a portrait is not simply an indexical or analogical likeness of a real face, but gathers into its representational ambit a host of conceptual ideas – social status, for example – that attach to the face to create a persona. Entering the virtual and AI realm, robotics and gaming practitioners contend with the 'uncanny valley', the perceptual–psychological interstice first described as *Bukimi no Tani Genshō* by Masahiro Mori (1970) as the almost abject process by which we elect whether to suspend our disbelief in the face of humanoid entities (automaton, dolls and social robots).[4]

When presented with the problem of reproducing likeness or similarity, which is by extension a question of understanding difference, the scientific imperative is to quantify. Knowledge is synonymous with particular kinds of measurement. Where faces are concerned, the scientific tendency is to act like a land surveyor, plotting facial landmarks and measuring the distances between them, comparing spatial configurations with the shape of individual features, interpolating missing data, noting deviations and ultimately producing geometrically elaborate 2D and 3D topographies of the face. Appearance becomes a technology of individuality and difference.

But what of the less tangible data that faces give us? Assuming the face is visible (and not covered for religious or other reasons), and we are able to see it (blindness and prosopagnosia excepted), its complex topography is what we attempt to read, albeit in different ways, when we are meeting someone for the first time. Following Levinas, the face as a space of empathetic encounter echoes the focus of our attention in our early infancy as we learnt the fundamentals of reciprocal communication. Recognition, Michael Podro reminds us, is 'the most basic relation we have to the world'.[5]

The innumerable permutations of the face's mechanical and expressive movements etch their effects on our appearance over time. We can probably recognize someone we know well even if they completely change their hairstyle, or grow or shave their facial hair, or gain or lose a great deal of weight, but perhaps not? Making decisions about these factors, individually and in relation to one another, in terms of featural specificity as well as the holistic gestalt of a face, is what is required when we are forced to rely on a facial image to identify or locate someone in the absence of the accepted scientific methods – fingerprinting or DNA. (It may not be possible to extract a DNA sample from a set of human remains due to their condition. A DNA sample is only useful if there is a matching sample recorded and stored on a database for comparison, or if leads produce a family member who agrees to their DNA being taken for comparative purposes.)

The range of recognized forensic facial depiction methods used to promote identification of the living, missing or the unidentified dead are, at their most basic, either two- or three-dimensional representations of a particular human face, produced with direct or indirect reference to either material (skeletal) remains and related written/visual case documentation, or visual documentation of the person concerned. In the case of eyewitness composites, the primary reference material is at least one memory-derived verbal description. Techniques range from the hand-drawn or sculpted to hybrid manual-digital techniques, digital compositing and fully computer-automated systems.

The other most 'visible' expression of forensic art outside of the police sketch are faces produced from skulls, a method for which the most robust scientific techniques exist.[6] Less visible techniques are post-mortem depiction and age progression, which rely on a combination of skills required by both skull-based reconstruction and memory-derived composites. Post-mortem depiction uses either drawing or digital photo-based compositing techniques (or a combination thereof) to sanitize photographic images of the unidentified dead to create publishable images, informed by a knowledge of post-mortem facial changes as a result of trauma or decomposition. Likewise, age progressions may be produced as sketches or more commonly, as digital photographic composites to produce impressions of what a missing child or adult (victim or fugitive) might look like years after their original disappearance. With the latter, the question of whether the individual is alive or dead must also remain open. Such images may be a lifeline for confused or bereaved families who have lost loved ones; we invest them with an enormous responsibility which they may fulfil if they are seen by the right person, even if they do not appear to achieve perfect verisimilitude with the target face.

Despite established data for predicting and depicting faces from skulls, marked differences in results are often seen in reconstructions of the same individual. Why does this happen? Is it an artefact of poorly applied scientific method or inadequate artistic skills, or both? Suspect composites show similar (even amplified) effects, but being derived from verbal description and memory, we should anticipate a far more fluid set of variables to be present.

It could be argued that reconstructions from the skull and composites are distinct processes, requiring different skill sets, so why would it make sense to draw this comparison at the outset? Because of their operational function to track down active criminals, and the fact that eyewitness identification holds significant legal weight, composites have been given the lion's share of research attention in the area of face perception/cognition. These processes are embedded in their construction via eyewitness recall, in both linguistic and visual terms. Seen differently, considering the composite allows us to spend time with the face as both an imaginative and a recollected object. The composite is a specific kind of image, one produced from language, and one which language may further mobilize as an evidentiary object. Composites tend to depict suspects, which signify an immediacy that depictions of victims do not possess in quite the same way. An urgency is certainly present in victim images – it is a legal imperative to identify the dead and certainly an ethical one to locate the missing – but such depictions denote a vulnerable subject (loss of agency) whereas suspect images connote potential threat.

We could set out their differences according to simple differences: reconstructions from the skull are intended to identify victims and composites, perpetrators; in other words, the former serves the dead and the latter targets the (presumed) living. The former uses material things – human remains and associated objects – and the latter relies on more immaterial traces and expressions of memory and language. They are informed by different research areas, with anatomy and pathology informing facial reconstruction methodology and cognitive psychology driving eyewitness interviewing and identification techniques and their products. Operationally, these identification tasks are usually undertaken by different units within law enforcement agencies, although different countries adopt different approaches.

Both areas of practice certainly benefit from specific consideration of their respective functions, techniques and challenges. But imagined as a Venn diagram, the space in which they intersect is a central, core function of both endeavours: an understanding of how such images may be received by the communities that will ultimately allow them to do the work of depiction. How facially literate are we? Do we understand how these images are read and interpreted; in other words, how they are likely to be processed both cognitively and culturally? Are we, broadly speaking, visually literate when it comes to forensic facial depictions?

The paradox between resemblance and recognition – that accuracy and likeness may not correlate as closely as we imagine they should – is *the* enduring conundrum of this work, and describes what animates my interest at a primary level. What might represent an artistic and aesthetic failure – poor anatomical accuracy or clumsy sketching, weak sculptural or Photoshop skills – may still be a forensic success *if* the depiction possesses sufficient gestalt to spark recognition. Visually different faces can be recognized as sufficiently similar to – if not precisely the same as – a missing loved one. Notwithstanding the practical fact of the image needing to be seen by the right person, the depiction may still be recognized, producing leads and – hopefully – a scientifically acceptable positive confirmation of biological identity (DNA and fingerprinting).

Structure

With reference to an interdisciplinary body of literature and practice-based experience (lab and field studies, criminal cases and recent contemporary art), this chapter explores the visual intelligence of forensic facial depictions, and the ways in which ideas about the face vary across the (often reluctantly) complementary practices of science and art. I consider these questions in relation to forensic art more broadly, its relationship to portraiture and concepts of accuracy and likeness by considering how the forensic composite is influencing the work of contemporary artists. Looking at forensic art 'awry' in this way (to borrow Slavoj Žižek's phrase) foregrounds the related but distinct ideas of 'identity' and 'identification', which inevitably entails a mention of the future of facial composites as bio-data models, and what the social, political and cultural implications of these new technologies and practices might be.[7]

I begin with a discussion of concepts of the face within forensic and visual art, exploring the languages we use to mobilize the face as a cultural and philosophical object and agent, from its ontology to its representational vocabulary. The discussion then moves to focus in on the composite, its functions and limits, and how this visual object has been used within cognitive psychology and computer vision research to develop knowledge about face perception.

Finally, the point about 'accuracy' is most clearly made where more than one reconstruction exists of a single, known individual. This principle guides my selection of examples, all of which explore correlations between multiple depictions, but also encompass the 'difference' in my title. The logic of the permutation is embedded in the ways in which composites are produced, and in turn how they have been studied within cognitive psychology, from the question of what to do with multiple witnesses to a single event, to the fact that caricatures in different visual styles can still be recognized as a certain person. The paradox animating the relationship between resemblance and recognition is extended by the idea that 'sketchy', 'incomplete' or cursorily drawn details, as well as distortion, can correct perceptual errors, and deliver a more convincing likeness than an image that attempts to slavishly mimic every detail of a face.

I conclude with a consideration of how concepts within visual art pertain to the (forensic) facial image, and in turn, how composites construct particular kinds of subjects. Considering time and language as significant contributing factors to their particular materiality, I understand composites as embodying the 'fugitive', in both literal and figurative terms. The uncanny spectres of faces created from pure data, including DNA, provide a further opportunity to consider whether, in the near future (or perhaps it is a moment we are already inhabiting?) notions of identification will float free of the facial image, and when, if ever, safety and security agencies and industries will liberate the face from their biometric-orientated, technophilic desires.

Thinking through forensic art

The very phrase 'forensic art' links two words with very different implications, intentions and practices, appearing to embody a paradoxical conjoining of applied science, on the one hand, and creative interpretation and invention, on the other.[8] The scientific literature, particularly that focusing on accuracy studies (the relationship between likeness and repeatability) in facial reconstruction from the skull, for example,[9] has attracted a range of critique from so-called hard scientists, with one even going so far as to dismiss the very notion of 'scientific art' as 'bizarre'.[10] Of course, this is historically and practically nonsensical. One only needs to cast the briefest glance at the rich and documented history of Western visual culture, from the Renaissance onwards, to find numerous examples of visual depiction – illustration, art, design, call it what you will – in the service of science, and artistic techniques that have scientific principles at their base, such as mathematical equations to achieve perspectival illusionism. A dismissive attitude towards the knowledge-producing potential of art also betrays a dangerously narrow view of what constitutes the 'scientific', which cannot be reduced to the purely quantitative. Art and science have more in common than not: both are informed by theory, driven by method, prompted by concept (innovative or not), executed by technique (skilled or not) and influenced by technology.[11]

Conventionally speaking, 'forensic art' is closely associated with facial images, reproduced with reference to human remains, testimony or visual documentation, although nomenclature within law enforcement may be confusing. The term 'forensic reconstruction' might be an operational term referring to spatial, rather than facial, visualization, as well as the production of demonstrative visual material for use in legal proceedings, such as sanitized injury depiction via computer-generated imagery (CGI) ('body mapping'), motivated by a regard for the dignity of the victim, and the emotional burden of those present. It may also include the work of the court sketch artist who visually documents legal proceedings where cameras are forbidden; a special genre of visual reportage. In the interests of disambiguation, the capitalized phrase 'Forensic Art' is generally used to refer to expert facial analysis and depiction, including methods of producing or comparing images of the human face and/or skull, according to scientifically assessed methods and standards. Such a proper compound noun suggests a discipline with agreed standards, principles and professional qualifications, which the practice cannot claim. I prefer the lowercase 'forensic art', which acknowledges the applied status of the 'forensic'; avoids any claims to 'Art' proper, and implicitly embodies an expanded field that includes spatial reconstructions and other forms of visual presentation used in court. Known as 'exhibits', these may also include material evidence, or visualizations of trace evidence interpreted through expert analysis and opinion.

Forensic art is usually practised by individuals working with law enforcement agencies, or those working alongside law enforcement in a freelance or consultative capacity. Over more than a century, the methodologies that have informed contemporary forensic facial depiction – which draws on studies in anatomy, facial growth and development, reconstructive surgery, dentistry, pathology, psychology and portraiture – have been developed and refined, but training and techniques are varied, so it stands to reason that the quality of forensic facial depictions can be described as wildly uneven. Practitioner experience may also be very varied; training within an academic environment may provide lab-based practice on historical remains but actual forensic field experience is limited, whereas working for law enforcement is entirely casework focused. Research and consultancy work might attract a range of face-based work, whereas law enforcement may separate the work of facial identification (composites and comparisons) and victim identification (reconstruction and superimposition). The South African Police Service, for example, is structured in this way, with little communication between the two units. Practitioners work across a range of media, including drawing, sculpture and digital montage, often in combination. Likewise, techniques may be traditional (manual), digital (or a combination thereof) or fully automated. It is understandable, given the range of techniques and available technology, that the accuracy and quality of depictions reflects these variables.

A visage by any other name

The ontology of images, particularly in disciplines which do not regard themselves as having visual cultures – science, law, geography and so on – has been given considerable attention by art historian James Elkins.[12] The deployment of facial images within those disciplines arguably serve a more instrumental purpose than in visual art, where conceptual and symbolic attachments and inferences of the facial image are central. But it is the paradoxical universality and elusiveness of the face as a known – visible, legible – object that is of particular interest:

> As adults, we are usually complacent and secure enough to define faces as we see them – I see you, and I know which part of you is your face. But of course that is not defining a face; it is identifying one. It's an odd situation we find ourselves

> in, not being able to say for sure what faces are, because faces must be among the most important objects that can be seen. They are indispensable for relations of any kind, and our own faces (and people's reactions to them) tell us what kind of people we are.[13]

Our ideas about faces tend to be normative and anthropocentric. This is significant because of the way in which we understand the face as functioning as both an icon and index of human interaction. We perform our identities through our faces and ideas about ourselves are projected onto us by others. We can construct and enhance our facial appearance (make-up or masks, surgery), or disease or trauma may alter our appearance temporarily or more permanently. As a biometric, it is a technology of identification. Deviations from the 'normal' face, the inability to make facial expressions (Moebius syndrome) or inability to recognize faces (prosopagnosia) operates as an injunction to normal' communication. Through the face, we relate to or reject one another.

Our assumptions about the co-ordinates of the human face are also reflected in our seemingly limitless capacity to project this idea of a face onto pretty much anything, animate or otherwise, even punctuation. Facial pareidolia describes the irresistible compulsion most of us have to project or read faces (and then expressive character), into all manner of inanimate objects from automobile design (headlights and front grills) to architectural structure to everyday objects.[14] Our use of emoticons in short-form text messaging and social media platforms might seem terribly contemporary but as Jon Calame (2013) reminds us, satirical US magazine *Puck* published set of sophisticated 'proto-emoticons' in 1881, demonstrating complex emotions – Joy, Melancholy, Indifference, Astonishment – with the simplest visual notation possible.[15]

We might be able to talk confidently about the shapes of faces from the perspective of anatomy, based on an understanding of the relationships between hard tissue (bone) and the soft tissues (muscles, skin and fat) of the face, and their particular structures and functions.[16] From the perspective of cognitive psychology, the relationship between resemblance and recognition, and the role that representational styles play in either encouraging or discouraging us to recognize similarities and differences between facial depictions, throws up a different, but related set of questions suggesting that the concept of 'likeness' is fairly fugitive or at least contingent. A mannerism or a 'look' can remind us of someone as much as a slavish reproduction of features. This throws into confusion assumptions we have about concepts of 'realism', 'naturalism' and 'verisimilitude' relative to the facial image.

Designations such as 'portrait', 'likeness' and 'composite' all suggest particular conceptual and technical parameters regarding depictions of the human face. Of these, 'portrait' is the most complex as it designates not just an object, but a genre that is historically and ideologically determined, and which has generated a vast raft of scholarship, predominantly in the fields of art history and visual culture.[17] It is after all a portrait, Leonardo Da Vinci's *Mona Lisa* (c. 1503–06), that is possibly the most famous (and most parodied) work of art in the world.

Within the field of human identification, questions of accuracy are played out in the terms of reference themselves. Accepted wisdom within the field is that forensic facial depictions do not (and should not) claim to function as a form of portraiture because of the inherent limits of what can either be predicted from the skull or depicted from memory.[18] The most common designation for building faces from skulls is 'facial reconstruction', which is also the term most commonly used in popular media. But related epistemological impasses have provoked objections to the claims implicit in the term 'reconstruction' – that an 'accurate' face can be reproduced from the skull. The term 'facial approximation' has therefore been proposed as a more precise reflection of process and result.[19] However, 'approximation' is more accurately used to describe fully computerized facial models that use fitting algorithms to match or 'wrap' a database-derived skin 'mask' to the target skull. Relying on statistical probability, and so eliminating highly characteristic features that we may associate with a specific individual face, such automated face models inevitably produce a facial 'type'. But forensic identification is specifically interested in unique features that identify a *particular* individual, and ideally aims to avoid misidentification wherever possible. As it is never possible to 'prove' identification with 100 per cent categorical accuracy (despite claims made about DNA), the principle of exclusion should predominate, as misidentification has significant consequences.

'Facial reconstruction' is often – reasonably – confused with reparative, corrective or cosmetic facial surgery. While forensic depiction owes a debt to maxillofacial and craniofacial surgical research, it is a different project. The alternative 'facial restoration' invokes the embalmer and their mortuary cosmetics which too, is a different project (although results can look uncomfortably similar). Further considerations concerning evidential weight (evidence vs. intelligence); circulation (among the appropriate community for recognition); and lack of widely agreed standards are more accurately represented by the relatively neutral term, 'depiction'. Hence my preference for the phrase 'forensic facial depiction', as it embraces the full range of facial images produced from human remains or from eyewitness memory, in the case of suspect composites. It acknowledges the anthropological, anatomical and representational (pictorial or sculptural) skills required to produce an individual or 'characteristic' face from information obtained either through analysis of the skull or verbal description. Pictorial and sculptural methods both involve attention to the qualitative and quantitative aspects of individual features within the holistic face 'complex'. Creative interpretation by the artist must remain sensitive to the limits of this knowledge. We are not *re*producing a face, but producing a 'new face' with little, limited or corrupted visual reference material for the target face.

'Depiction' therefore cautiously acknowledges the possibilities and the limits inherent to this work, echoing what Michael Podro recognizes from within visual art: 'Depiction has two main conditions: first our capacity to recognize through difference, and second, the intention to use the object that is materially present – the painting or drawing – to imagine what we recognize within it.'[20]

Faces from memory

In 2009, a story about the unlikely success of an eyewitness sketch went viral. The image, the most notional of drawings depicting the alleged murderer of a Bolivian taxi driver, has since developed the anecdotal reputation of one of the worst suspect images ever produced, and as such, makes for enduring 'listicle' fodder on websites like Buzzfeed.[21] It is necessary to stress that this image was not made by a professional police artist, but by the eyewitness themselves.[22] Nonetheless, a suspect was identified, arrested and convicted with little else to go on other than a thatch of thick, straight hair parted down the middle, which I am hazarding was sufficiently characteristic to pick him out from a crowd. So how and by whom should these facial images be produced? How are they expected to work? What forensic authority do they possess and how is this constructed?

Police composites (or 'artist's impressions') are facial depictions of criminal suspects, based on interviews with the relevant witness(es). There is variation in both composite and interviewing techniques; practitioners may use pencils or computer software, relying more or less on photographic reference catalogues or databases of features to compile a likeness. Recommended best practice is the cognitive interview, designed to retrieve stored memories via holistic context reinstatement, versus the brusque, 'just the facts, ma'am' investigative approach.[23]

Identikit and Photofit are two of the earliest manual, feature-based systems that enjoyed widespread use on both sides of the Atlantic since their introduction some four decades ago. The images produced by these systems possess a cultural endurance that is visual shorthand for phrases like 'suspect', 'fugitive' and 'person of interest'. Photofit replaced Identikit's catalogue of drawn features printed on acetate strips with photographic samples printed on rigid card, presumably to enhance the realism and by implication the accuracy, of the resulting representation. A witness would select features and assemble them like a Dada collage. Both systems offered the possibility of customizing the final image (manually adding facial marks or other characteristic features not offered by the stock images), but it remained difficult, if not impossible, to remove the demarcation lines that exposed the piecemeal, feature-based construction of the image. And the catalogue of available features was limited.

The problem is that with feature-based systems a fundamental error is introduced right from the start: it is a broadly accepted view that we do not encode faces feature for feature, and external features – hairstyle, facial hair – dominate our recall of unfamiliar faces (see Jane Draycott's chapter in this book for the inseparability of hair from the face as a whole). Rather, we encode faces holistically, as a gestalt or organized whole.[24] Only through the process of learning faces and becoming familiar with them, do internal features begin to dominate what we remember about a face. This is why when a friend or relative changes their hairstyle or shaves their beard, we will most likely be able to pick them out from a crowd.[25]

Subjected to controlled experiments, facial images produced using feature-based systems did not produce a good match with the target face, 'even under the favourable (and unrealistic) condition where the target's face was visible during construction'.[26] Cognitive psychologists have made a strong case for the wholesale replacement of feature-based systems with a sophisticated, 'holistic face' approach, which has effected policy changes in the UK as to how such images are produced and forensically understood: according to Association of Chief Police Officers guidelines (2009), developed in line with the evidence base produced by laboratory research, initial composites are regarded as 'primary' exhibits, with any further modifications (morphed, animated) as 'secondary' exhibits.[27] Technology has responded accordingly with computerized compositing systems such as

EvoFIT using evolutionary algorithms to 'breed' increasingly individualistic faces based on user selections from sets of similar faces that all conform, more or less, to a basic physical description. They can also be animated.[28]

Holistic composite systems enfold mathematical shape models with biological principles of sexual selection, described in terms of 'genetic algorithms'. As EvoFIT developer Charlie Frowd explains, variations on a set of reference faces can be generated by applying random coefficient values, conceptualized as 'face genes', to the reference face data, with the result being 'a search of the space of possible faces and, ideally, evolution towards the relevant identity':

> Genetic mutation can also be applied, an operation that replaces genes with a random value, the aim of which is to maintain variability in the population of faces. The resulting face has characteristic of both parents, with some variation. … The breeding process is iterated using faces that witness have selected from the (evolved) array. Note that this approach inherently involved chance due to the random nature of selecting (a) breeding pairs and (b) individual genes taken from each parent. The consequence is that sometimes a good likeness emerges early on, but at other times the evolution takes longer.

The final face can be further refined using a set of in-built 'holistic scales', which in addition to adjusting for age-related facial changes (slackening the jawline, for example), also allow for more qualitative value-based judgements such as 'health' and 'honesty': 'The accumulated effect of these 14 or so holistic scales can substantially improve the likeness of an evolved face.'[29]

Holistic composite systems are designed to address what artists have known for long: we don't encode faces according to individual features, but we read them as a whole. Certainly, some people may have a single feature that is highly distinctive, which a skilled caricaturist would make function as emblematic of that individual, but we essentially read faces as a complex spatial topography, with features occupying the area in relative proportions to each other.

Given their conditions of production, composites are heavily contingent things, subject to the vagaries of eyewitness recall, the skill and experience of the officer conducting the interview and producing the image, the technology and the *time* s/he has available to do the job. Time figures as a feature from event through to artefact: the speed of the event witnessed; the time of day possibly affecting visibility; how soon after the event the witness provides a description; the length of time available for an interview; the time available to the forensic artist to produce an image; how soon (and where) the image is circulated and whether there is any media investment in keeping the story alive will all have an influence on the resulting facial image(s).

Suspect renderings: The conviction of Norman Simons

The functions and limits under which composites operate are relevant for any facial depiction (reconstruction from the skull, age progression) produced as part of a forensic investigation. They are a specific kind of image, a pictorial statement designed to aid identification though generating leads and keeping an event alive in public memory. They are commonly referred to as secondary methods which may indeed *lend support* to primary methods, which in human identification terms is currently limited to scientifically supported methods like DNA and fingerprinting. Since 1993, the Daubert standards, a set of principles that determine whether evidence is based on scientifically valid reasoning appropriate to the matter at hand, have governed the admissibility of any evidence presented in court.[30] As expert practitioners, these standards apply equally to the forensic artist or facial identification officer.

The troublesome case of Norman Avzal (Sarfaraaz) Simons, convicted in Cape Town in 1995 of being the notorious Station Strangler, has the dubious reputation of flouting these fundamental principles. It is one example where a police composite – and visual identification methods in general – carried specious evidential weight in a case that was already deeply fraught in sociopolitical terms, and remains controversial. The process that identified Simons as a suspect, and then the ways in which he was 'made into an image' remains doubly suspect, in both senses of the word.

The hunt for the Station Strangler – 'the largest hunt for a serial killer in South African criminal history' – began at the height of Apartheid South Africa's State of Emergency in 1986, spanned the release of Nelson Mandela from prison in 1990 and culminated in an arrest in April 1994, only days before that country's first democratic national election.[31] Over this period the decomposing bodies of twenty-two pre-teen and teenage boys who had gone missing from in and around the sprawling settlement of Mitchell's Plain, a visually bleak, impoverished and crime-ridden area known as the Cape Flats, had been found dumped, some in shallow graves, in the surrounding dunes. The stress this case placed on the communities directly affected by it was considerably amplified by the acutely febrile political situation in South Africa at that time. Traumatized, disenfranchized and angry people were pushed to breaking point, and there was intense pressure to make an arrest. The hunt gained real momentum when a young boy, Ryno van Rooyen, reported his brother, ten-year-old Elroy van Rooyen, missing after they'd helped a man carry boxes from a local shopping centre towards a nearby train station on 11 March, 1994. Van Rooyen's remains were found shortly thereafter in similar circumstances. An eyewitness, Fouzia Hercules, was found who corroborated Ryno's account, which allowed police to identify a viable suspect: ex-Mitchell's Plain schoolteacher Norman Simons.

Although twenty-two boys were thought to be victims of a single serial predator, Simons would only be convicted of Elroy's abduction and murder. He had eventually confessed to this murder during questioning, but later retracted. The case against Simons otherwise rested on visual identification, a composite construction to which both witnesses contributed, and an identification line-up, where Ryno was unable to positively identify Simons, but Fouzia did. Simons received a 25-year sentence for murder and a further 10 years for kidnapping, effectively sentenced to life in prison. His 2015 parole application was unsuccessful.

Three different composites were produced with the participation of both witnesses, which according to Hercules' evidence took place eleven days after Elroy's abduction, on 22 March. The trial judge recounts Ryno's testimony in his judgement: 'The police showed them a lot of pictures of ears, noses, mouths and so on. She and he looked at them together because they had to say how the man's eyes, his nose and his mouth looked. The police officer then showed them a picture that looked like the man who took Elroy away.'[32] From this account, we learn that the facial identification officer produced three 'drawings', which were entered into evidence as Exhibits J, K and H. At the judge's request, these were shown to Ryno during his testimony, and 'he immediately pointed out Exhibit K as that of the man who took Elroy away'.[33]

Exhibit K received special mention in the trial judgement, with more than a full page's worth of commentary relating to the 'clear similarities in facial features, eyes, nose and mouth between the face on the identikit … and the accused's face', an unprecedented move that confers undue evidentiary weight to what is, at best, an average composite produced under questionable circumstances, but ultimately confers responsibility onto the defence to take issue with admissibility or by extension, any prejudice that may have arisen as a result thereof:

> Exhibit K is largely in line with the accused's face, has not been contested by the defense or refuted by any other evidence [or witness].
>
> viz. Exhibit K and the accused's face, where similarities are apparent from a comparison of the accused's photograph (Exhibit L) with Exhibit K, except for the scar, which is prominent on the identikit and *looks like a cut*. However, the accused has visible dark-coloured scars or other lesions [marks] under his right eye, as can be seen clearly on the pictures Exhibits L and Dl and D2. As can be seen in the photograph, Evidence CCl, the accused's hair may be 'combed up' higher than it was when Evidence L was taken.
>
> It is also common cause that no photo or identikit of the accused – or the suspect – was published before Fouzia Hercules made a statement to the police on 22 March and the identikit, Evidence K, was drafted.[34]

Three years later, the judgement in Simons' unsuccessful appeal against his guilty verdict also makes reference to the 'striking resemblance' the Identikit bears with 'the face of the appellant':

> The only material difference is the style in which the appellant's hair is combed, but [the fact] that he combed it like that on occasion emerged via another witness. … And as far as the reliability of her identification is concerned, *it is substantiated by the identikit*, and the confession to which I will shortly refer.[35]

Both judgements make reference to both a distinctive facial mark – a large, permanent and pigmented scar on Simons' right cheek which is referred to by one judge as a 'cut' (quite a different thing) and an easily alterable external feature (Simons' hairstyle) as key points of comparison. Simons in fact objected to being the only person included in the identification parade with a facial scar. Recalling the salience of external versus internal features in the context of unfamiliar faces, Frowd points out that external features are important in the naming process: it is difficult to *identify* faces with only internal feature information, but including external features at the early stages of *producing a composite* can be a distraction.[36]

Judicial verdicts in the Simons case are written in Afrikaans, the administrative language of the Apartheid Nationalist government which understood the power of language as a tool of exclusion and control. The quotes are my own translations, and in doing this work, I was reminded of the difficulty of the word *getuienis*, which means both 'testimony' and 'evidence', in the sense of the act of providing it. 'Evidence' as an object is *bewysstuk*; *bewys* can also be translated as 'proof', to demonstrate something to be truthful. It is in this act of translation that the difficulty of imposing forensic responsibilities on an object of intelligence might gain a dark import, and demonstrates how open objects are to mobilization.

Inhabiting the eye of a perfect storm of social and political upheaval, the legal outcome of the case is regarded as having severely limited integrity. A facial identification research group in the Department of Psychology at the University of Cape Town has reconstructed the visual identification practices used by the police in two experiments in a controlled, laboratory context.[37] The first experiment set out to test whether the line-up was obviously biased against Simons. It asked participants to identify the guilty suspect from the (admittedly poor quality) video. Played silently, participants only had low-resolution visuals to work with, making judgements based on behaviour and body language; out of eighty participants, not one picked Simons. A second experiment tested resemblance between the composite and fifty randomly chosen faces matching Simons' ethnic appearance (one photograph in the array was of Simons). To prevent participants from choosing Simons merely because of the distinct scar (or 'cut') visible on the composite, researchers placed a plaster in the same position on all faces in the array. Simons was the third most frequently chosen face, but he was more likely to be selected out of a larger group of faces than an array including fewer choices.

However likely Simons is as the perpetrator, there has been sufficient post-conviction revision to suggest that his guilt was not proven beyond a reasonable doubt in the terms of hard evidence.[38] In lieu of this, we are left with the disturbing precedent that a high court judge, as well as a team of three appeals court judges, considered a composite image to have adequate authority to act in lieu of any physical evidence linking Simons to the case, and use this to support a murder conviction and life imprisonment.

Quantifying likeness: Questions of materiality

An ideal accuracy study should set out to measure the range of variables that determine facial appearance – morphology, proportion, textural resemblance – in relation to method and technique. This is frequently a challenge: forensic cases based on skeletal remains may never be identified, and if they are, the quality of an available ante-mortem photograph is often poor, and only lends itself to two-dimensional visual (proportional and morphological) comparison. Medical imaging technology now permits in vivo 3D analysis of facial reconstruction methods, where morphological deviation between subject and model can be mapped.[39] Reliable photographic images will not, however, exist for historical individuals prior to the mid-nineteenth century. Other visual depictions, such as portraits, etchings, honorific sculpture and possibly a life/death mask, may exist for notable individuals post-seventeenth century. The Manchester method, so-called because it was developed within the now-closed Unit of Art in Medicine at the University of Manchester, combines facial anatomy and mean tissue depths, developed by Richard Neave after the Russian (Gerasimov) and American (Gatliff) methods. Photographic superimposition was then used to compare reconstruction with portrait, demonstrating a range from extreme accuracy to identifying an instance of the wrong portrait being associated with a particular mummy. This method was first introduced in the Buck Ruxton case and more recently given credit for producing the 'forensic aesthetics' of popular culture via the images produced by Richard Helmer in support of the forensic identification of Joseph Mengele.[40]

Where no reliable comparison image exists to assess accuracy and likeness, suspect composites may offer at least an image-to-image comparison if the target individual is eventually arrested. In 2013, a group of researchers led by Scott Klum published a study that set out to test two different modalities – 'sketches' (drawn) versus 'composites' (using software) based on eyewitness accounts – to mugshots of the perpetrators. The team gathered seventy-five forensic sketches, fifty of which were drawn by 'the world's most successful forensic artist' Lois Gibson,[41] who serves the Houston Police Department, with the balance produced by forensic artists employed by the Michigan State Police. Arrest photos (mugshots), presumably resulting from the arrest of these individuals based on the sketches, were then used as experimental stimuli, and shown to a group of volunteers. Two days after viewing the mugshot, the group of volunteers used a facial composite software to produce images of these 'suspects' from memory. Memory of the *photograph*, that is, not a memory made in a real-life scenario. The research team then subjected the forensic sketches and the computer composites to a 'recognition accuracy' analysis using two different automated face recognition systems.[42]

This the only study I have found that attempts a quantified comparison of different representational registers. However, inherent problems may render the findings not very useful; for example, it is questionable whether the role of the volunteers within the controlled study really did '[mimic] a witness of an actual crime scene' as the researchers claim; and some of the composite drawings by Gibson contain details – angle of head, lighting across the facial surface – that are uncannily similar to the mugshot taken at arrest. This visual similarity prompts the question of whether she adjusts the published versions of her sketches after having had sight of the mugshots, or whether she has developed a drawing style that deliberately mimics the unforgiving frontal view of the mugshot, in which case an argument could be made for introducing implicit prejudice at the stage of the composite.

Two years prior to Klum et al.'s study, Susan Hayes and Nick Milne, representing an interdisciplinary collaboration across visual art and human biology, published the results of a study which attempted to quantify the differences between a photographic reference image and a portrait drawing of the same subject, across a cohort of thirty individuals. Their specific contribution was the novel application of geometric morphometrics (GM) to facial depictions, in addition to the more conventional methods of visual assessment and anthropometry. Of the three analytical methods used, they express confidence in GM's ability to quantify tendencies within an individual artist's style, and suggest that the method has wider application across a range of facial depictions.[43]

In simple terms, GM offers a set of analytical tools to talk about the differences between shapes in mathematical space in statistical terms. It is typically applied to the study of variation in biological forms; here the authors treat portrait drawings as biological forms in themselves, 'to find out how an individual artist changes the shapes of a sitter's face' expressing an interest in what they call intended versus 'unintentional' distortions that may be present in the drawn portraits. Intended distortions are those choices we make about highlighting characteristic features. As they observe,

> While both exaggeration and generalization do not constitute a mimetic rendering of an individual's facial appearance, they are not entirely inaccurate either. The more tangible transformations that occur in the translation of a living person into a traditional 2D portrait tend to be somewhat less than desirable; that is, largely unintentional manipulations of the sitter's facial shapes which more often than not are unintended side-effects of the processes involved in visual perception.[44]

In their experimental design, all the drawings were made by one artist (Hayes), and were initiated in the presence of the volunteer subject. In other words, there was a 'life drawing' session of varying length that allowed for direct observation at first. At the session's conclusion, a reference photograph of the sitter was taken from which the drawing was later completed. Of course, this also means that multiple processes of translation are taking place here – from direct observation, to photographic representation, and then the drawing being a further (hybrid) translation of both.

Calling on canonical figures in art history and theory – Richard Brilliant, Ernst Gombrich and Roger Fry among others – to address issues of accuracy and likeness in portraiture, Hayes and Milne frame a position that prioritizes portraiture's mimetic function. They work to separate the genre from its aesthetic values, stressing its artisanal and mechanical aspects. Their findings are, in their words, 'at best partial': the drawings were not 'pure outlines', and GM cannot measure degrees of texture and shading that we *translate* into shape information. The portraits that did demonstrate better shape accuracy according to GM principles were judged better likenesses, 'but not significantly so'.[45]

Hayes and Milne acknowledge that absolute mimesis (whatever this might be) cannot be achieved, acknowledging Gombrich on the inevitable influence of factors extrinsic to pure optical recording that will affect representational 'translation' of a real human face into an image of one on a page: the choice of medium (conté crayon in this case); the sociopolitical and cultural contexts of the production and reception of images; and that qualitative factors such as time, and the nature of interaction between artist and subject, are complex to quantify.

Artists are well versed in the paradox that slavish pictorial mimesis can in fact work against the effect of likeness. Podro notes that providing too much detail – representing each facial detail with equal attention – is less convincing than a depiction which balances salient details well described with more open or 'provisional' areas and allowing our imaginations to interpolate between these to 'produce' the face (or whatever object we are contemplating).[46] This finds a parallel in studies which suggest that photo-realism may be operationally detrimental in a forensic context; facial depictions produced with a photographic level of 'accuracy' may in fact *reduce* the possibility of the individual being recognized. Theoretically, any mimetic slip may cause us to reject the image, even if it mostly resembles the target face. 'A good likeness, therefore, is not an accurate likeness, but one that has an *illusion* of accuracy.'[47]

Criticisms levelled against artistic subjectivity usually attend to repeatability. However, the significant difference in context between controlled studies undertaken in laboratory contexts and the contingencies of field operations should be taken seriously. A 'gold-standard protocol' for testing composite systems in the lab suggests that the issue of subjectivity and error is less an artistic problem and more fundamentally a human one:

> As constructors typically produce rather different-looking images for a given target … the protocol also recommends that at least eight constructors be recruited per system. Similarly, evaluators vary in recognition ability and so at least eight evaluators should attempt to name each composite.[48]

Distorting the image corrects perceptual errors

Within visual arts practice, it is a basic rule of thumb that considering an image from different angles, whether in the studio or in a gallery, is essential to assessing formal strengths and weaknesses, and overall compositional integrity. It is as true for non-figurative (abstract) work as it is for figurative representations. It is necessary to shift one's perspective in relation to the object, by stepping away and regarding it from a distance and from different angles, masking sections to view areas in isolation and then in relation to the whole. Details may look convincing close up and in isolation, but may get lost in the composition as a whole. I have always found it useful to photograph work in progress, and assess it via a single-generation mediation. This process reintroduces an objective perspective that can get lost in the haptic intimacy of making. Uncannily, criminalistics pioneer Hans Gross described something very similar in the late nineteenth century:

> The photograph is the image reflected by a mirror but it is a fixed image; this definition itself proves that photography, however paradoxical the assertion may appear, shows us more than the eye, even when it shows us no more than the eye can see. … A painter, … after having worked for a certain time, places his portrait before a mirror and considers the image which the latter reflects; he often discovers great faults which he was incapable of seeing upon the portrait itself. The reason is that when one looks for a long time at an object, … one always sees it under the same aspect, which prevents certain defects being noticed; but when the image is reflected by the mirror one sees the object under lateral inversion and in consequence under another aspect; details may then perhaps be discovered which have formerly escaped notice. In photography exactly the same may be

> said; an object has been observed with great minuteness and application; a whole series of observations have been made regarding it; nothing striking has been noticed about it because one has become accustomed to its appearance; but if it be photographed, the new colour, the new situation, and the new aspect enable us to see it from another point of view and reveal fresh details which have not yet been discovered.[49]

The knowledge gained by 'looking awry', is well-known within visual arts practice. That it now has an evidence base via cognitive psychology is very encouraging. Recognition studies have shown that subjecting a facial image to a 'physical linear stretch' (doubling its height or width) has been found to 'substantially' improve correct naming, despite a concession that 'stretched images look distorted and inappropriate for the serious application for which they are used'.[50] This can be mitigated, they suggest, by looking at an image side-on, which creates a similar effect that is perceptual rather than physical. The future of improved composites performance therefore appears to be with multiple viewpoints, if not full three-dimensional composites: 'A multi-view model should be able to accurately render differences in perspective, so that the image at construction is a better match to the image seen at the crime scene … . The same principle may be applicable to other factors such as lighting and facial expression.'[51]

Looking awry: *Super Noi* and *Portrait Machine*

While research aimed at understanding visual cognitive processing can be instructive, looking at how professional contemporary artists co-opt facial recall techniques offers another way of 'looking awry'. In the hands of those with both technical and conceptual expertise, and an unwillingness to accept given forms, committed instead to literal and figurative acts of deconstruction, the deadening effect of metric evaluation opens up to often tongue-in-cheek poetics of chance and intention, presenting parallel enquiries that thicken and enrich our understanding of the complexities of depicting the face.

In the 1990s, Italian artist Maurizio Cattelan, a conceptual artist who perpetually claims to 'have no ideas', embarked on a project he dubbed *Super Noi* ('Super Us'), collections of composite drawings of the artist constructed by police artists in different cities based on verbal descriptions provided by close friends and acquaintances, which capitalize on Cattelan's distinctive physiognomy. The various iterations of the work – sometimes referred to as *Il Super Noi* or suffixed with a specific city name like Bologna, Torino, New York – comprise fifty drawings per set on A4-size sheets of acetate.[52] They are wall-mounted, cloud-like arrangements, which sidestep the temptation to foreground the modernist–bureaucratic associations of such images by displaying them in a rationalist grid. Representing 'a network of other people's appraisals', the effect is 'a kaleidoscopic representation of an individual', playfully calling into question the fundamentally contingent nature of a profession intended to be forensically precise.[53] The title itself is suggestive of a conceptual portmanteau, linking the psychoanalytic theory of the Super Ego, by which we internalize social rules and behavioural norms, with the personality 'disorders' of schizophrenia or multiple personality disorder.

Performing the paradox of identity as simultaneously that which defines the 'self' and which one seeks to escape, Cattelan's fragmentary and serial rogue's gallery is a humorous but ultimately complex mediation on perception, representation and judgement. He entertains his criminal fantasies in the visual language of suspect apprehension. He has said of the work: 'That piece was really about how people around you perceive you in different ways than how you really are. So I was thinking about visualizing the idea of the self. The drawings really looked like me, but at the same time they were like cartoons. They were terrific. I don't know if it was a fluke.'[54]

The 'composite-Cattelan' advanced by this work clearly demonstrates what the artist recognizes about the simultaneous likeness and caricaturish 'un-likeness' that emerges when a facial image is produced from verbal description. UK-based South African artist Trasi Henen's *Portrait Machine* (2011–ongoing), takes this even further.[55]

Henen describes *Portrait Machine* as 'an ongoing, participatory drawing project that is concerned with interpersonal exchange, subjectivity, perception, collaboration and conversation'. An experiment in perception, practice and pedagogy, the project triangulates the relationship between sitter (subject), describer and artist, performed as a one-to-one reciprocal exchange, with Henen acting as describer and artist and sitter switching roles after the first drawing is completed, and then one-to-many, performed with groups of art students in real time but not necessarily in real space – some sessions were conducted with Henen describing a subject via Skype. While spatial arrangements have been varied, all participants have consistently comprised either artists, designer/illustrators or art students; in other words, all have a measure of visual training.

Portrait Machine anticipated cosmetic brand Dove's 2013 US campaign film, *Real Beauty Sketches (You're More Beautiful Than You Think)* featuring forensic artist Gil Zamora. Here forensic art skills are harnessed to demonstrate the difference in self-perception (generally heavily self-critical) and perception by others (more generous) among a diverse group of female participants. Waiting-room-style encounters were set up between women who were strangers to one another. Entering Zamora's studio individually and with the artist screened from view, Zamora asks each woman to describe herself. In a separate session, he asks them to describe the woman they had just met. The resulting finely detailed pencil sketches were exhibited side-by-side, with emotional commentary by the participating women.[56]

With Henen's consent, I staged *Portrait Machine* with a group of undergraduate (second year) Fine Art students at the University of Stellenbosch in 2014. I had been on sabbatical from the university for the previous eighteen months, so this group of students would not yet have encountered me within their teaching context. Hidden from view behind a screen, I listened as a teaching assistant, who would also act as 'witness-constructor', and introduced the exercise to the class. The assistant proceeded to describe my appearance, avoiding words which might prematurely reveal my identity (gender pronouns, for example), and the class attempted to translate her words into a facial image. The class had an opportunity to request any further salient information before completing their drawings. I stepped out from behind the screen to a mixture of laughs, cries of dismay, and 'Oh, I thought it was you!'

Having had rigorous cognitive interview and compositing training myself, it was this experience which really brought home the significance of descriptive language, the effects, as Henen writes of 'the speaker [acting] as the scribe's eyes'. What words and analogies are shared or missed between witness and artist? The possibilities and failure of language is ultimately what *Portrait Machine* foregrounds: hearing yourself described, and the unspoken self-censorship that might very well be occurring if the describer doesn't want to offend is an uncanny and acutely self-conscious scenario, which is amplified by then being faced with multiple versions of oneself, produced simultaneously and in real time, by a diverse group of people many of whom are equally self-conscious about 'getting it wrong' (despite there being no way of getting it right, although I think one got very close indeed).

Constructing fugitive subjects: *Prosopopoeia* and the counter-forensic

Translating the living face to facial image with incontrovertible fidelity eludes human ability. An essential factor, some fundamental animating spirit, escapes synthetic reconstruction. Paradoxically, this is amplified when we strive for mimetic realism without the necessary exceptional technical skill, and more tolerated when our depictions attend only to salient or characteristic details, or when our abilities can only manage the most rudimentary of sketches. As the various case studies discussed here show, representational style and technique, along with technical facility and dimensional translation, have a material effect on how the image does its work. As for the rest, it is a question that continues to vex and exercise those invested in understanding the social, cultural and legal authority of the facial image. Considering time and language as significant contributing factors to their particular materiality, composites are presented as embodying the 'fugitive', in both literal and figurative terms, exploring the limits of portraiture, the relationship between verisimilitude and the 'real', the image as record and performing the self. Subjecting the ineffable qualities of portraiture to various metric analyses may get us some way, but it feels a little too much like Narcissus staring at his own reflection in a pond, an exercise in hubris.

In our current climate, where the relationship between identity and identification is increasingly contested and constructed (via selfies and other online avatars), the facial image and the self are related yet discrete conceptual entities. Within the humanities, interest in the face as an object of study can be understood as an extension of identity and body politics into 'critical subjectivities' that coalesce in different periods, aligning with various disciplines and discourses that frequently overlap (cultural studies, anthropology and philosophy, or psychoanalysis and criminology, for example). Yet despite inhabiting a world made of faces – or certainly, pictorial representations of faces – only a limited number of contemporary scholars in the humanities and social sciences – have begun to focus closely on the face itself.[57] Within an expanded art history, the work of Hans Belting is instructive.[58] This wave of recent scholarship suggests that the fugitive nature of the face (temporary and surgical cosmetic procedures being a rather literal case in point) is being recognized in newly complex ways. Deleuze and Guattari's complex and rather abstracted notion of 'faciality', which offers a machinic counterpoint to Emmanuel Levinas' empathic concept of 'the face of the other', seems to resonate with the politics of 'facelessness', as well as our current digital culture where images are being made by and for machines, which means that a large part of what we understand as 'visual culture' is in fact invisible to humans.[59]

So, in the spirit of looking awry, I conclude with an introduction to two concepts I will be taking forward in my future work that considers forensic art in an expanded sense, reproducing the faces of the vulnerable or the violent in the pursuit of justice, while also attending to the more fugitive aspects of the roles faces play in our contemporary lives. The arguments advanced here pivot on a central idea about the function of forensic art in the context of wider practices of scientific methods of human identification: their basic function is to facilitate and enact the translation of object to subject, which they achieve in inherently narrative and metaphorical ways. They need stories to be told about them, stories that are most often told in reverse, reconstructed from debris and data collated post hoc.

The Forensic Architecture group's advancement of the concept of 'forensis' retrieves the public, and by extension political, imperatives of the term 'forensic' as embodied in the etymology of the word, the *forum*, a public space that foregrounds courageous critique and challenge, not simply immutable authority.[60] The language and forms of *forensis* are performative. The objects that bear the responsibility of truth telling are mute witnesses until someone demonstrates their significance to us. In other words, they require interpretation and translation. These objects have to be made to speak, or rather, conditions have to be designed to enable them to do so.

Thomas Keenan has identified and transposed the concept of prosopopoeia – a concept innately tied to the face – as a way of understanding how objects function within a forensic context, where absent, imagined or dead persons or things are represented as 'speaking'.[61] As performative technology, collected evidence and court exhibits are mobilized to do the work of testimony, made to speak via the interpretative skills of an expert witness, making representations on behalf of those who cannot, and giving a face and voice to inanimate things including abstract concepts like 'city' and 'state', 'enabling claims to be made, positions to be taken, justice to be pursued'.[62] Here, material and immaterial things are crafted as objects of authority and truth.

The Simons' case demonstrates the challenge to individual liberty than can result when a number of intensely contingent factors come together to differentiate someone as exceptional, based not only on their facial appearance but also how this notionally translates into an image produced under stressful circumstances. Here, investigators and legal representatives chose to ignore a basic ground truth, which led to a person's conviction and imprisonment: composites should not be used in isolation. They can be very valuable, but their contingent nature requires us to be alert to their very real limits, both at the point of their construction and then how they might be understood later. As Frowd reminds us, their legibility is not stable:

> Human observers make errors when constructing composites and making judgments based on identity … the likeness of a composite to a defendant is related to whether (participants acting as) jurors *believe the defendant to be guilty or not,* suggesting that composites may not provide independent evidence and that their use in court is questionable.[63]

Where the safety and security industries are concerned (global conflict providing a justifiable if double-edged cause for real concern), the project to tame and control the fugitive face has extended to producing faces from DNA trace evidence, an endeavour in fact anticipated by artist Heather Dewey-Hagborg with her project *Stranger Visions*.[64] More recently, scientists have reported success in deriving almost-identical facial images from the brain activity of macaque monkeys from those previously shown to the monkeys. As researcher Doris Tsao commented, 'A face is impossible to describe in words … . One can imagine applications in forensics where one could reconstruct the face of a criminal by analysing a witness's brain activity … you could imagine reading out what face a person is imagining.'[65]

Forensic art, I suggest, produces prosopopoeial objects in a very literal sense, and in the hands of visual artists, we see how these methods may perform counter-forensic roles.[66] Attending to the human face in all its complexity, not just as an indexical likeness, which is expected to do particular kinds of work, but also as an icon or emblem of individual personhood onto which we project various fears and desires, forensic art produces culturally complex objects that are in excess of its forensic imperative, crafting 'subjects' in very particular ways that are, in turn, embodied in images which possess a certain performative agency. Acting as our focalizing lens, the composite sketch also serves to illustrate that all forensic art represents a compositing process: we create faces from a range of data that is observed, heard, read and gathered through material interaction and touch. It represents a similar hybrid practice of producing knowledge through both haptic- and optic-based observation that Joanna Sofaer describes in her bioarchaeological work.[67] Further, these visual depictions open up conversations between artist, witness and community. In facilitating the transfer of valuable information from one context to another in the interests of social cohesion and justice, they arguably represent a particular genre of community art.

Critical interdisciplinary practices question received knowledge – how it is made and how it is used – in very particular ways. Any considering of the applications of science entails a consideration of its dual-use potential and by extension, its political implications. As such, it represents a specific intersection of art, science and legal agency, placing specific demands on how we understand – and practise – the relationship between ethics and aesthetics. In its work with the living and the dead, attending to vulnerable people subjected to the worst aspects of human behaviour – neglect, exploitation, violence and death – forensic art troubles, in critical and productive ways, the illusion of empirical distance that is so much part of the construction of the scientific gaze.

Notes

1 Wittgenstein cited in M. Podro, *Depiction* (New Haven: Yale University Press, 1998), 180, note 1.

2 See The Innocence Project (https://www.innocenceproject.org/) which is committed to overturning wrongful convictions based on eyewitness identification.

3 I am particularly interested in recent projects aimed at recuperating the analogous rather than indexical nature of photography's verisimilitude or those which make a powerful case for the indexicality of the digital photographic image: Roland Barthes, *Camera Lucida: Reflections on Photography* (New York: Hill and Wang, 1981); Susan Sontag, *Regarding the Pain of Others* (London: Hamish Hamilton, 2003); K. Silverman, *The Miracle of Analogy: Or the History of Photography, Part 1* (Palo Alto, CA: Stanford University Press, 2015); Kris Paulsen, 'The Index and the Interface', *Representations* 122.1 (Spring 2013): 83–109; Errol Morris, *Believing is Seeing* (New York: Penguin, 2014) and see Jason Bate's chapter in this book for the telescoping of past with present.

4 M. Mori, 'The Uncanny Valley (trans. K. F. MacDorman and Norri Kageki)', *IEEE Robotics and Automation* 19.2 (2012): 98–100 (Original work published in 1970). doi:10.1109/MRA.2012.2192811. Online: http://spectrum.ieee.org/automaton/robotics/humanoids/the-uncanny-valley (accessed 7 September 2016). A. Tinwell, *The Uncanny Valley In Games and Animation* (Boca Raton: CRC Press, 2015), extends this into the sophisticated CGI creations of gaming and animation, and interrogates that affective quality of such bodies and faces, offering a productive application of developmental psychologist Edward Tronick's Still-Face Effect: E. Z. Tronick, 'Things Still To Be Done on the Still-Face Effect', *Infancy* 4.4 (2003): 475–82.

5 Podro, *Depiction*, Preface, n.p.; E. Levinas, 'Exteriority and The Face', in *id.*, *Totality and Infinity* (The Hague, Boston and London: Martinus Nijhoff Publishers, 1979); *id.*, 'The Face' in *Ethics and Infinity: Conversations with Philippe Nemo* (Pittsburgh: Duquesne University Press, 1985).

6 Due in large part to media exposure via television shows like BBC's 'History Cold Case' (2010–11) and any number of forensic procedural dramas. For an illustrated history of how this method has developed scientifically and artistically, see the virtual exhibition 'The Evolution of Facial Depiction from Human Remains', curated by Face Lab (LJMU) for the Arts & Humanities Research Council (UK) website. Online at http://www.ahrc.ac.uk/research/readwatchlisten/imagegallery/facialdepiction/ (accessed 19 September 2017).

7 Slavoj Žižek, *Looking Awry: An Introduction to Jacques Lacan through Popular Culture* (Cambridge, MA: MIT Press, 1991); H. Aldersley-Williams, K. Arnold, M. Gordon, N. Kotsopulos, J. Peto and C. Wilkinson, eds, *Identity and Identification* (London: Black Dog Publishing, 2009).

8 C. M. Wilkinson, 'A Review of Forensic Art', *Research and Reports in Forensic Medical Science* 5 (2015): 17–24.

9 For example, W. D. Haglund and D. T. Reay, 'Use of Facial Approximation Techniques in Identification of Green River Serial Murder Victims', *The American Journal of Forensic Medicine and Pathology* 12.2 (1991): 132; Richard P. Helmer, S. Rohricht, D. Petersen and F. Mohr, 'Assessment of the Reliability of Facial Reconstruction', in *Forensic Analysis of the Skull*, ed. M. Y. Iscan and R. P. Helmer (New York: Wiley-Liss, 1993); C. N. Stephan and M. Henneberg, 'Building Faces from Dry Skulls: Are They Recognized above Chance Rates?' *Journal of Forensic Sciences* 46.3 (2001): 432–40.

10 C. N. Stephan, 'Facial Approximation – From Facial Reconstruction Synonym to Face Prediction Paradigm', *Journal of Forensic Sciences* 60.3 (2015): 566–71.

11 The current interest in so-called SciArt and STEAM (as an extension of STEM, adding Art to the acronym) suggests that there is committed interest in recuperating aspects of this historical relationship from an educational perspective, despite a lack of clarity about the broadness of the terms of reference and power dynamics that may inform

interdisciplinary collaborative work, which is rich in breadth, depth and international presence.

12 J. Elkins, 'What is a Face?' in *id.*, *The Object Stares Back* (San Diego, New York and London: Simon and Schuster, 1996, repr. Harvest, 1997); *id.*, *Picturing the Body: Pain and Metamorphosis* (Stanford: Stanford University Press, 1999); *id.*, *Visual Practices Across the University* (Paderborn: Wilhelm Fink, 2007); see also W. J. T. Mitchell, *Image Science: Iconology, Visual Culture, and Media Aesthetics* (Chicago: University of Chicago Press, 2015).

13 Elkins, 'What is a Face?' 161.

14 Facial pareidolia even enjoys its own Twitter account *Faces In Things* @FacesPics.

15 J. Calame, 'Saving Face', *Cabinet*, 49 (2013): 28–31. This nascent 'typographical art' would find its digital maturity decades later with ASCII art, but what is really noteworthy is the correspondence between the emotional expression depicted here and those of the Facial Action Coding System, based on the work of Carl-Herman Hjortsjö, *Man's Face and Mimic Language*, trans. W. F. Salisbury (Lund: Studentlitteratur, 1969/70) http://diglib.uibk.ac.at/download/pdf/782346?name=Man's%20face%20and%20mimic%20language (accessed 30 March 2017) and adapted by Paul Ekman and Wallace V. Friesen, with a significant updated published in 2002: P. Ekman and W. C. Friesen, *Facial Action Coding System: A Technique for the Measurement of Facial Movement* (Palo Alto: Consulting Psychologists Press, 1978); *id.* and J. C. Hager, *Facial Action Coding System: The Manual on CD ROM* (Salt Lake City: A Human Face, 2002). FACS is used in psychology and animation to quantify and analyse the mechanics and affects of facial expression.

16 B. A. Fedosyutkin and J. V. Nainys, 'The Relationship of Skull Morphology to Facial Features', in *Forensic Analysis of the Skull*, ed. Iscan and Helmer, 199–214.

17 Podro, *Depiction*; Richard Brilliant, *Portraiture* (London: Reaktion, 1991); Marcia Pointon, *Portrayal and the Search for Identity* (London: Reaktion, 2013).

18 C. Wilkinson, 'Facial Reconstruction – Anatomical Art or Artistic Anatomy?' *Journal of Anatomy* 216 (2010): 235–50. Increasing attention is being paid to best-practice methods to optimize the possibility of identification (an already highly contingent process) and minimize the introduction of error into the depiction itself: S. Davy-Jow, 'The Devil is in the Details: A Synthesis of Psychology of Facial Perception and its Applications in Forensic Facial Reconstruction', *Science & Justice* 53 (2013): 230–5; C. M. Wilkinson, 'A Review of Forensic Art', *Research and Reports in Forensic Medical Science* 5 (2015): 17–24.

19 J. S. Rhine, 'Coming to Terms with Facial Reproduction', *Journal of Forensic Sciences* 35.4 (1990): 960–63; Stephan, 'Facial Approximation'; *id.* and M. Henneberg, 'Recognition by Forensic Facial Approximation: Case Specific Examples and Empirical Tests', *Forensic Science International* 156 (2006): 182–91; Alan G. Morris, *Missing and Murdered: A Personal Adventure in Forensic Anthropology* (Cape Town: Zebra Press, 2011), 87.

20 Podro, *Depiction*, 5.

21 Patrick Smith, 'The 21 Worst Police Sketches Of All Time', *BuzzFeed* https://www.buzzfeed.com/patricksmith/the-27-worst-police-sketches-of-all-time (accessed 21 May 2017).

22 Rory Carroll and Latin America correspondent, 'Childlike Photo-Fit Leads Bolivian Police to Murderer', *The Guardian*, 24 November 2009, section World news https://www.theguardian.com/world/2009/nov/24/bolivia-photo-fit-murderer (accessed 21 May 2017).

23 On composite techniques, see S. Mancusi. *The Police Composite Sketch* (New York: Humana Press, 2010); D. P. Hinckle. *Mug Shots: A Police Artist's Guide to Remembering Faces* (Colorado: Paladin Press, 1990); L. Gibson *Forensic Art Essentials: A Manual for Law Enforcement Artists* (Cambridge, MA: Academic Press, 2007); K. T. Taylor, *Forensic Art and Illustration* (Boca Raton: CRC, 2001). On the cognitive interview, see R. P. Fisher and R. E. Geiselman *Memory Enhancing Techniques for Investigative Interviewing: The Cognitive Interview* (Springfield, IL: Charles C. Thomas, 1992).

24 J. W. Tanaka and M. J. Farah, 'Parts and Wholes in Face Recognition', *The Quarterly Journal of Experimental Psychology* 46.2 (1993): 225–45; G. M. Davies, J. W. Shepherd and H. D. Ellis. 'Remembering Faces: Acknowledging Our Limitations', *Journal of the Forensic Science Society* 18.1–2 (1978): 19–24.

25 C. Frowd. 'Facial Composites and Techniques to Improve Image Recognizability', in *Forensic Facial Identification: Theory and Practice of Identification from Eyewitnesses, Composites and CCTV*, ed. T. Valentine and J. P. Davis (West Sussex: Wiley Blackwell, 2015), 48. Of course, a significant point about the limitations of doing facial recognition research in laboratory environments versus 'in the wild', is that different cognitive processes are at work in processing still facial images versus faces in motion, within a specific context where a host of other distractors may be present. Having contextual information improves recognition rates.

26 Ibid., 46–7. The interview process, in which a facial description is elicited, plays a formative role in producing an image. Asking witnesses to make seven whole-face judgements regarding health, masculinity, pleasantness, honesty, distinctiveness, intelligence and likeability showed a striking improvement in accurate naming for composites produced using a feature-based system (a leap from 9 per cent to 41 per cent, tested with PRO-Fit) as well as a holistic system (24 per cent became 39 per cent, testing EvoFIT): ibid., 56. The marked improvement reported for the feature-based system seems to demonstrate the double effectiveness, and therefore necessity, of applying a holistic approach in both processes of descriptive recall and visual construction.

27 Ibid., 58–9, 63–4.

28 The three commercial systems currently in existence are South Africa's ID, EFIT-V and EvoFIT: C. Tredoux, D. Nunez, O. Oxtoby and B. Prag, 'An Evaluation of ID: An Eigenface Based Construction System', *South African Computer Journal* 37 (2006): 90–7; 2006 (37): 90–7; C. J. Solomon, S. J. Gibson and J. J. Mist, 'Interactive Evolutionary Generation of Facial Composites for Locating Suspects in Criminal Investigations', *Applied Soft Computing* 13.7 (2013): 3298–306; C. D. Frowd, P. J. Hancock and D. Carson, 'EvoFIT: A Holistic, Evolutionary Facial Imaging Technique for Creating Composites', *ACM Transactions on Applied Perception (TAP)* 1.1 (2004): 19–39; C. D. Frowd, P. J. Hancock, V. Bruce, F. C. Skelton, C. J. Atherton, L. Nelson, A. H. McIntyre, M. Pitchford, R. Atkins, D. C. I. Webster and J. Pollard, 'Catching more Offenders with EvoFIT Facial Composites: Lab Research and Police Field Trials', *Global Journal of Human-Social Science* 11.3 (2011): 35–46. EvoFIT's website offers comprehensive information about the system, including links to published research. See http://www.evofit.co.uk/ (accessed 15 May 2017).

29 Frowd 'Facial Composites', 51–2. Susan Hayes and Cameron Tullberg, 'Police Witness Identification Images: A Geometric Morphometric Analysis', *Journal of Forensic Sciences* 57 (2012): 1487–94, note that the results indicate that compared with arrest photographs, witness identification images systematically depict suspects with lowered and medially located eyebrows.

30 *Daubert Standard* (definition). Legal Information Institute, Cornell University Law School. Available at https://www.law.cornell.edu/wex/daubert_standard (accessed 15 January 2017).

31 Candice Bailey, 'Revisiting the Station Strangler Cases', *IOL (Independent Online)*, 16 August 2005. http://www.iol.co.za/news/south-africa/revisiting-the-station-strangler-cases-250970 (accessed 15 January 2017).

32 *The State versus Norman Sarfaraaz Simons* (High Court of South Africa, Good Hope Provincial Division, 1995) SS.104/94: 1404–5.

33 Ibid.

34 State v. Simons 1995: 1407, my emphasis.

35 *Avzal Sarfaraaz Simons versus The State* (South African High Court of Appeal 1998), SS104/1994: 4–5.

36 Frowd 'Facial Composites': 54.

37 A. Nortje, C. G. Tredoux, K. Kempen and A. Vredeveldt, 'Applying Laboratory Techniques to a Real-Life Case: What Insight can we Provide about the Station Strangler Case?' Poster presentation at the American Psychology-Law Society Annual Conference, Portland, United States, March 2013.

38 Simons himself and his lawyer have maintained this position from the beginning. This is not particularly surprising. More striking is the way the conversation recently re-entered the public imaginary in the context of a procedural parole hearing: https://traveltrailtony.wordpress.com/category/informative-articles/ http://www.iol.co.za/news/crime-courts/shofer-was-the-station-strangler-2053684

39 See W. J. Lee, C. M. Wilkinson and H. S. Hwang, 'An Accuracy Assessment of Forensic Computerized Facial Reconstruction Employing Cone-Beam Computed Tomography from Live Subjects', *Journal of Forensic Science* 57.2 (2012): 318–27; W. Lee, S. Mackenzie and C. M. Wilkinson, 'Facial Identification of the Dead', in *Forensic Anthropology 2000-2010*, ed. S. Black and E. Ferguson (Boca Raton: Taylor and Francis, 2011), 363–94; S. Decker, J. Ford, S. Davy-Jow, P. Faraut, W. Neville and D. Hilbelink, 'Who is this Person? A Comparison Study of Current Three-Dimensional Facial Approximation Methods', *Forensic Science International* 229 (2013): 161.e1–161.e8; K. Smith, 'The Doppelgänger Effect? A Comparative Analysis of Forensic Facial Approximation Methods'. Unpublished thesis, University of Dundee, 2013.

40 See Caroline Wilkinson's chapter in this book, and B. Brier and C. Wilkinson, 'A Preliminary Study on the Accuracy of Mummy Portraits', *ZÄS* 132 (2005): 111; J. Prag and R. Neave, *Making Faces* (London: British Museum Press, 1999); C. Wilkinson, *Forensic Facial Reconstruction* (Cambridge: Cambridge University Press, 2004–8); J. Glaister and J. C. Brash, 'Medico-legal Aspects of the Ruxton Case', volume 33 (Edinburgh: Livingstone, 1937); Thomas Keenan and Eyal Weizman, *Mengele's Skull: The Advent of a Forensic Aesthetic* (Berlin: Frankfurt Sternberg Press/Portikus, 2012).

41 'The World's Most Successful Forensic Artist: How Lois Gibson's Incredible Sketches Have Helped Solve Thousands of Crimes', *Guinness World Records*, 2017 http://www.guinnessworldrecords.com/news/2017/1/lois-gibson-facing-crime-with-world-records-458852 (accessed 21 May 2017).

42 S. Klum, H. Han, A. K. Jain and B. Klare, 'Sketch Based Face Recognition: Forensic vs Composite Sketches', *International Conference on Biometrics* IEEE (2013): 1–8. See also G. Davies, H. D. Ellis and J. Shepherd, 'Face Recognition Accuracy as a Function of Mode of Representation', *Journal of Applied Psychology* 63.2 (1978): 180–7.

43 Susan Hayes and Nick Milne, 'What's Wrong with This Picture? An Experiment in Quantifying Accuracy in 2D Portrait Drawing', *Journal of Visual Communication* 10 (2011): 149–74.
44 Ibid., 149.
45 Ibid., 150, 169.
46 Podro, *Depiction*, 28.
47 Hayes and Milne, 'What's Wrong?', 151, my emphasis.
48 Frowd 'Facial Composites', 49. 'Constructors' are shown faces with which they are unfamiliar, and after a specific delay, recall the face within an interview context with a trained professional. 'Evaluators' are those familiar with the target face who are asked to perform a recognition task with very basic contextual information.
49 Hans Gross, *Criminal Investigation: A Practical Handbook for Magistrates, Police Officers and Lawyers* (1893, English translation 1907), cited in Greg Siegel, 'The Similtude of the Wound', *Cabinet* (2011), 99.
50 Frowd 'Facial Composites', 61, but contradicted/nuanced by ibid., 65, citing C. H. Liu and A. Chaudhuri, 'Reassessing the 3/4 view effect in face recognition'. *Cognition* 83.1 (2002): 31–48.
51 Ibid., 65.
52 Maurizio Cattelan *Il Super Noi*, http://www.laboralcentrodearte.org/en/recursos/obras/super-noi-torino-1996.
53 Guggenheim Education Teacher Resource Unit, 'Maurizio Cattelan *All*', 2011 Online: https://www.guggenheim.org/arts-curriculum/resource-unit/maurizio-cattelan-all (accessed 24 May 2017).
54 A great fan of gangster and heist films, Cattelan's practice is peppered with pranks and thefts, as well as a host of self-portraits of varying scale and bodily fragmentation, 'a battery of lookalikes, mini-mes, doubles, and surrogates'. He toys with the obsession the world of art has with the authentic signature (the mark of the self); celebrity culture's 'self as brand' (Cattelan occupies both worlds); and the agency of reinvention: the most famous person you'll never see: Guggenheim '*All*', 2011; V. Nussbaum, 'Serial Künstler: Portrait of the Artist as a Malefactor', in *Elective Affinities: Testing Word and Image Relationships*, ed. C. MacLeod, V. Plesch and C. Schoell-Glass (Amsterdam: Rodopi, 2009), 212.
55 T. Henen, http://portraitmachine.blogspot.co.uk/
56 Dove Real Beauty Sketches https://www.youtube.com/watch?v=litXW91UauE
57 Apart from the authors in the present book, see Suzannah Biernoff, 'The Rhetoric of Disfigurement in First World War Britain', *Social History of Medicine* 24.3 (2011): 666–85.; *ead.*, 'Medical Archives and Digital Culture', *Photographies* 5.2 (2012): 179–202; *ead.*, *Portraits of Violence: War and the Aesthetics of Disfigurement* (Ann Arbor: University of Michigan Press, 2017); Heather Laine Talley, *Saving Face: Disfigurement and the Politics of Appearance* (New York and London: New York University Press, 2014); Jenny Edkins, 'Forensic Identification', in *Missing: Persons and Politics* (Ithaca and London: Cornell University Press, 2011); *ead.*, *Face Politics* (London and New York: Routledge, 2015); Sharonna Pearl, *About Faces: Physiognomy in Nineteenth-Century Britain* (Cambridge, MA: Harvard University Press, 2010); *ead.*, *Face/On: Face transplants and the Ethics of the Other* (Chicago and London: University of Chicago Press, 2017).
58 Hans Belting, *Face and Mask: A Double History*, trans. Thomas S. Hansen and Abby J. Hansen (Princeton and Oxford: Princeton University Press, 2017).
59 Gilles Deleuze and Felix Guattari, 'Year Zero: Faciality', in *id. A Thousand Plateaus – Capitalism and Schizophrenia*, trans. Brian Massumi (Minneapolis and London: University of Minnesota Press, 1987); Levinas, 'Exteriority'; *id.*, 'The Face'; T. Paglen.

'Invisible Images (Your Pictures are Looking at You)', *The New Inquiry*, 8 December 2016. Online https://thenewinquiry.com/essays/invisible-images-your-pictures-are-looking-at-you/ (accessed 8 December 2016).

60 Eyal Weizman, 'Introduction: Forensis', in *Forensis: The Architecture of Public Truth*, ed. Forensic Architecture (Berlin: Sternberg Press, 2014).

61 Thomas Keenan, 'Getting the Dead to Tell Me what Happened: Justice, Prosopopoeia, and Forensic Afterlives', in Forensic Architecture, ed., *Forensis*, 35–55.

62 Ibid., 38.

63 Frowd, 'Facial Composites', 46, my emphasis.

64 P. Claes, D. K. Liberton, K. Daniels, K. M. Rosana, E. E. Quillen et al., 'Modeling 3D Facial Shape from DNA', *PLoS Genetics* 10.3 (2014): e1004224. doi:10.1371/journal.pgen.1004224; H. Dewey-Hagborg, *Stranger Visions*, http://deweyhagborg.com/projects/stranger-visions

65 L. Chang and D. Tsao 'The Code for Facial Identity in the Primate Brain', *Cell* 169.6 (2017): 1013–28.e14. Cited in H. Devlin, 'Scientists Discover How the Brain Recognises Faces – by Reading Monkey's Minds', *The Guardian*, 1 June 2017. Online https://www.theguardian.com/science/2017/jun/01/scientists-discover-how-the-brain-recognises-faces-by-reading-monkeys-minds (accessed 5 June 2017).

66 Thomas Keenan, 'Counter-forensics and Photography', *Grey Room* 55 (2014): 58–77, taking up Allan Sekula's phrase in the context of ordinary citizens taking on role that state representatives have abnegated, such as legal identification and serving justice, in order to identify individuals and events that have been wilfully and violently anonymized or otherwise silenced by the state or its operatives.

67 J. Sofaer, 'Touching the Body: The Living and the Dead in Osteaoarchaeology and the Performance Art of Marina Abramović', *Norwegian Archaeological Review*, iFirst (2012): 1–16; and see *ead., The Body as Material Culture: A Theoretical Osteology* (Cambridge: Cambridge University Press, 2006).

7

From 'Staring' to 'Not Caring': Development of Psychological Growth and Well-Being among Adults with Cleft Lip and Palate

Patricia Neville, Andrea Waylen and Aidan Searle

According to Goffman's typology of stigma, cleft lip and palate (CLP) registers as an example of stigma of the body, a mark of difference that is physical in nature and immediately apparent on meeting. The social and psychological burden of living with Goffman's 'discredited identity' has been well documented within the academic record, and in turn has served to maintain the stigmatizing effects of CLP.[1] However, there have been recent calls from some clinicians and academics challenging the negative conceptualization of CLP. One way to counter the hegemony of stigma in CLP studies is through qualitative research and exploring the self-representation of adults with CLP. This chapter discusses qualitative findings from face-to-face interviews with adults with CLP. These data acknowledge that living with CLP registers as an experience of feeling different and that the experience is negotiated through a number of key transitional periods as people grow into adulthood. Progression through each stage can be either helped or hindered by the actions of some key institutional players – the immediate family, peer group and the school and work settings. Overall, this chapter contends that CLP can be a life-affirming experience, albeit with challenging moments at times.

It is widely agreed that we live in a society with an aesthetic preference for facial symmetry and regularity. Conformity to this societal ideal resonates as a particularly potent social process and any one whose bodily traits differ from this 'ideal' are rendered different and atypical as a result. For those born with some degree of facial difference we can expect their lives to be shaped by an active process of 'othering', an experience that undermines their sense of self and self-identity as well as singling them out for a lifelong struggle against the forces of stigma, social exclusion and discrimination. Many theorists, from Goffman to Garland-Thomson, have documented the phases and stages of this corrosive social process.[2] As a result, the life experiences of all people with facial difference are assumed to fall in, uncontested, with these theoretical observations.

However, it is worth noting that stigma is a 'negatively defined condition' and so using it as a conceptual tool to analyse and interpret the lived experiences of people with facial difference may adversely bias the nature of the research, priming researchers to attend to instances of self-blame and negative self-concept.[3] This tendency of stigma and difference research to foreground the pathological is an inherent limitation of facial difference research, one that is not often reflected upon in the literature. As researchers working with adults with CLP we are acutely aware of how current scholarship is skewed in its negative conceptualization of CLP. Moreover, our experience of working with people with CLP and listening to their life experiences and testimonies reminds us that an alternative, more constructive discourse about living with CLP also exists, though underreported.

The aim of this chapter is to challenge the long-held assumption that all people with facial difference are perpetually perplexed by the 'otherness' of their lives, that the effects of facial stigma impose very real limits on their life and that the label of 'difference' is the constant medium through which they experience and encounter the social world. We plan to contest this by attending to the life experience of adults with CLP through a qualitative lens. We do not deny that a mechanism of difference and stigmatization is experienced by people with CLP. Indeed the pathological effect of being different in a society that values the 'norm' is well documented. However, we contend that by pursing an interpretative approach to understanding people with CLP, inviting and listening to their personal testimonies, an alternative discourse based on their first-person experience will emerge. By enabling people with CLP to share their own biographical narratives, we create the possibility for an understanding about facial difference to be informed by self-authority and self-representation rather than the self-perpetuating cycle of stigma and difference. In this way we strive to circuit break the stigmatizing effect of CLP.

In this chapter we will begin by defining what is meant by cleft lip and/or palate. Next we will outline the current scholarship on CLP and the centrality of the notion of stigma found there. From this we will present our argument for the merits of an alternative paradigm to researching CLP through the interpretivist approach. Qualitative research with fifteen adults with CLP will be presented revealing an alternative, strengths-based narrative which will challenge existing canon as well as our own preconceptions about people living with facial difference.

What is CLP?

CLP is a congenital condition affecting a child's facial structures which also has functional consequences for feeding, chewing, breathing and impaired dental, facial, speech and language development.[4] CLP occurs due to fusion failure in the structures forming the different parts of the mouth during the first trimester of pregnancy and is among the most common congenital malformations. The global incidence of CLP is approximately 1/700 live births.[5] Cleft lip and/or palate is generally categorized into three types: (1) unilateral or bilateral cleft lip and/or alveolus (approximately 25 per cent of CLP cases); (2) unilateral or bilateral cleft lip, alveolus and palate (approximately 25 per cent of CLP cases) and (3) isolated cleft palate which is the most common form of CLP accounting for approximately 40 per cent of all affected children.

Visible scarring, an underdeveloped maxilla or a flat asymmetric nose may characterize the look of a person born with CLP and thus have significant implications for appearance. A child born with CLP may also experience functional difficulties when feeding, chewing and breathing as well as having impaired dental, facial, speech and language development.[6] Typical speech deviations in CLP are hypernasality, audible nasal air leakage, decreased pressure on consonants and deviant articulation patterns. For these reasons, a child with CLP is likely to begin corrective surgery within the first few months of life. The cleft of the lip is the most visible part of the cleft and is usually closed in the first year of life. In the UK lip closure surgery is usually undertaken at about three months with palate closure following at about six months, and additional surgery and intervention may continue into early adulthood. In most Western countries, a person with CLP is cared for by a multidisciplinary cleft team, usually including a plastic surgeon, orthodontist, speech and language therapist and audiologist. Increasingly, psychological services are being included in cleft care teams in the UK in response to a body of evidence concerning difficulties faced by children who are visibly different and who may have communication deficits.[7] Most cleft teams follow their patients routinely until their late teens. Clearly, CLP is an intense medicalized process for children and families alike until their early adulthood in which they face ongoing involvement with multidisciplinary healthcare teams to ensure the child's optimal physical and psychosocial development.

Stigma and CLP: A literature review

It has been long assumed that living with facial difference has a psychosocial impact for children born with CLP. According to Goffman's typology of stigma, CLP registers as an example of stigma of the body, a mark of difference that is physical in nature and immediately apparent on meeting. Though all stigma gives rise to 'a sense of undesired differentness', the 'hare' or cleft lip does register as a 'stigma symbol', something that 'break(s) up an otherwise coherent picture' of a human face. In this way people born with CLP may acquire a 'discredited identity'.[8] There are three consequences for being 'visibly conspicuous' in society.[9] First is the perception that one is different from the majority that society manifests. This is called felt stigma. Second is enacted stigma, or being explicitly treated differently from the majority of society (either in the form of discrimination or ostracization). Last, self-stigma, or where 'internalised feelings of guilt, shame, inferiority' overwhelm the internal dialogue of the person.[10]

Stigma theorists posit that the experience of being stigmatized is an incredibly painful process and an untenable state. This position has been confirmed by some CLP studies. In quantitative studies differences have been established between young people born with CLP and non-cleft peers that may support this position. For example, those with CLP are reported to drop out of school more frequently and be less likely to belong to clubs and societies.[11] In addition, young adults with CLP are less likely to aspire to further education compared with their non-cleft peers, and are more frequently unemployed and have a lower income than their non-cleft peers.[12] More recent research suggests that children and young adults with CLP have fewer friends than non-cleft peers.[13] Children with CLP are also believed to marry less often and later and childless marriages are more likely to occur.[14]

Other studies confirm instances of felt stigma and enacted stigma among children with CLP. Hunt et al. conducted a cross-sectional study of 160 children (8–21 years) to determine the psychosocial effects of living with CLP.[15] The study also included a comparison group of children without CLP. Children born with CLP reported greater behavioural problems, greater depression and teasing and were less satisfied with their facial appearance than controls. Teasing was a significant predictor of poorer social functioning and it was concluded that children with CLP require psychological assessment as part of routine cleft care. Hunt et al. conducted a further study with 129 parents of children with CLP to determine the level of agreement of perceived teasing between children born with CLP and their parents.[16] Their experiences were compared to those of ninety-six parents of children without CLP. Parents of children born with CLP reported that their child was teased more often than their non-CLP peers and parents were less satisfied with their child's speech. Children who had been teased were more anxious, less happy with their appearance and had greater behavioural problems.

Interestingly, family members are not impartial bystanders to the stigmatization that people with CLP feel and experience. Goffman contends that family members assume a 'courtesy stigma' by virtue of having a family member with a stigma.[17] Research on parents of children with CLP attests to the psychosocial impact of having a child with a 'visible difference'. For instance, research shows that both the antenatal diagnosis of cleft and the months following diagnosis may be a traumatic period for parents and their family.[18] While most of the research on children with CLP and their families documents this unsettling time as a reaction to their movement through the cleft treatment pathway, a recurrent theme in this literature is the role that perceptions of facial appearance and difference plays in the decision-making process of families. Forth recognized that coping with a CLP diagnosis has similarities to the process of grieving due to the loss of a 'perfect' child.[19] Farrimond and Morris also discovered that parents' primary motivation for CLP repair was to make their baby acceptable to others even though they themselves loved their babies as they were.[20] It has been suggested that the mechanics of this desire for 'normalizing' surgeries is also driven by a moralistic component. Nelson et al.'s research into the beliefs and motivations in relation to treatment decision-making of families of children born with CLP reported the fulfilment of a moral obligation to be 'good' parents by pursuing 'normalizing' treatments.

Thirty-five in-depth interviews were conducted with parents of children with CLP between the ages of 20 weeks to 21 years to explore experiences across the treatment programme.[21] Findings showed that, throughout the child's childhood and adolescence, parents experienced conflicting emotions about their child's treatment and the stigmatizing attitudes of others. Uncertainty about treatment/interventions was a potential source of conflict for parents. At diagnosis, either prenatally or at birth, parents sometimes reported a simultaneous mixture of grief about the impairment and delight about having a newborn. There was confusion around simplistic categorizations of 'normality' and 'difference' and there was some ambivalence about labelling a child as 'normal' or 'abnormal'. Conflicting emotions were particularly evident with regard to parents' views of surgery. As much as parents believed in the possibility of surgery to improve a child's function or 'normalize' his or her appearance, there was a strong emotional urge to protect their child from discomfort and distress in relation to 'surrendering' the child to the surgical team. Experiences of treatment were sometimes reported to become more arduous for both children and their parents as children grew older. Post-surgery discomfort such as bleeding, swelling, infection and nausea was also challenging for parents pursuing repeated surgical treatments. It was reported to bring about a profound sense of guilt and the best way to resolve such guilt was to place their trust in the specialist clinicians involved in cleft care. Such treatments were viewed as a way of facilitating their child's social inclusion and helping them reach their full potential.

The ambivalence experienced by parents of infants and children with CLP towards their child's changing face is an important finding often underreported in the literature. However, the parental discourses produced by those whose children were born with CLP serve as a potent reminder of the hegemony of facial appearance and of a parent's ardent desire to protect their child from the social and psychological consequences of being visibly different. Here, we find evidence of people 'opting out' of the process of stigmatization and creating their own terms of reference for how they engage with facial difference. Social expectations of 'normal' and 'difference' are rewritten as inversions of each other and a new, alternative ideological space is created for the family to inhabit and maintain through their interactions with their child. Though operating at a micro level, this interrogation and negotiation of social norms and pressures is encouraging as a localized response against the process of stigmatization. This confirms Goffman's claim that family offers 'a protective capsule for its young' as a key site where the negative stereotypes and ideologies associated with the stigma are not allowed to penetrate or influence their home life.[22]

Proposing a new paradigm for CLP

The stigma associated with CLP appears to popularize the notion that CLP resonates as a negative social experience and significant psychological burden. However, there have been regular calls from some clinicians and academics challenging this negative conceptualization of CLP.[23] Advocates of this paradigm shift have been inspired by a critical appraisal of the psychological research conducted on the self-concept and self-esteem of people with CLP, a reassessment of the role that families play in the dealing with stigmatization, the emerging literature on resilience and health and a resurgence in qualitative research with people with CLP.[24] Together, these methodological reflections and empirical studies point to the possibility of an alternative paradigm in conceptualizing the CLP experience that doesn't take stigma as its starting point.

The exploration of the psychological impact of living with CLP brings into question the methodology of many of the studies which pronounce CLP as a negative experience. In part, this is because there are methodological differences between these studies, and some are cross-sectional, based on a small sample size and often deal with different age groupings.[25] It should also be noted that previous research has used a wide variety of quantitative assessment methods. For the most part, parental self-report questionnaires have been utilized and there is a possibility of poor agreement between child and parent pairs. In addition, many studies neglect to document the age of infants or children and findings are based at one time point.

There appears to be conflicting evidence as to whether children born with CLP experience major psychological problems throughout adolescence and into adulthood. For instance, while children and adolescents with CLP undergo periods of psychological adjustment, especially during adolescence, they generally have been found to have a normal self-concept.[26] There is also evidence that those with acquired rather than congenital facial difference (e.g. changes due to trauma or accident) experience more difficult psychological adjustment.[27] In their study of the stigma experiences of eleven to eighteen-year-olds with facial difference, Strauss et al. found that young people with congenital conditions reported fewer instances of being stared at and were less likely to talk about their appearance to others than those with an acquired facial difference. This finding led them to conclude that young people with congenital conditions have better coping strategies than those with acquired facial difference because they have lived with the condition longer than others and so have become less vulnerable to the opinions of others.[28] Moreover, as surgical techniques for the repair of CLP have become more refined, the potentially visible 'deformity' may not be as pronounced as it has been in the past and may help to explain why adults with CLP seem relatively satisfied with their body image.[29]

Affective influence of families

Previously, it was reported how families struggle with the CLP diagnosis, with many fearful about the social rejection of their infants. However, this parent research has also revealed an interesting contradiction. Farrimond and Morris suggest that parents had adjusted their 'internal' picture of their baby so well during the initial weeks following birth that they did not want to change their child's facial appearance by the time the surgery was scheduled.[30] This finding was most salient when the 'defect' was less severe. Furthermore, in their evaluation of specialized cleft counselling, Ray-Bellett and Hohfield reported that some mothers expressed distress following lip closure as they had to readapt to the baby's new facial features and expressions.[31] This phenomenon was also described by Hammond et al. They reported that some parents adjust well to the appearance of their baby and the cleft raising doubts about the perceived pressures from society to conform to the idea of 'perfection'.[32] However, the initial joy of a safe delivery may be quickly replaced by a parental concern for the infant to be healthy and perceptions about the management of the facial difference through corrective surgery. However, the research suggests that the path to achieving a 'normal' appearance for the infant is more complicated than initially implied. Clearly, the mechanics of stigmatization can be challenged and resisted within the family context and through the transition from childhood to adulthood. Indeed, parents and siblings can potentially be powerful agents against stigmatization and can act as catalysts in the development of emotional resilience in individuals born with CLP.

Resilience and CLP

Emotional resilience refers to one's ability to adapt to stressful situations or crises. In an effort to shift the focus away from the potentially negative effects of CLP, the concept of emotional 'resilience' has been introduced to determine why some individuals with CLP do better than others in the context of psychosocial development and functioning.[33] Resilience was initially presented as a useful way to interpret this formational aspect of CLP because it is concerned with positive human development in spite of identifiable risks.[34] However, there is much debate as to how to define resilience: at its most basic, resilience refers to a dynamic process of adaptation within the context of severe adversity.[35] Since the 1970s a growing body of literature has identified other protective factors and competencies necessary to respond positively to adversity.[36] These include the impact that a sense of self, self-determination and prosocial attitude can have in protecting against adversity.[37] Although resilience has contributed to our understanding of child development some theoretical issues raise doubts about its applicability to the study of individuals born with CLP. First, the resilience literature is based on the experiences of 'at-risk' children and those who have experienced childhood neglect, abuse and deprivation.[38] As a result, resilience research is linked to the study of psychopathology.[39] Secondly, the model of psychological growth proposed in resilience theory posits that environmental factors play an important part in the development of resilience. Although resilience has been demonstrated in individuals born with CLP it is unclear which aspects of an individual's environment promote the development of a resilient identity. Indeed, most resilience research considers the impact of external risks like poverty, abuse and neglect on child development.[40] However, CLP is not associated with psychopathology and it is not considered a clear risk to child development.[41] Finally, resilience theory has paid little or no attention to the role that intrinsic factors, such as personality, may play in helping people adapt to challenging circumstances. As a result, how individuals appraise, internalize and respond to external events is missing from the framework.[42] Such an omission detracts from the value of the theory since it fails to acknowledge that life experiences and their meanings are forged iteratively and dialectically, through the continual interchange between our external and internal worlds. For these reasons, the concept of resilience may not be appropriate in the context of understanding individuals' experiences and their response to CLP. However, this raises the issue of the need for a new theoretical framework that can explain their experiences with CLP.[43]

Qualitative research and CLP

Despite this critique of resilience theory, there is value in putting forward a theory that promotes a positive understanding of living with CLP. While living with a 'spoiled identity' does come with challenges, being born with CLP may also provide opportunities for self-growth that go beyond the terms of the negative label.[44] A growing number of qualitative research studies with people with CLP confirm this.[45]

Nelson has postulated that qualitative research could make a contribution to understanding the factors that influence adjustment in young people and adults born with CLP and their satisfaction and adherence with treatment.[46] Recent research studies show a growing interest in the application of qualitative research methods to help in this.[47] In addition, other researchers have investigated the application of quality of life measures with CLP.[48] These qualitative studies provide a rich narrative from which we can describe the experiences of individuals born with CLP. Qualitative methods can help in understanding the lived experience of people born with CLP and their families in context and may also help to clarify the previous contradictory and inconsistent findings that are present in the existing literature. Havstam et al. conducted a qualitative study with thirteen young adults (25–34 years) born with CLP in Sweden.[49] Their aim was to explore the experience of growing up with CLP and associated deviant speech. This research emphasized the important role that parents can play by speaking openly with their child about how to present to other people. Experiences of teasing were reported by all participants and, although it was not always a dominant feature of childhood, teasing was perceived to be the most negative consequence of CLP. The extent of identification with the 'world' of CLP varied widely – some young adults expressed the benefits of joining a cleft palate association as it was, potentially, an excellent opportunity to learn more about cleft and meet others with similar experiences; however, others did not want to be grouped with other people on the basis of one feature they believed to be fairly insignificant to their identity. Moreover, the authors found that some participants felt special because of their CLP due to the extra attention received from parents and classmates during difficult times of treatment and also because of the ability of many others to see qualities in people beyond their appearance. These findings resonate with previous research demonstrating that high self-esteem is common among persons with CLP.[50] Such findings serve to reinforce a move to investigating the experience of living with CLP from a positive perspective.

The forgoing has revealed some limitations and assumptions regarding CLP research and in turn validates the argument that a new approach to the study of the experiences of CLP is warranted. We contend that researchers need to pursue more qualitative research with people with CLP. The rich data that qualitative research generates will allow in-depth exploration of life experiences, and allows individuals born with CLP to describe their construction of reality and meanings. Accordingly, we will be able to better understand these experiences and in time develop participant-led theories to challenge stigmatization by exploring how people with facial difference negotiate their life path in an aesthetically normative world and the role that families play as mediators and role models in this process. What follows is a record of fifteen qualitative interviews that were conducted with the purpose of exploring the lived experiences of living with CLP.

Method

The research was funded by a National Institute for Health Research (NIHR) Research for Patient Benefit Grant (PB-PG-0110-21049). This study is nested within a programme of qualitative research undertaken in the development of the Cleft Lip and Palate module for Healthtalk.org. Healthtalk.org is a website that provides 'free, reliable information about health issues', by sharing individuals' experiences of a wide variety of health conditions.[51] Qualitative researchers with expertise in the patient experience conduct the recruitment and interviewing of participants and analyse data.

In-depth interviews were conducted in two parts.[52] First, individuals were invited to tell their story of living as an individual born with CLP. Second, a topic guide was used to further probe and prompt participants, to generate richer data about their experiences. An advisory panel of lay people affected by CLP, health professionals and academics assisted in the development of the research and development of the topic guides. The topic guide included topics pertinent to living with a cleft lip and/or palate and engagement with cleft services in the UK and life for young adults and older adults that are no longer engaged with specialist cleft services.

Participants were recruited through a variety of methods, including personal networking, social media including the Facebook pages of organizations such as Cleft Lip and Palate Association (CLAPA), Healthtalk.org and Adult Voices (an online forum for adults born with CLP). All potential participants were sent an information sheet via email. Participants were given a period of one to two weeks before they were re-contacted for their agreement to be interviewed either by video or audio only. Those agreeing to participate signed a consent form prior to interview. Ethical approval for the study was granted to Healthtalk.org/Health Experiences Research Group, University of Oxford by Berkshire Research Ethics Committee for health research.

Participants and interviews

Interviews with fifteen adults aged 17–62 years born with CLP were conducted between April 2013 and April 2014 (nine female, six male). All were of white British origin although one male was born in Canada and one female was born and living in Northern Ireland. Of the fifteen interviewees all had been born with CLP except one female who was born with an isolated cleft palate. Three adults interviewed were also parents of children born with CLP. Further details can be found in Table 7.1. Interviews were conducted by Aidan Searle in participants' homes across the UK. Twelve interviews were conducted face to face and video recorded and three interviews were audio recorded via telephone. The interviews ranged from 22 minutes to 66 minutes duration and the mean was 55 minutes.

Table 7.1 Individuals born with cleft lip and/or palate (N = 15)

ID	Interviewee	Cleft	Children	Marital Status	Occupation	Ethnicity
Mally	Individual 62 years, with cleft / father and grandfather of cleft child	Born with cleft lip, son and grandson also born with cleft	4	Single	Docker	White British
Michael	Father 23 years, (cleft lip and palate)	Son with cleft lip and palate	2	Cohabiting	Mother/Lifeguard	White British
Lizzie	Female, 25 years	Born with cleft lip and palate	0	Married	Call centre	White British
Kendal	Female, 24 years	Daughter born with cleft palate	1	Single	Teacher	White British
Hannah	Female, 22 years	Born with cleft lip and palate	0	Single	Student	White British
Ryerson	Male, 41 years	Born with cleft lip and palate	0	Married	Lecturer	Canadian
Karan	Female, 47 years	Born with cleft lip and palate	1	Married	Self-employed/Street Pastor	White British
Jon	Male, 38 years	Born with cleft lip and palate	0	Single	Arts promoter	White British
Iona	Female, 19 years	Born with cleft lip and palate	0	Single	Nursing student	White British
Mary Z.	Female, 39 years	Born with cleft lip and palate/skeletal dysplasia	0	Single	Teacher	White Irish
Josh	Male, 17 years	Born with cleft lip and palate	0	Single	Student	White British
Alex	Female, 19 Years	Born with cleft lip and palate	0	Single	Student	White British
Mary W.	Mother, 53 years	Born with cleft lip and palate. Daughter born with cleft lip and palate	3	Single	Teacher	White British
Gemma	Female, 18 years	Born with cleft lip and palate	0	Single	Student/Childcare worker	White British
Elliott	Male, 20 years	Born with cleft lip and palate	0	Single	Retail Manager	White British

Results

Five key themes emerged from the thematic analysis:

- Support from others (family, peers and professionals)
- Perception of self (appearance and speech)
- Public awareness and social interaction
- Experience of treatment and outcomes
- Treatment and health professional communication

In addition to these key themes the findings are presented within three transitional life stages:

- Transition to primary education
- Transition to adolescence
- Transition to higher education and/or employment.

The transition to primary education

The experience of going to primary school was where individuals born with CLP reported that they first developed or acquired a sense of difference, through their interactions with their peers. As a result, the notion of difference was extrinsic in origin, a consequence of their new social context:

> I can't remember exactly when I remember kind of thinking, 'Oh I've got a cleft.' But it was more probably around primary school age, which probably about 6-ish kind of was first kind of memories. And it wasn't so much school wasn't so much of an issue as in like bullying and kind of name calling. I just remember like going to school and people saying, 'Crikey, what a difference from us' or, 'Crikey, you speak funny' or, 'What's wrong with your face?' It was quite a lot of that. (Iona)

While the experience of attending primary school confronted these children with a sense of their difference on a regular basis, this difficult social context also challenged their parents as well.

> I don't actually recollect much about, you know, being made aware that I was different until I actually went to school. I started, you know, at about age 4. I remember little comments from the other kids. Kids are naturally quite curious: if they see something different they want to know. And my mum remembers after the first day, my first day of school coming back and I said, 'Mummy, what's wrong with my face?' And that was her first sort of experience of what was … to come basically, all the questions and everything like that. (Lizzie)

Individuals recalled turning to their parents for advice and support to help them understand and cope with these peer judgements. All the participants said that their family background was overwhelmingly positive and one in which having CLP did not single them out for differential treatment. As a result, the family context emerged as a supportive environment, one that nurtured the whole child, rather than dwelling on CLP. In this respect, the family context not only contrasts sharply with the challenging encounters of primary school, but also provides the building blocks for the development of psychological growth and well-being.

> I wasn't really aware that I had a visible difference until I went to school. And then I started to notice that people were staring at me, and then that, that is really when I became aware that I, I have this. But before that I was quite sheltered at home so you know, I wasn't made to feel any different than anyone else. So that has kind of been one of my values, you know, to get me through school and everything else, that you know, I don't, it doesn't make me a different person; it's just I look different. (Hannah)

> My father was very ill from the [participant's] age of 10 and so as a family we financially struggled a lot. But our values were strong, so the values were right and my parents did the best they could and that's, you know, that's all you can hope for, isn't it? So I think as a family we were very, we were a very strong unit. Yeah I have no qualms over my childhood. I mean every family has its problems and skeletons in the cupboard. But, no, my father did his best, as my mother did, so I was very lucky really. (Karan)

> I don't know, I think just my parents were very secure in the fact that I was no different to my siblings or you know, I was still a, a person inside, and that was more important to them than the way that I looked. So, I think because they were comfortable with it, I was comfortable with it. So, you know, it's like all the operations and stuff, it never seemed to overwhelm them. It was just, 'We'll do what we have to do', and, you know, for me that was a really strong upbringing. So you know, right now I owe everything to my parents, you know, bringing me up to be confident. (Lizzie)

However, one individual felt that the family support he received came more from his mother than his father;

> I don't know about guidance, but quite strong confidence, kind of getting confidence from my mother, who I guess mothers tend to be more emotive and fathers don't. I guess that's probably a terrible generalisation but broadly true. So I guess quite a bit of support then from my mother, I always felt that I was able to chat any kind of anything through really. Although I guess with father it felt like a bit more of a formal relationship, but less so as I got older perhaps. (Jon)

Transition to adolescence

In adolescence individuals born with CLP became more autonomous in their decisions and competence in life. While the primary school environment invoked a sense of difference in children, as adolescents they did not passively succumb to these negative experiences. Buoyed by a supportive family environment, they developed a repertoire of coping strategies which enabled them to take these negative experiences and turn them into opportunities for positive self-growth.

One strategy was to reframe the negative social interactions, and recategorize them into 'standard' bullying and 'CLP specific' bullying.

> But yeah mainly, mainly it was just kind of … you know, teasing and sort of … snide comments, I guess. But actually there was always a level of bullying which I think actually quite a lot of which wasn't related to having a cleft; a lot of it was just … it happened to everyone as well. So I guess one has, you know, one has to pick apart the stuff that was cleft related and the stuff that was just teenage boys being nasty to each other [laughs]. I reckon there was probably more of the latter than the former actually. I think the kind of cleft, as I say, might occasionally have been a catalyst, might have set you apart. (Jon)

Another strategy was being open about the condition and prepared to talk to others which served to increase their connectedness to others in their environment. For example, Iona was also proactive in seeking out and creating a support network.

> It does take time to develop ways to deal with it. At first it's very hard and you just want to run away from it all but that's not the answer. I personally found people that I could talk to, whether it be people at school or family and you just kind of you kind of you learn to accept that you're born with a cleft, and that's not going to change. And over time you learn to think, 'That is what makes me different and that's actually a good thing.' But until you reach that point, you need to find a good like support network to kind of talk over what you're thinking because if you don't voice what you're thinking, nothing's going to get done. I feel it's quite important to talk about it. Yeah, positive things just having a good support network and helping the helping you to come to terms with, as I said, helping you to come to terms with the fact that having a cleft is a good thing. It's not everyone's going to have something that makes them different, and you don't want to be like everybody else, and having a cleft gives you a story, it makes you who you are and it shouldn't be seen as a negative. People might treat you like it's a negative, and they might say things, but you know, I'm sure they're not perfect, and their imperfection might not be noticeable as having a cleft is, so yeah. (Iona)

One male born with CLP reported that there was a lot of interest in his visible difference and people were interested in knowing more about the condition.

> Most people know. It's kind of you can see it pretty well, that's kind of the first thing that most people notice is the cleft lip. And then they go, 'Oh do you have a cleft lip?' 'Yes, and palate'. Yeah and they're always they're always kind of interested actually. Interest is the main reaction actually. (Josh)

Furthermore, the experience of supportive family environments may positively impact on psychological growth so that strength of character could emerge in adolescence.

Indeed, those who had experienced adverse conditions in childhood developed the confidence to fully engage with life in adolescence.

> My school years were … bit of a roller-coaster. Of course kids can be very cruel, especially when they don't understand what they don't, you know, with something different they don't understand it and they can be very, very cruel. So, yeah, I got bullied, I got teased. But that kind of spurred me on to … basically be the person that I am today, you know, I am very determined, and if someone says, 'You can't do this because of this, that and the other', I try and set out to prove that I can. So it's made me the person I am. I think my parents definitely helped with that, because they're both very, very driven, very determined people anyway, and they kind of made me … the sort of saying of 'don't let them get you down' sort of was very well … very, that basically summed up what they stood for, and what I stand for is the same: don't let them get you down. (Lizzie)

Engagement in creative pursuits and sports suggests that individuals with CLP act as agents within their supportive environments. They experimented with new behaviours and habits and do not let their CLP set the limits of their abilities. Creative pursuits included singing, writing music and performing in public. Physical activities included rock climbing, swimming and fencing. The participants found that engaging in these pursuits enabled new ways of dealing or challenging the negative peer interactions.

> Yeah. I think the what's, what's shaped me is a supportive family, good friends and you know, the influence of music and the space to be creative and to allow yourself to express yourself. And I think that's so important, and anything creative is, is going to get that out of you. And I think if I didn't have that I can't imagine what [laughs] I'd be like: probably clinically insane. But yeah, I'd definitely say the music is the core thing that brought me out of my shell. (Hannah)

Perhaps not surprisingly, both physical and creative activities appeared to provide an outlet for self-expression enabling the individual with CLP to enhance their supportive environment, accommodate their developing psychological growth and pursue a sense of connectedness with others despite their birth condition.

> It's been really good. I mean obviously I go clubbing, as a young adult. I have been clubbing once or twice, and you do meet people that look at you a bit strangely because you look different from what, you know, normal people would look like without a cleft palate. But when you're out people don't really look and think of me any different. I mean they might look at me and give me a few strange looks, but they won't treat me any differently. (Gemma)

Transition to higher education and employment

The final transition among adults born with CLP was from education to employment. Participants were either employed or, if in higher education, had work experience.

The working environment was generally found to be supportive and for many it was a transformative experience, liberating them from the negative attitudes they experienced at school and allowing them to develop psychological growth in the adult world.

> It was kind of a last minute decision to take a year out, and it was just kind of just a regular job in a restaurant, nothing it was definitely beneficial and yeah it was in terms of kind of coming out your shell a wee bit more, kind of going from school where everything about school was remembering about being bullied, and not being happy as I was meant to be, and to then kind of going somewhere new and meeting new people, and nobody knows anything about that, and you don't really have any memories of, like any bad memories, no. It was it was good to go somewhere where people didn't point and ask questions. (Iona)

> I found it (college) was a lot more enjoyable for me. I was studying something I really enjoyed (public services). I found the whole attitude in college was a lot more mature than school, I'd say, especially the course I was on. I don't think you really get too many time wasters or people who bully people on the course. Yeah, you're at the place you want to be because you chose to do that, rather than school where everything's just a bit of a laugh with your mates and you study in the meantime. College for me was a great experience because I think when I got to college I was very self-conscious and not very confident at all, and then the course was very confidence boosting. (Elliott)

Career paths taken by individuals born with CLP included education, healthcare and other caring professions – career pathways that require good interpersonal skills and where communication has a key role. Taking these career paths emphasizes individuals' sense of autonomy, and an ability to engage with society, rather than withdraw from it and succumb to its normative judgements.

> I have never felt, though I said I'm self-conscious about the way I look, I have never, since I've come here, ever felt that anyone's looked at me and said, 'Hey, you know, that guy, he's just a little bit too ugly for the post', no, not at all. Oh it affects my confidence in sort of funny areas. Yeah, no, I … if any, in a way … hmm I think it's probably contributed to me being more outgoing in some ways because I had to be. (Ryerson)

> In the occupation I am, I do have to be very open and very talkative to new people I don't know, because we're always going to have new parents come in to visit, to look at [school name], and new children come in. So I guess part of what's made me very confident is working where I do, because I have had to come out of my shell, I have had to meet new people and …. Yeah, yeah I think that's been the best thing for me that I've ever done really, is going there. Because it's, it's really helped me in, in my personality and it's helped me grow a lot more. And … I've always kind of known since I was 10 really that I wanted to work with children. I've got it mapped out, and I know what I want to do. And just, I mean knowing what I want to do has helped me as well. I've developed that confidence to just get out there and do what I want, sort of thing. (Gemma)

Discussion and conclusions

This chapter has reviewed the CLP literature with regard to living with CLP and presented data from in-depth interviews with individuals born with CLP. While the existing academic record would propose that living with CLP is a negative experience, contemporary qualitative research reveals that people with CLP are not passive bystanders to the processes of stigmatization but actively engage with and test the assumed normative limits of their lives. The thematic analysis of the data contributes to a greater understanding of the transitional periods implicated in the development of psychological growth and well-being in individuals born with CLP from childhood to adulthood. These include transition to school, transition to adolescence and the transition to further education/work. These lifespan events register as 'normative transitions', widely acknowledged transitions that occur in the lifespan and tend to coincide with 'developmental change'.[53] As a result, our findings confirm the general development of psychological growth and well-being from adolescence into young adulthood. These transitions may also be coupled with periods of potential vulnerability and uncertainty, as well as opportunities for psychological growth and well-being. Key to these transitions is how individuals internalize the influence and response of family and peers, as well as those of the wider social institutional context. Individuals born with CLP encouraged questions and embraced 'open' dialogues with others in their communities and this dialogue served in the development of psychological growth and well-being.

With regard to stigma theory, many individuals found that starting school can be a transition where individuals encounter their first and sustained instances of 'stigma learning', being taunted and treated differently by their peers. Other transitional points for stigma learning are believed to emerge in adolescence, and later when entering the employment market. Goffman refers to these transitional points as 'occasions of stigma learning' that can be incidental in nature as well as systematic or planned. The family appears to be the bedrock upon which people with CLP build their lives. Goffman contended that a family can offer 'a protective capsule for its young' where negative stereotypes and ideologies associated with stigma are resisted by process of 'normalization' and instruction on how people should interact with a stigmatized individual.[54] In the qualitative research reported here we have noted that the role of the family can support the development of a positive self-identity in the early and teenage years.[55] While the family context and the schooling context are independent of each other, they do converge, with events happening in the school environment being mitigated by the family and vice versa. The family emerges as a pivotal developmental setting where children feel valued and cherished as 'children' and not as 'children with CLP'. However, the school environment may emerge as posing a particular threat to the development of individuals born with CLP, who may require support from families in knowing how to act with those who provoke stigma.

Relationships with parents, peers, schools and employers remain important at times of transition and in future-orientated domains such as education and career. However, it is also not clear to what extent families are implicated in autonomy-supportive environments once an individual with CLP has left the family home. Indeed, over time most teenagers come to prefer the companionship of peers to that of their family: peers are less likely than parents to coerce, criticize and lecture and are more willing to give each other personal validity, social status and shared interests.[56] The centrality of peers in adolescents' social lives means that there is scope to introduce peer-based autonomy-supportive environments. One approach is school pupil-mentoring programmes in which peer-led support could enhance the development of psychological growth and well-being at school to counter sustained instances of 'stigma learning' in the school environment. The findings also suggest that the transition from school to further/higher education and employment were key periods when individuals were liberated from the 'spoiled identities' of earlier years and developed an enhanced relatedness to others.[57]

This research exemplifies the benefits that qualitative research offers to the study of CLP. By using interviewing as our main research method we discovered people who had not let CLP define them but who, instead, had taken ownership and attributed their strength of character to their cleft. Individuals were encouraged to tell their 'story' and did so with great candour often recalling experiences from childhood and adolescence that could be considered to be implicated in the development of 'stigmatized identities'. We learnt that they were able to consider their own interests and emotions which in turn fostered volition with engaging in educational, occupational and personal lifestyle choices. These authoritative and empowered voices are too often missing from the research. Their exclusion from the canon lessens it as a result.[58]

Notes

1 Erving Goffman, *Stigma: Notes on the management of Spoiled Identity* (Harmondsworth: Penguin, 1963), 14.
2 Goffman, *Stigma*; Rosemarie Garland-Thomson, 'The Politics of Staring: Visual rhetorics of Disability in Popular Photography', in *Disability Studies: Enabling the Humanities*, ed. Sharon L. Snyder et al. (New York: Modern Language Association of America, 2002), 56–75; Rosemarie Garland-Thomson, *Staring: How we Look* (Oxford and New York: Oxford University Press, 2009); and see chapter by Jane Frances in this book on how academic research has been slow to identify its own exclusionary practices and language.
3 Lee F. Monaghan and Simon J Williams, 'Stigma', in *Key Concepts in Medical Sociology*, ed. Jonathan Gabe and Lee F. Monaghan (London: Sage, 2013), 58.
4 Orlagh Hunt, Donald Burden, Peter Hepper and Chris Johnston, 'The Psychosocial Effects of Cleft Lip and Palate: A Systematic Review', *European Journal of Orthodontics* 27 (2005): 274–85.
5 Peter A. Mossey, Julian Little, Ron G. Munger, Mike J. Dixon and William C. Shaw, 'Cleft Lip and Palate', *The Lancet* 379 (9703) (9 November 2009): 1173–8.
6 Orlagh Hunt, Donald Burden, Peter Hepper, Mike Stevenson and Chris Johnston, 'Self-reports of Psychosocial Functioning among Children and Young Adults with Cleft Lip and Palate', *Cleft Palate-Craniofacial Journal* 43 (2006): 598–605.
7 Elaine Lockhart, 'Mental Health Needs of Children and Adolescents with Cleft Lip and/or Palate', *Clinical Child Psychology and Psychiatry* 8 (2003): 7–16.
8 Goffman, *Stigma*, 14, 15, 59, 60, 14 respectively.

9 Garland-Thomson, 'The Politics of Staring', 576.

10 Ann Jacoby, 'Felt versus Enacted Stigma: A Concept Revisited: Evidence from a Study of People with Epilepsy in Remission', *Social Science and Medicine* 38.2 (1994): 269–74; R. G. Block, '"Is It Just Me?" Experiences of HIV-Related Stigma', *Journal of HIV/AIDS and Social Services* 8 (2009): 1–19; Debbie Patterson, Alex Barnes and Chloe Duncan, *Fighting Shadows. Self-stigma and Mental Illness* (Auckland: Mental Health Foundation of New Zealand, 2009), 14.

11 John Peter and Rosalie R. Chinsky, 'Sociological Aspects of Cleft Lip and Palate Adults. 1. Marriage', *Cleft Palate Journal* 11 (1974): 295–309; John Peter and Rosalie R. Chinsky, 'Sociological Aspects of Cleft Lip and Palate Adults. 2. Social Integration', *Cleft Palate Journal* 11 (1974): 304–10; John Peter and Rosalie R. Chinsky, 'Sociological Aspects of Cleft Lip and Palate Adults. 3. Education', *Cleft Palate Journal* 11 (1974): 443–9.

12 John Peter, Rosalie R. Chinsky and Mary J. Fisher, 'Sociological Aspects of Cleft Lip and Palate Adults. 4. Vocational and Economic Aspects', *Cleft Palate Journal* 12 (1975): 193–9.

13 Joe Noar, 'Questionnaire Survey of Attitudes and Concerns of Patients with Cleft Lip and Palate and their Parents', *Cleft Palate-Craniofacial Journal* 28 (1991): 279–84.

14 Tore Ramstad, Ernst Ottem and William C. Shaw, 'Psychosocial Adjustment in Norwegian Adults who had Undergone Standardised Treatment of Complete Cleft Lip and Palate. Part I. Education, Employment and Marriage', *Scandinavian Journal of Reconstructive Surgery and Hand Surgery* 29 (1995): 251–7.

15 Hunt et al., 'Self-reports of Psychosocial Functioning'.

16 Orlagh Hunt, Donald Burden, Peter Hepper, Mike Stevenson and Chris Johnston, 'Parents' Reports of Psychosocial Functioning in Children with Cleft Lip and/or Palate', *Cleft Palate-Craniofacial Journal* 44.3 (2007): 304–11.

17 Goffman, *Stigma*, 44.

18 F.-J. Kramer, C. Baethge, B. Sinikovic and H. Schliephake, 'An Analysis of Quality of Life in 130 Families having Small Children with Cleft Lip/palate using the Impact on Family Scale', *International Journal of Oral and Maxillofacial Surgery* 36.12 (2007): 1146–52.

19 S. Forth, 'Antenatal Diagnosis of a Cleft Lip and Palate: Supporting the Parents', *Journal of Neonatal Nursing* 9 (2003): 57–61.

20 J. Farrimond and M. Morris, 'Knowing or not Knowing before Birth: Parents' Experiences of having a Baby with a Cleft Lip, with or without Cleft Palate' (Unpublished BSc. Diss., University of the West of England, Bristol, UK, 2004).

21 Pauline A. Nelson, Anne Louise Caress, Anne Marie Glenny and Sue A. Kirk, '"Doing the Right Thing": How Parents Experience and Manage Decision-making for Children's "Normalising" Surgeries', *Social Science and Medicine* 74 (2012): 796–804; Pauline A. Nelson, Susan A. Kirk, Anne Louise Caress and Anne Marie Glenny, 'Parents' Emotional and Social Experiences of Caring for a Child through Cleft Treatment', *Qualitative Health Research* 22.3 (2012): 346–59.

22 Goffman, *Stigma*, 46.

23 Tomoko Omiya, Mikko Ito and Yoshihiko Yamzaki, 'The Process Leading to Affirmation of Life with Cleft Lip and Cleft Palate: The Importance of Acquiring Coherence', *Japan Journal of Nursing Science* 9 (2012): 127–35; Ronald P. Strauss, '"Only Skin Deep": Health, Resilience and Craniofacial Care', *Cleft Palate-Craniofacial Journal* 38.30 (2001): 226–30.

24 Strauss, 'Only Skin Deep'.

25 Hilary L. Broder, Maureen Wilson-Genderson and Lacey Sischo, 'Examination of a Theoretical Model for Oral Health Related Quality of Life Among Youths with Cleft', *American Journal of Public Health* 104.5 (2014): 865.

26 Ronald P. Strauss, Barry L. Ramsey, Todd C. Edwards, Tari D. Topolski, Kathy A. Kapp-Simon, Christopher D. Thomas, Carla Fenson and Donald I. Patrick, 'Stigma Experiences in Youth with Facial Differences: A Multi-site Study of Adolescents and their Mothers', *Orthodontic Craniofacial Research* 10 (2007): 96–103; Strauss, 'Only Skin Deep', 227; Ornella Masnari, Clemens Schiestl, Jochen Rossler, Stefanie K. Gutlein, Katrin Neuhaus, Lisa Weibel, Martin Meuli and Markus A. Landolt, 'Stigmatization Predicts Psychological Adjustment and Quality of Life in Children and Adolescents With a Facial Difference', *Journal of Pediatric Psychology* 38.2 (2013): 162–72.

27 Masnari et al., 'Stigmatization', 163.

28 Strauss et al., 'Stigma Experiences in Youth', 100–1.

29 Tore Ramstad, Ernst Ottem and William C. Shaw, 'Psychosocial Adjustment in Norwegian Adults who had undergone Standardised Treatment of Complete Cleft Lip and Palate. Part II. Self-reported Problems and Concerns with Appearance', *Scandinavian Journal of Reconstructive Surgery and Hand Surgery* 29 (1995): 329–6.

30 Farrimond and Morris, 'Knowing or not Knowing before Birth'.

31 Celine Ray-Bellett and Judith Hohfield, 'Parental Diagnosis of Facial Clefts: Evaluation of Specialised Counselling', *Swiss Medical Weekly* 134 (2004): 640–4.

32 V. Hammond, Amy Fletcher, Jenny Hunt and Sue Channon, 'Mothers' Experiences of Cleft Repair Surgery for their Babies born with a Cleft Lip and/or Palate', paper presented at the Cranio-Facial Society of Great Britain and Ireland, Annual Conference, Bristol, 25–27 April 2012.

33 Kirsten B. Feragen, Ingela L. Kvalem, Nicola Rumsey and Anne I. H. Borge, 'Adolescents with and without a Facial Difference: The Role of Friendships and Social Acceptance in Perceptions of Appearance and Emotional Resilience', *Body Image* 7 (2010): 271–9.

34 Staci M. Zolkowski and Lyndal M. Bullock, 'Resilience in Children and Youth: A Review', *Children and Youth Services Review* 34.12 (2012): 2295–303.

35 Michael Rutter, 'Resilience, Competence and Coping', *Child Abuse and Neglect* 31.3 (2007): 205; Janice G. Dyer and Teena Minton McGuiness, 'Resilience: Analysis of the Concept', *Archives of Psychiatric Nursing* 5 (1996): 276–82; Suniya S. Luthar, Dante Cicchetti and Bronwyn Becker, 'The Construct of Resilience: A Critical Evaluation and Guidelines for Future Work', *Child Development* 71.3 (2000): 543–62.

36 Dyer and Minton McGuiness, 'Resilience: Analysis of a Concept'; Zolkowski and Bullock, 'Resilience in Children and Youth'.

37 Dyer and Minton McGuiness, 'Resilience: Analysis of a Concept'.

38 Zolkowski and Bullock, 'Resilience in Children and Youth'.

39 Ibid.

40 Rutter, 'Resilience, Competence and Coping'.

41 Zolkowski and Bullock, 'Resilience in Children and Youth'.

42 Rutter, 'Resilence, Competence and Coping'.

43 For further discussion of this issue Aidan Searle, Patricia Neville and Andrea Waylen, 'Psychological Growth and Well-being in Individuals born with Cleft: An Application of Self-Determination Theory', *Psychology and Health* 32.4 (2017): 459–82.

44 'Spoiled identity': Goffman, *Stigma*, 14.

45 Christina Havstam, Katja Laakso and Karin C. Ringsberg, 'Making Sense of the Cleft: Young Adults' Accounts of Growing Up with a Cleft and Deviant Speech', *Journal of Health Psychology* 16.1 (2011): 22–30; Omiya et al., 'The Process leading to Affirmation of Life'; Pauline A. Nelson, 'Qualitative Approaches in Craniofacial Research', *Cleft Palate and Craniofacial Journal* 46.3 (2008): 245–51.

46 Nelson, 'Qualitative Approaches'.
47 Havstam et al., 'Making Sense of the Cleft'; Omiya et al., 'The Process leading to Affirmation of Life'.
48 Masnari et al., 'Stigmatization'; Broder et al., 'Examination of a Theoretical Model'.
49 Havstam et al., 'Making Sense of the Cleft'.
50 Martin Persson, Gustaf Aniansson, Magnus Becker and Henry Svensson, 'Self-Concept and Introversion in Adolescents with Cleft Lip and Palate', *Scandinavian Journal of Plastic and Reconstructive Surgery and Hand Surgery* 36 (2002): 24–7.
51 http://www.healthtalk.org/about/overview
52 Health Experiences Research Group, *Researcher's handbook Healthtalkonline & Youthtalk modules, Version 27* (Oxford: University of Oxford, 2012).
53 Urie Brofenbrenner, 'Ecology of the Family as Context for Human Development: Research Perspectives', *Developmental Psychology* 22.6 (1986): 723–42.
54 Goffman, *Stigma*, 46.
55 Jennifer G. La Guardia, 'Developing Who I am: Self-Determination Theory Approach to the Establishment of Healthy Identities', *Educational Psychologist* 44.2 (2009): 90–104.
56 Reed Larson, Maryse Richards, Giovanni Moneta, Grayson Holmbeck and Elena Duckett, 'Changes in Adolescents' Daily Interactions with their Families from Ages 10–18: Disengagement and Transformation', *Developmental Psychology* 32 (1996): 744–5.
57 'Stigma learning': Goffman, *Stigma*, 46; 'spoiled identity': ibid., 59. Mel Bartley, David Blane and Scott Montgomery, 'Health and the Life Course: Why Safety Nets Matter', *British Medical Journal* 314 (19 April 1997): 1194–6.
58 Dawn Snape and Liz Spencer, 'The Foundations of Qualitative Research', in *Qualitative Research Practice: A Guide for Social Scientists and Researchers*, ed. Jane Ritchie and Jane Lewis (London: Sage Publications Ltd., 2003), 1–23.

8

Making up the Female Face: Pain and Imagination in the Music Videos of CocoRosie

Morna Laing

The surrealist movement revered imagination as a vehicle for social change, with André Breton claiming 'Imagination alone offers me some intimation of what can be.'[1] As well as offering a vision of what can be, we may also turn to imagination to envision what might have been. By virtue of their music and performances, Freak-Folk musicians CocoRosie articulate pain in the present by ventriloquizing marginalized figures from the past (such as witches), spectres repressed by the heterosexual matrix, and their own past and future selves. The face lies at the heart of CocoRosie's performances. It conveys intense emotion to the viewer as well as being a canvas upon which layer upon layer of gendered meanings can be muddied through the application of make-up and addition of false hair. Such facial interventions open up 'geometric possibilities'[2] of meaning, that is, multifaceted identities beyond the tired male/female binary that haunts the human subject. By presenting selves that are not theirs they permit the skeletons of repressed or outlawed selves to shine through: selves made visible through make-up, adornment and intertextuality. In so doing they locate contemporary identity struggles in the context of long-standing discourses on marginalized femininities, sexualities and modes of being.

This chapter is located in the Cultural Studies tradition and presents visual analysis of stills from two CocoRosie videos: 'Lemonade' (2010) and 'We Are on Fire' (2012). In the first section I lay down a framework for understanding gender through the work of American philosopher Judith Butler. From there, I turn to the female-gendered face, arguing that facial hair is key to policing the boundary that separates male from female faces in the West. Through analysis of stills from 'Lemonade' I argue that cosmetics and faux facial hair allow not only for the visualization of repressed male selves but also serve as a stand-in for trauma or nostalgic imagining. The idea of the ageing female face, replete with wrinkles and tufts of hair, is explored and linked to the visual iconography of witchcraft: a theme that carries over into the final part of the chapter where I analyse stills from 'We Are on Fire'. In this final section I explore visual discourses on witchcraft, with particular emphasis on metamorphosis, pain and the symbolic language of female sexuality. The magic of witchcraft is connected with the perception that make-up has magical properties which can transform the face.

Woven throughout the chapter is reference to Donna Haraway's cyborg politics and, in particular, her call to move away from gender binaries and other reductive oppositions that serve to categorize and therefore 'master' the social world. Haraway's metaphor of the cyborg lends a vocabulary for making sense of the categories and binary codes that structure – and feel resistance from – the multiple selves that appear in the music videos of CocoRosie. I turn to Haraway not to theorize the body through technoscience but rather to find a vocabulary for the 'noise' and 'pollution' advocated by cyborg politics[3] and visualized through the forms created by CocoRosie and the myths they reorient.

CocoRosie

CocoRosie was formed in 2003 by two sisters, Bianca and Sierra Casady. The sisters were born and raised in North America, and had a relatively unconventional upbringing, frequently moving to different regions of the United States. Their mother was an artist and teacher while their father was devoted to Shamanism and the Native American religion Peyote Church.[4] The sisters became estranged during childhood, reuniting later as adults, in Paris, where Sierra was training at the Conservatoire. It was in a bathtub in Montmartre that they recorded their debut album, Maison de Mon Rêve (2004). Their music tends to be located within the 'Freak Folk' genre,[5] and might itself be considered 'queer' in its resistance to categorization and its amalgamation of influences: from pop, blues, electronica, hip hop, beat boxing, classical opera and so on. They sample an eclectic range of sounds, from children's toys to mewing kittens.[6] One of their most notable collaborations has been with transgender musician Anohni, formerly known as Antony Hegarty of Antony and the Johnsons. As well as collaborating musically, CocoRosie and Anohni have worked together on the Future Feminism project, comprised of a feminist manifesto, exhibition, performance and lecture series.[7] One half of the sisterly duo identifies as queer – Bianca – and she has commented on how this informs the way the sisters fashion themselves, with drag and same-sex drag being one strategy they employ.[8]

Although CocoRosie are musicians, the face, the body and fashion are central to the ideas and feelings they convey to the viewer. This might be true of musicians more generally in that Janice Miller notes 'the fundamental part that appearance can play in communicating and completing music performances, particularly for women, for whom fashion has been made an endemic part of their lives for hundreds of years'.[9] The power vested in the musician's face when performing is brought to the fore in instances where the singer's face is obscured, to disconcerting effect: such is the case with Sia, who in performances tends to '[conceal] herself behind wigs, masks and surrogates'.[10] CocoRosie have ideal female faces by contemporary standards in the West: they are both white, slim and relatively young women (perhaps in their mid-thirties), with classical features and clear skin. Yet the sisters use their faces as one element in a larger strategy that challenges normative ways of thinking about gender, sexuality and identity. In this chapter I consider the way the sisters 'disassemble' and 'reassemble' their faces, to borrow the words of Haraway.[11] This is a strategy in which 'noise' and 'pollution' are nurtured through the use of make-up, false hair, and past, present and future selves.

Gender binaries and abjection

Before proceeding further it is worth considering what it means for a gender or sexuality to be normative, and thus by definition what a normative female face might look like. American philosopher Judith Butler has theorized sex, gender and sexuality as constructed through exclusion, and regulated by the heterosexual matrix:[12]

> The heterosexualization of desire requires and institutes the production of discrete and asymmetrical oppositions between 'feminine' and 'masculine', where these are understood as expressive attributes of 'male' and 'female'. The cultural matrix through which gender identity has become intelligible requires that certain kinds of 'identities' cannot 'exist' – that is, those in which gender does not follow from sex and those in which practices of desire do not 'follow' from either sex or gender.[13]

Butler is here outlining the idea that 'male' and 'female' are positions constructed in binary opposition. Binaries are the simplest way of marking social difference in language, the simplest way of categorizing people.[14] However, such simplicity is also reductive; it limits the range of behaviours, appearances and personality attributes that can legitimately be adopted by men and women. Furthermore, the sides of the binary are not equal; the male side of the binary, and its associated attributes, are nearly always pitched higher: active versus passive; rationality versus emotion; culture versus nature; human versus animal; adult versus child.[15] In this way, the male/female binary intersects with a whole host of other binary oppositions, limiting the expectations and experiences of each gender.

The heterosexual matrix regulates the kind of identities that count as intelligible, and those that fail to be recognized. Cultural norms prescribe that a person whose sex is female should be feminine and, in turn, her object of desire should be asymmetrical: a man. This identity will count as 'intelligible' and normative under social and psychic conditions as they exist in the West. A gender will thus be intelligible to the extent that it maintains coherence along the lines of sex and gender, sexual practice and desire. Butler's argument is that this 'unity', while seeming like the 'cause' of gendered behaviours and appearances, is in fact the effect of regulatory discourses, which construct masculinity and femininity as asymmetrical and mutually exclusive. In this way, 'every [living and desiring] being is constrained by not only what is difficult to imagine, but what remains radically unthinkable.'[16]

Yet, as Butler points out, 'Heterosexuality does not have a monopoly on exclusionary logics.'[17] Any form of identity politics tends to rely on oppositions in order to establish coherent subject-positions, meaning that binaries persist within the gay community. For example, Butler highlights the exclusionary logic of 'butch' versus 'femme', or the punishing idea that bisexuality somehow involves a 'failure of loyalty or lack of commitment – two cruel strategies of erasure'.[18] The subject's quest for coherence therefore comes 'at the cost of its own complexity'.[19] The subject's assumption of a coherent position 'is produced through the production, exclusion and repudiation of abjected spectres that threaten those very subject-positions'.[20] These 'abjected spectres' are both psychical and physical – an example of the latter being the facial and body hair that women now routinely remove in order to exclude it from their sense of self.[21]

The female face: Make-up and magic

Make-up and adornment of the face play a key role in elaborating gender differences between men and women.[22] These can be used to communicate one's conformity with gender norms as well as resistance to them. One need not necessarily be a woman to construct a feminine face through make-up, as RuPaul's Drag Race (2009–present) has emphatically shown. As well as building a picture of femininity, make-up can be used to emphasize female sexuality through reddened lips and flushed cheeks.[23] While powder and skin preparations have been used for centuries (as Theresa Tyers' chapter in this book demonstrates), the mainstream use of visible make-up is a modern phenomenon. The mainstreaming of visible make-up can be traced back to the late 1910s when it was used to create a sense of glamour on the faces of Hollywood starlets, becoming central to women's beauty routines in the 1930s.[24] Prior to this, visible make-up had been associated with the theatre and with prostitution and tended to evoke suspicion when used outside this ambit.[25] For instance, in the late seventeenth century the use of cosmetics was seen as '[usurping] the power of God to determine appearance' as well as the ability of men to determine a woman's beauty (and, therefore, her worth).[26] Made-up faces could conceal signs of ageing or disfigurement from disease, potentially leading men to marry under 'false pretences'. Significantly for this chapter, the potential of 'scents, paints, cosmetic washes, artificial teeth [and] false hair' to 'seduce' and 'betray' was linked explicitly to witchcraft in an act passed by the English parliament in 1770.[27] Marriages made on the basis of such 'seductions' would stand null and void upon conviction. The prevailing ideal for the female face at this time was therefore 'ascetic'.[28]

Visible make-up has been normalized in the contemporary context but the association between make-up and magic lives on. For instance, the paraphernalia of witchcraft – powders, ointments, lotions and potions – ties in with the products used in beauty rituals of women (and men) today. In the early twentieth century, Max Factor was hailed a magician for his ability to 'conjure images'.[29] His cosmetics could be used 'to overcome imbalances in shape, size or angle and to disguise "defects"'.[30]

Furthermore, Brian Moeran notes the language of magic and enchantment used in contemporary beauty advertisements and fashion magazines. He argues that advertising involves 'a heightened use of language that aims to combine word and deed (the persuasion to purchase and make use of a product) by using spells especially constructed to effect a magical transfer'.[31] He draws a further parallel between beauty advertisements and ritual, with language being used in the former to '[connote] their power (over beauty) to exorcize the demons of unattractiveness'.[32] In spite of advertising's potential to manipulate, there is nevertheless agency to be had in the application of make-up. For those living with facial disfigurement in the present, cosmetics play a protective role; make-up can 'provide a defensive barrier against intrusive stares and comments as they go about their day-to-day lives'.[33] More generally speaking, make-up can also be used as a tool in the everyday expression and creation of the self.[34]

Facial hair and shame

In their music videos, CocoRosie use make-up to communicate a sense of loss vis-à-vis repressed versions of themselves – those masculine qualities not permitted to show through on account of their having assumed a female subject-position. For instance, 'Lemonade' (2010) sees CocoRosie appear with exaggerated facial hair: Bianca's beard is drawn in dark kohl (Figure 8.1) while Sierra's is made of light grey fluff attached in wispy clouds around her mouth (Figure 8.2).

This visualizes Butler's idea that 'if identity is constructed through opposition, it is also constructed through rejection',[35] and this rejection entails a loss of other possibilities: possibilities CocoRosie imagine via their faces in 'Lemonade'. The young girl's realization that she cannot be both male and female entails a profound sense of loss vis-à-vis a whole set of attributes that she is 'forbidden' from assuming. Inhabiting a female subject-position not only excludes characteristics coded as male, but also means that there will always be a gap between one's body as experienced and the position mapped out in discourse. One can never quite manage to fully and completely identify with the female (or male) subject-position one feels compelled to inhabit. And this gap between the is and the ought represents another source of pain.

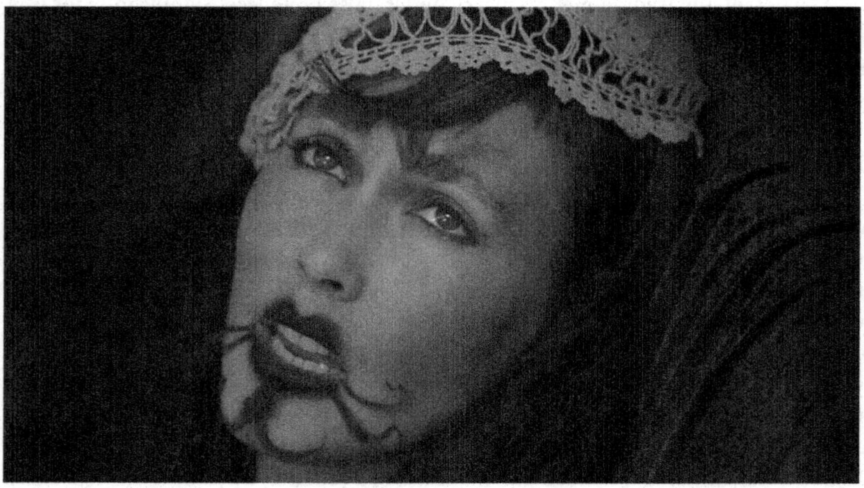

Figure 8.1 Bianca in 'Lemonade' by CocoRosie (2010). CocoRosie, 'Lemonade' directed by Emma Freeman, 2010, used by permission of CocoRosie.

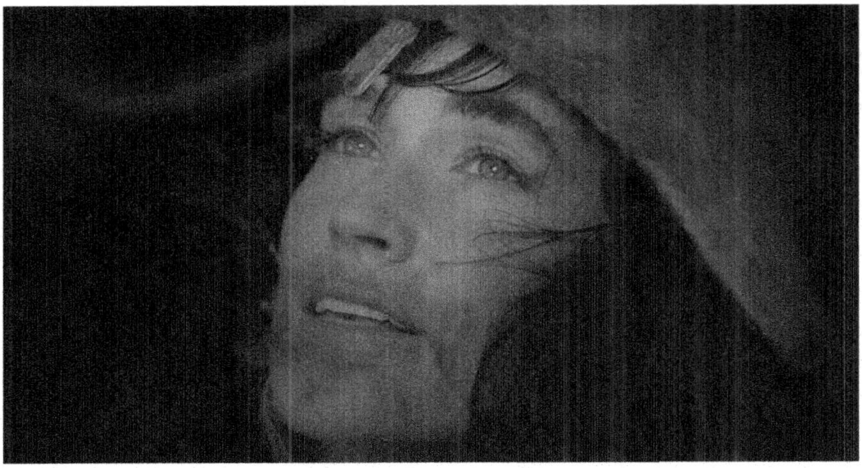

Figure 8.2 Sierra in 'Lemonade' by CocoRosie (2010). CocoRosie, 'Lemonade' directed by Emma Freeman, 2010, used by permission of CocoRosie.

One facial attribute banished from the female ought is dark and visible facial hair. Karin Lesnik-Oberstein has described body hair on women in the West as 'something not to be seen or mentioned; prohibited and circumscribed by rules of avoidance; surrounded by shame, disgust, and censure'.[36] Thanks to feminism, there is more open discussion of female sexuality and female bodies, yet the censure of body hair stubbornly hangs on. Body hair, in this sense, is 'the last taboo';[37] depilation is 'feminism's lost battle'.[38] Detecting the presence or absence of facial hair thus becomes key to policing the boundary that separates 'male' from 'female'. Presence of this element signals and bolsters masculinity. By contrast, absence of this element comes to signify femininity, leading Lesnik-Oberstein to posit women's body hair as 'an area of silence and blankness', harking back to Freudian definitions of femininity founded on 'lack'.[39] This blankness can be further connected to the 'blankness' of pain, as discussed below.

Facial hair on women therefore involves a collective disavowal. It is generally known that dark hair grows on the female face, but there is a discursive demand that this be minimized and managed, through practices of bleaching, waxing, threading, electrolysis or, more recently, laser therapies. Many of these processes are painful and all of them are party to the 'labour' of femininity. Such practices are arguably motivated by shame,[40] and constitute self-imposed attempts to approximate the female ideal: a classic example of Foucault's subject who self-regulates, having internalized the rules of society and culture. For as Bartky has argued: 'No one is marched off for electrolysis at gunpoint.'[41] These practices of the self can be understood as 'purifying' the category 'female' in opposition to 'male', in line with the famous passage from Mary Douglas' *Purity and Danger*:

> I believe that ideas about separating, purifying, demarcating and punishing transgressions have as their main function to impose system on an inherently untidy experience. It is only by exaggerating the difference between within and

> without, about and below, male and female, with and against, that a semblance of order is created.[42]

Woman is defined without, below and against facial hair. The presence of dark hair on the face of a woman therefore constitutes 'matter out of place'[43] or, as Lesnik-Oberstein puts it, 'Its visibility does indeed reveal femininity as that which hides within itself the potentially masculine.'[44]

Yet, the removal of facial hair is not just about bolstering the difference between femininity and masculinity. The black smudge on Bianca's face (Figure 8.1) overflows the boundaries of her lips, streaming outwards, like inky roots reaching to murky waters below. These roots form a beard on and around her mouth. The resemblance to roots is significant, given the liminal status of hair in relation to the body: 'Being both in and on the skin body hair polices a boundary: its follicle is inside the body, while the shaft is visible outside of it.'[45] Smelik argues that the disgust evoked by female body hair stems from its status as the abject. Smelik, reading Julia Kristeva, defines the abject as 'that which has been ejected out of the social order as dirty, impure, improper or unclean', particularly that which 'signifies an unstable boundary between the inside and outside of the body'.[46] Body hair, therefore, must be rendered 'radically other' in order to allay disgust and be excluded from one's ladylike sense of self.[47] Bianca's kohl beard, therefore, serves not to bolster her femininity, as is typical of make-up in the West,[48] but rather to cut through it. Bianca's beard looks painful in its bruised incarnation, suggesting trauma or damage in the gendering process. Beards are unwelcome on young feminine faces.

Where facial hair is present on a woman, it tends to set her apart from the mainstream – it signals her difference or 'otherness'. A notable example is the Mexican artist Frida Kahlo, whose many self-portraits include pronounced facial hair both on her upper lip and between her eyebrows (the latter being pejoratively termed a 'monobrow'). Yet Kahlo was a painter, and in some ways the values and mythology of the art world – its bohemianism, subversion of the mainstream values and the myth of the 'tortured artist'[49] – permit women a certain degree of freedom from punishing norms of appearance.[50] Like Kahlo in her role as bohemian artist, CocoRosie perhaps also have the creative and 'subcultural capital' to carry off such transgressions with limited rebuke.[51] Also noteworthy is that Kahlo presented herself alongside animals, such as her pet monkey.[52] This is significant because some scholars have argued that the desire for a hairless self is an attempt to reinforce our nature as humans rather than animals, given our evolutionary path from furry primates to humans.[53] Removing hair can therefore be understood as an act that neutralizes the lurking threat of beastliness, '[reasserting] a strict boundary between human and animal'.[54] One of the fears that propelled discourses on witchcraft, as explored below, was the idea that these women were capable of crossing the human/animal divide, whether through practices of bestiality or through metamorphosis into animal form.[55]

Yet, in contrast to Kahlo, whose facial hair is downy and appears in soft dark wisps above her fulsome lips,[56] CocoRosie's facial hair is perhaps simply too 'made-up'. The hair fails to border on animality, or even masculinity, in the way that Kahlo's does. We know too easily that the hair is make-believe. The sense that they are just dressing up, temporarily, saves the sisters from abjection, as do their classical features, their youth and their whiteness. Furthermore, their exaggerated facial hair might actually act as a foil for their beautifully feminine features. In this sense, might it not be more subversive for the sisters to wear no make-up at all,[57] or to appear with their own facial hair darkened?

The subversiveness of the sisters' faux facial hair depends, in a sense, on what they are trying to convey. 'Lemonade' can be read as a reflection on their childhood past, both nostalgically and through the lens of trauma. The childlike 'dressing-up-box' nature of their beards, therefore, is perhaps what the performance demands. Adornment furnishes the sisters with the capacity for transformation, which ties in with the association between make-up and magic. After all, 'Make-up promises transformations. Surface blemishes are hidden, youth is regained, skin becomes smooth.'[58] The very word 'make-up' acts as a double-entendre for made-up, meaning something that is invented or untrue or that springs from the imagination. CocoRosie seem to be using make-up as a means of communicating inner emotions, trauma and pain. In turn, this ties in with figures from the collective consciousness – such as witches or their own personal narratives. Or it might simply signal what would be different if we could simply 'ungender our discourse'.[59] Nevertheless, when it comes to facial hair in 'Lemonade' the sisters are quite far away from the '"hairy" women [who are] monstrous in being like men, or masculine'.[60] As such, the facial hair poses less of a challenge to gender norms, perhaps standing in, instead, for childhood experience.

The ageing female face: Hags, sorcery and trauma

Childhood is a theme that recurs in the music of CocoRosie, with the name of the group itself comprised of pet names given to the girls by their mother. Furthermore, the sisters featured in a documentary entitled The Eternal Children (2006) alongside Devendra Banhart and Antony and the Johnsons (directed by David Kleijwegt).[61] In 'Lemonade' (2010), one of their more autobiographical songs,[62] the Casady sisters appear in three incarnations: as children, as elderly women, and as their current selves: young women. This triple presentation of the female self harks back to paintings such as The Ages and Death (1541–4) by Hans Baldung Grien. As an artist, Baldung was instrumental in the development of a visual iconography for witchcraft in sixteenth-century Europe.[63] In this particular painting, the young woman is illuminated as the most desirable of the three whereas the elderly woman appears sallow and wrinkled: envious of the younger woman's beauty. This contrasts with 'Lemonade' where judgement on the elderly women is not passed in the same way.

'Lemonade' opens with two elderly women, crowned with bejewelled headdresses and slowly descending staircases, in parallel, on their way to visit past selves (Figure 8.3).

Their crown-like headdresses and ornate gowns recall medieval visions of the classical sorceress Circe,[64] whose spells transformed men into beasts.[65] A flash of a classical bust featuring woman and child seems to support this interpretation as does a ship in a bottle and a taxidermic stag, fox and raven. Yet while Circe has tended to be depicted as a very beautiful young woman, capable of enchanting men,[66] the women

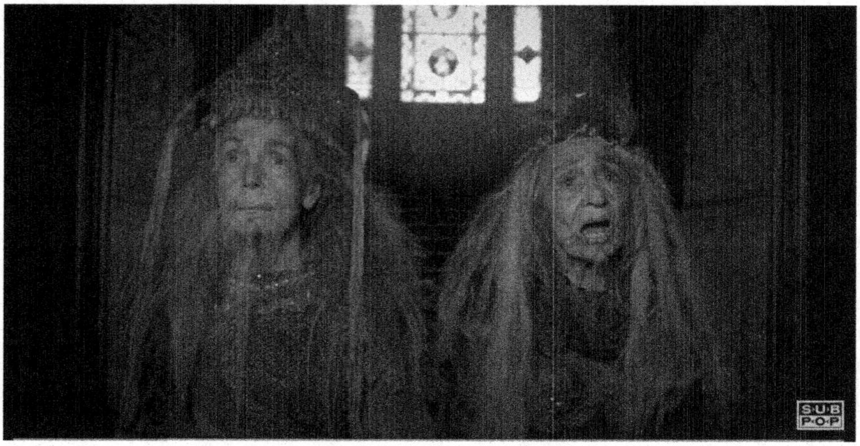

Figure 8.3 The elderly selves. Still from 'Lemonade' by CocoRosie (2010). CocoRosie, 'Lemonade' directed by Emma Freeman, 2010, used by permission of CocoRosie.

in 'Lemonade' are elderly with wrinkled faces sprouting tufts of hair. Their faces more closely fit descriptions of the witch-as-hag in the medieval and early modern periods. For instance, in 1646 John Gaule, a critic of the witch prosecutions, wrote the following: 'Every old woman with a wrinkled face, a furred brow, a hairy lip, a gobber tooth, a squint eye, a squeaking voice, or a scolding tongue … is not only suspected but pronounced for a witch.'[67] Reference is made to sorcery through the introduction of tarot cards, glass bottles, amethyst crystals and tea – all suggestive of alchemy. At one point the elderly women sit facing a lightning storm, with a tea set, suggesting weather sorcery or astrological powers. Meanwhile Bianca sings of her childhood 'search for magic powers to heal our mother's pain'. The relationship between Bianca and Sierra as sisters, lends itself well to activating historical discourse on witchcraft such as that in Shakespeare's Macbeth: 'The Weird Sisters, hand in hand, Posters of the sea and land'.[68]

The figure of the witch emerged in the first half of the fifteenth century in European cultural discourse.[69] Press reportage was often sensationalized and pamphlets, treatises and news reportage were circulated widely thanks to the recently established printing press. This allowed for a discursive elaboration of the practices and potency of witches as 'evil-doers … who inflict malefic (maleficium)', usually in association with the devil.[70] Charles Zika has presented a rich overview and analysis of visual iconography from the fifteenth and sixteenth centuries, demonstrating how symbolically loaded the trope of the witch is. The figure of the witch, as she emerged in visual discourses during this period, brought together strands of discourse on the classical witch as well as biblical scriptures.[71] Most importantly for this chapter, the witch represented the breakdown of the social order and the categories that structure human experience.

Visual culture in the late-medieval period was central to fostering belief in the figure of the witch: 'Visual images helped embody the witch; they helped make her more immediate, recognisable and credible.'[72] Crucially, the majority of these images were male authored and served a number of functions, including providing a pretext for indulgence in fantasies and fascinations vis-à-vis female sensuality.[73] The Casady sisters rework the misogynistic myths bound up in the figure of the witch, which have structured and limited female experience. They expose these myths not only to portray pain but also to indicate the potential for different visions in the future. For if a figure as potent as the witch can emerge out of discourse – and recede again – then surely there is room for movement in our definitions of gender, sexuality and subjectivity as they stand. The idea of reworking historical myths through the moving image can be understood as one tool in the 'recrafting' of the body: 'The boundary is permeable between tool and myth, instrument and concept, historical systems of social relations and historical anatomies of possible bodies, including objects of knowledge.'[74] Historical repositories of myth, Haraway argues, inform our imaginations and we might think of the cyborg as 'a kind of disassembled and reassembled, postmodern collective and personal self. This is the self feminists must code.'[75] CocoRosie are engaged in this process of coding and recoding the face and body, working and reworking the myths and ideas that haunt our personal and collective sense of self – whether we like those myths and memories or not. They are making certain selves visible to us, and certain bodies possible, through the 'communications technologies' that Haraway identifies as 'the crucial tools recrafting our bodies'.[76]

By combining elements from two different discourses on witchcraft – the classical Circe combined with the 'old hag' – CocoRosie frame these women so they can be read as beautiful in their wisdom and alive in their power. For magic had been practised by individuals prior to the emergence of the newfound figure of the witch in the early fifteenth century. What changed, at this point in time, was that magic and sorcery came to be demonized in discourse.[77] As such, 'Men and women who had traditionally specialized in a range of magical practices – love and fertility magic, fortune telling, treasure seeking and healing – as well as those who had a reputation for harmful sorcery directed against fertility, bodily health or community well-being, now stood accused of witchcraft.'[78] The women in 'Lemonade' seem to belong to an older discourse on sorcery and magic, which was not necessarily malicious or aligned with the work of the devil. Thus, while in medieval discourse, frizzy hair was a sign of evil,[79] in 'Lemonade' the long frizzy locks that hang from the women's heads seem to add a sense of mystery or wisdom to the women. Furthermore, there is no sign of the envious nature attributed to the witch-as-hag, such as the eating or sacrificing of children,[80] or the envy directed to younger women.

Although their costume signals that these are women, their facial hair casts doubt on this judgement; one cannot be sure that these are women and not men. The woman on the left has a long, wispy beard, yet unlike the clouds on the face of Sierra (Figure 8.2), it is no longer possible to tell whether these wisps are natural or artificial – rooted on the face or simply stuck on. This recalls the scene in Macbeth (1623) where Banquo encounters the three witches:

> What are these,
> So wither'd and so wild in their attire,
> That look not like th' inhabitants o' th' earth,
> And yet are on't? Live you? or are you aught
> That man may question? You seem to understand me

> By each at once her choppy finger laying
> Upon her skinny lips: you should be women,
> And yet your beards forbid me to interpret
> That you are so.[81]

Banquo's final lines – 'you should be women, / And yet your beards forbid me to interpret / That you are so' – resonate with the white facial hair that covers the faces of CocoRosie's future selves (Figure 8.3). The facial hair throws their identity as women into question, reinforcing the sense in which they, and the witches upon whom Banquo rests his eyes, cannot possibly be of this world. Yet, in some ways the hair on their chins (in Figure 8.3) goes with the grain of discourses on ageing femininity, whereby elderly women are posited outside the ambit of the sexualizing male gaze. As such they are no longer expected to conform to the normative standards of femininity. Facial hair then becomes of little consequence: no longer 'matter out of place'.[82]

The house the women of 'Lemonade' inhabit is old and dimly lit, suggesting not only that dusk has fallen but also that these are women in the twilight of their lives. The sense of decrepitude that reaches out from the house is mirrored in the crumpled faces of the women, signalling wisdom, experience and lives well lived. This is a positive interpretation of female ageing, quite at odds with the way women have historically been condemned for growing old, sometimes to the point where the charge of witchcraft was applied. In the eighteenth century Mary Wollstonecraft argued that because female perfection was premised largely on appearance, it meant ideal selfhood was short-lived, and women were 'made ridiculous and useless when the short-lived bloom of beauty [was] over' – a sentiment she attributed to a male writer who wondered of what use women 'turned of forty' were to this world.[83] As the elderly women in 'Lemonade' descend the staircase, their movements are in sync, and their faces look similar, conveying a sense of twinning, and reminding us that the musicians are sisters.

The light changes when the elderly women enter a room downstairs, encountering their childhood selves, in a brightly lit room, playing in a rowing boat with a Native American man: a reference to the sisters' father and his spiritual beliefs, perhaps. Here, the brunette children sport wispy facial hair. This is reminiscent of comments made by CocoRosie in an interview:

> We started to fashion ourselves after elemental beings – fairies and gnomes; sort of a post-human kind of style where gender roles don't apply in the same way necessarily; where facial hair isn't explanatory of one's gender. It's more of an adornment or just sort of evident that we're turning more and more to nature and becoming more a part of the landscape.[84]

This desire to separate facial hair, as signifier, from masculinity, as signified, is an interesting one. If one follows this train of thought, in the context of 'Lemonade' the facial hair appears without judgement for all three sets of selves: the children, the young women and the elderly women. The idea of 'elemental beings' suggests primitive and powerful creatures that are close to nature, and whose frame of reference means that hair on one's chin is of little consequence; it no longer signifies failed femininity or the failure to keep masculinity at bay.

In one sense this seems to contradict Haraway's cyborg politics and her assertion that there is no way back to nature, only endless overlap and confusion between the natural and the artificial.[85] Yet, on the other hand, the sisters' 'recoding' of an element of the face strongly tied to gender is reminiscent of Haraway's suggestion that we can, however, escape 'the maze of dualisms' through new constellations of meaning.[86] There 'is no "place" for women in these networks, only geometrics of difference and contradiction crucial to women's cyborg identities. If we learn how to read these webs of power and social life, we might learn new couplings, new coalitions.'[87] Haraway's word choice when she states there is 'no "place" for women in these networks' is interesting in that the words 'no place' hark back to the Greek for 'utopia'. The idea that the category of 'women' can be done away with completely might be seen as a sort of utopian thinking, tying in with critiques of cyborg imagery: for as Kirkup writes, its usefulness 'as a tool for material change is yet to be proved'.[88] Similarly, within the realm of representation, facial hair can pass without judgement but when one enters the social world, with its attendant critical gaze, one's experience is likely to be different.

Nevertheless, gender is not the only element deliberately mistranslated in 'Lemonade'. Another contradiction is realized through competing definitions of childhood: marking a shift away from the binary that defines childhood in opposition to adulthood, towards a multiplication of possible 'childhoods'.[89] The vision of childhood as plural is mirrored in the way vocals are split between the two Casady sisters, with each singing a different version of the past. Sierra sings the choruses, which cast the past as dreamily nostalgic:

> Shot a rabbit from the back seat window
> Sat and watched the summer corn grow
> Ate ice cream in a desert dream
> And got lost in father's singing
> Too hot inside
> Too hot outside
> Lazy days when I said let's go for a ride
> We'll sail on Spirit Lake
> Me, my pappy, and his lemonade

This conjures her childhood past as pastoral and sweet, resonating with the Romantic idealization of childhood as a dreamy and imaginative space, free from the vicissitudes of adult life.[90] By contrast, Bianca's verses sing of childhood trauma: 'no more breathing time, the ambulance sped'; 'we seven kids, we almost died'; 'in search of magic powers, to heal our mother's pain'. The two versions of childhood sung by the sisters can be understood as belonging to two different discourses on childhood: the former stems from eighteenth-century Romanticism, and posits childhood as a site of innocence, curiosity, joie-de-vivre and proximity to nature, while the latter is rooted in Freudian notions of childhood as a site of trauma.[91]

Sierra's rose-tinted take on childhood is reflected in her face, which smiles contentedly, looking dreamily upwards towards what one imagines to be a blue sky. Sierra does not have drawn-on facial hair like her sister; instead her face is adorned with grey woolly matter, almost in the manner of clouds. The 'woolly' texture here is significant, as the term can also be used to describe a vague, indistinct thought or articulation. Sierra's face, along with the words she sings, might suggest an account that lacks credence: it is dreamy, insubstantial and distant from what Freud would consider the 'truth'. Her blue eyes sparkle as if representing sunny days of lemonade, a symbol of optimism. When she sings, the room lights up, bright and clear. While Sierra represents the Romantic view of childhood as Edenic space, this can be contrasted with Freud's conceptualization of childhood as a space involving 'much more complex productions of repression, redaction and fantasy'.[92] The sisters' faces communicate competing visions of the past thus portraying visually the 'modern suspicion of memory'[93] as well as the central role memory and childhood occupy in the founding of adult selves.

Bianca's face is etched in thick black kohl, which is far more pronounced than her sister's nostalgic fluff. Her eyes are dark and deep in colour, with this emphasized by smoky make-up, messily applied so as not to appear sultry but instead to emphasize their depth. The blackness of her lips overflows into a purple smudge. It looks dark and painful, almost like a bruise – the lasting impression of a blow. These markings can be read through a psychoanalytic lens, symbolizing childhood as the site of fear and absolute dependence, resulting in literally unspeakable trauma.[94] This carries over into a later scene in 'Lemonade' where one of the elderly women looks in the mirror and sees her younger self reflected back (Figure 8.4). Her younger self, Bianca, has a furrowed brow; her eyes are pleading under the weight of her brow. These facial expressions play a key role in communicating her pain: such expressions are normative and learnt from collectively elaborated discourse and experience.[95] The mirror into which she looks is clouded, with a dark, jagged edge, cast by an object which obscures the straight lines of the glass. The silhouette of a raven is only just visible in the cloudy mirror, suggesting that one of these selves might be an alchemic illusion. Draped around her face are her black locks, which drip down onto her throat, slicing her pale skin and echoing her dark moustache.

In an interview Sierra explains: 'We were estranged by our childhood … . Our work's a going-through-time-in-reverse journey, recovering memories that come from black areas in the mind.'[96] Of course, memories can never recover the truth of the past; they are always being rewritten in the present. But there is a sense of digging up the self – both personal and collective – perhaps represented in the title of another of CocoRosie's songs, 'Gravediggress' (2013). The intergenerational spirit of 'Lemonade' draws our attention to the normative dividing lines between active adulthood and decrepitude as well as between child and adult. The 'adult' side of the binary tends to exclude some of the very human emotions associated with childhood such as wonder, curiosity and a sense of magic, which some argue have been expelled from the category of 'adulthood' since the Enlightenment.[97] Yet here they find reincorporation in the sisters' elderly selves thanks to discourses on alchemy in which wisdom and expertise are valued. The wrinkles and hair on the women's faces does not mark these women as 'past their best' but instead marks them out as learned, kind and worthy of respect. Bianca, in the same interview, stated, 'I'm just discovering what a fairy tale really is. Everything's symbolic in our work. The bloody twins, the scarecrow; these are dark

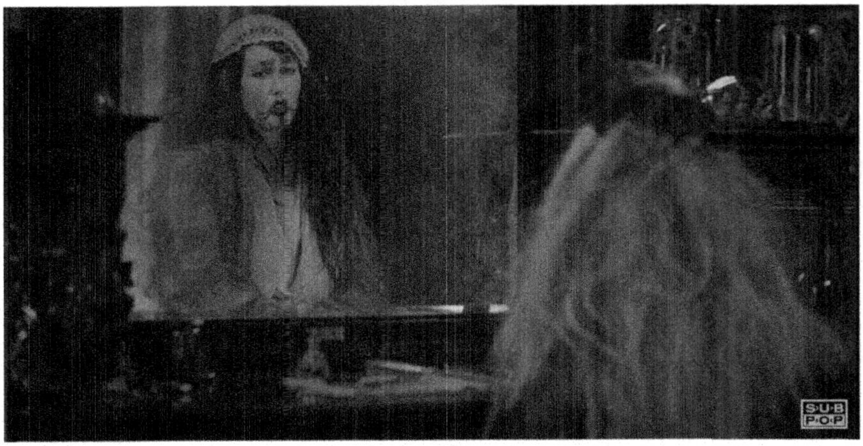

Figure 8.4 Still from 'Lemonade' by CocoRosie (2010). CocoRosie, 'Lemonade' directed by Emma Freeman, 2010, used by permission of CocoRosie.

figures from your own psyche.'[98] The sisters' exploration of the psyche, along with reference to symbolic figures and fairy tale, is reminiscent of André Breton's notion that 'there are fairy tales to be written for adults, fairy tales still almost blue'.[99]

In summary, CocoRosie employ surrealist strategies in relation to the face, using it as a site to articulate aspects of dreams and the unconscious: both personal and collective. In 'Lemonade' the face is used to signify pain and loss on a number of levels: the loss of masculine aspects of the self; the loss of outward beauty through ageing; and the loss involved in shifting from childhood to adulthood. Commenting on surrealist approaches to photography, Bancroft suggests that the photographic medium can articulate 'the dilemmas of one's own subjectivity, rendered in aesthetic form'.[100] It is not just assuming a sexed subject-position that entails loss; if we follow Lacan, then every subject suffers the loss of coherence and completeness in the coming forward as a separate subject during and after the mirror stage.[101] Identity is founded in alienation. That alienation does not happen just once; it happens again and again in order, paradoxically, to maintain the subject's boundary. As Bancroft writes, 'The continuing tension between fragmentation and unity … takes place on the surface of the body.'[102] Fashion, broadly construed to include make-up and adornment, can serve to visualize the alienation of the self, dispelling the myth of a coherent subject. Make-up thus holds imaginative potential, and made-up faces can remind us of what we have lost, who we have been, and who we might have been under different conditions. Make-up is a means of visualizing otherwise invisible conflicts.

Witches, fire and metamorphosis

'Lemonade' (2010) is an introspective song that explores the pain of ageing, childhood and familial relationships, with allusion to magic and alchemy. By contrast, 'We Are on Fire' (2012) makes more explicit the links to witchcraft, with particular emphasis on the human suffering that stemmed from persecutions. The video introduces several elements from discourses on witchcraft incorporated into the facial appearance of the performers, such as animal metamorphosis, active sexuality and envy as 'black bile'.

The video opens with a black mass emerging from the sea. This suggests transmutation, the force of the elements, or maybe even the return of the repressed. The next shot sees the two sisters bolt upright in an iron-framed bed: awoken from slumber, garbed in grubby nightdresses, their faces obscured by feathered animal skulls. The skulls mask the women's humanity while simultaneously connoting sacrifice, death and necromancy.[103] The skulls might even suggest the 'animal-masking rituals' of carnival and the inversion of the normal social order this entailed.[104] Bundles of hair are threaded through the eye sockets, perhaps from a horse or a goat (used by witches to inflict sickness).[105] These non-human elements – the skulls, the feathers, the animal hair – point to the breakdown of boundaries between the human and animal world. Witches had the power to metamorphose into animal form allowing them to 'negotiate and transgress the borders of the human realm; they had the power to move into the world of animals, just as they did into the world of the dead'.[106] Such shape-shifting powers were seen as woman's attempt to 'play God' and deceive onlookers: a more extreme version of the condemnation of make-up in the late seventeenth century already mentioned above.

As the vocals commence, Bianca appears, complete with facial markings: a sharp long line beneath each eye, etching the descent of her tears; three black lines pointing downwards from her lips; and thick chains surrounding her neck. Some of her teeth are blackened, chiming with Ovid's visual description of Envy, below. Fluffy white hair swells out from beneath her chin, calling from her throat like smoke from the pyre. The smoke symbolizes the punishment meted out to those convicted of witchcraft – burning – as well as the perceived threat of active female sexuality and lust.[107] Yet

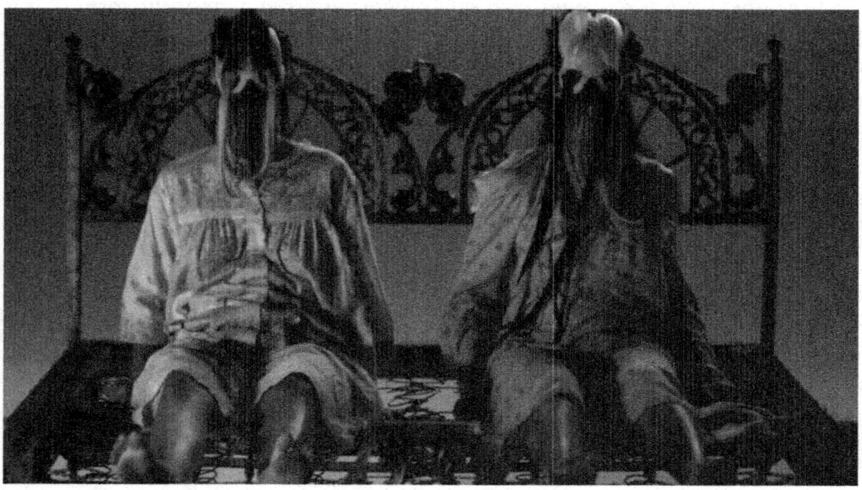

Figure 8.5 Still from 'We are on Fire' by CocoRosie (2012). CocoRosie, 'We Are on Fire' directed by Emma Freeman, 2012, used by permission of CocoRosie.

unlike the many images of witches with wild and billowing hair, signalling unbridled sexuality,[108] Bianca's hair is contained: she wears a black hairnet, suspending her hair and her femininity. This may be the image of a woman whose hair has been shorn in search of the devil's mark: 'Evidence' of witchcraft.[109] Everything is hard.

As the vocals commence, Bianca sings of 'tales of a grass widow':[110]

When I was young I thought I'd be more than just a fantasy
I wanna be this, I wanna be that
A big black dog with the soul of a cat
The blue-eyed devil inside of me
He won't leave, he's buried deep

The mention of fantasy reminds us of the witch's origin as an object in discourse. Taken at face value, 'The blue-eyed devil inside of me' could represent the sealing of a witch's pact through intercourse with the devil.[111] However, the devil mentioned by Bianca could also stand for internalized discourses on gender and sexuality that delimit the trajectories taken by male and female selves: in this case extinguishing of life on account of impossible acts. Mention of the devil is also noteworthy in that theological accounts of witchcraft posited the devil as the ultimate source of the witch's powers. Thus, even the idea of witchcraft as a 'female grab for power'[112] leaves a man (the devil) pulling the strings. In turn, this evokes Butler's notion of citationality whereby gender performativity is likened to a judge citing a law that precedes him. Yet, as Butler points out, 'From where does that law draw its authority? … Is there an original authority, a primary source, or is it, rather, in the very practice of citation, potentially infinite in its regression, that the ground of authority is constituted as perpetual deferral?'[113] Or is gender, and the power bestowed on men, to be taken as God-given?

Having been exposed to the fire, the sisters' legs are charred as we see them fly through the air, releasing white powder from their hands as they go – powder being something supplied to witches for the purposes of malefice.[114] Yet, it might be more accurate to say that the sisters are 'flung' – like objects in discourse – they do not seem in control of their movements, reminding us that witches were manipulated by the devil or, more accurately, by the discourses that brought them into being. The act of flying has been a metaphor for female sexuality since antiquity.[115] This is reinforced by the sisters' hair which flows out behind them (Bianca's hair now being released from the hair net). Not only did billowing hair signify the witch's active sexuality, it also 'connected her to the wild and to the bestial'.[116] Witches were thought to derive pleasure from a range of non-reproductive sexual acts including bestiality; riding broomsticks or animals through the sky; cunnilingus; sexual acts with other witches; and sensual exploration of 'the ambiguities of bodily orifices and vapours'.[117] The witch therefore posed a challenge to traditional definitions of domestic femininity, symbolized by the broken china cup in 'We Are on Fire'. Thanks to that humble household implement, the broom, the witch could escape the confines of the home to meet with other women.[118] The individual accused of witchcraft need not have engaged in any such sexual practices, but the myth applied to them through the charge of witchcraft activated belief that this was so. While this discourse was used to punishing effect in the historical persecution of witches, feminists have latterly revelled in the figure of the witch as a symbol of revolt: 'Witches who fly in hordes to attend Sabbaths are seen as being part of a female collectivity or coven that threatens the social order.'[119]

The idea of sexual indeterminacy ties in with what Butler has termed 'unintelligible' subject-positions: that is, those subject-positions that cannot be placed squarely within the categories outlined in the heterosexual matrix.[120] What happens when one fails to decode a face and body as male or female, homosexual or heterosexual? Furthermore, the stigmatization of witches was not just about gender and sexuality. Flying alongside Bianca and Sierra is a man wearing a feathered Native American headdress, which obscures his face – and humanity. The inclusion of the indigenous North American man harks back to the man in 'Lemonade' and also reminds us of 'the conflation of witchcraft with Amerindian cannibalism through the agency of the figure of Saturn'.[121] Both witches and Native American people were posited as uncivilized and therefore a threat to the moral and social values of Christian Europe in the late sixteenth century.

As they fly, black venom spills from the sisters' mouths. This symbolizes the 'black bile' associated with a melancholic temperament, chiming with Bianca's dripping hair and beard when she appears in pain Figure 8.4.[122] The black liquid is also reminiscent of Invidia in Ovid's Metamorphoses where, in the words of Petherbridge, 'Envy tramples flowering meadows and pollutes cities with her tainted breath, and turns people to stone by infusing them with her pitch-black venom.'[123] Not only was Envy venomous, she was also vile in appearance, as Ovid writes:

> [Her] face was sickly pale, her whole body lean and wasted, and she squinted horribly; her teeth were discoloured and decayed, her poisonous breast of a greenish hue, and her tongue dripped venom … . Gnawing at others, and being gnawed, she was herself her own torment.[124]

The emotion of envy was picked up upon in medieval portrayals of the witch.[125] It could be cited as a witch's motivation for maleficence, particularly in the elderly, post-menopausal witch – or 'hag' – who was thought to envy the fertility and beauty of younger women, as in Baldung's The Ages and Death (1541–4). The black venom mentioned by Ovid, and spilling from the sisters' mouths in the video, conveys a sense of putrefaction and dirt. The 'old hag' is not only socially spent – 'dirt' on the margins of society – but in her uselessness has become toxic to herself and to others. Ovid here draws a symmetry between outward appearance and inward character: a moralistic discourses on beauty-as-goodness and ugliness-as-badness that continues into the present.[126] Yet such discourses were, of course, contradictory. Witches were not uniformly ugly; the witch was also thought to reside in the enchantingly beautiful femme-fatale, such as Circe.[127] The beautiful witch is represented by Sierra whose long dark locks flow onto her shoulders, framing her classical features, as she burns at the pyre. Both repellant and attractive, the witch could not win.

The song reaches its climax in the operatic wails of Sierra while she burns at the stake, her voice adrift over the flames. The human suffering that stemmed from discourses on witchcraft is here rendered powerfully visible and visceral through Sierra's face and vocals. For a moment this overwrites the playful or sanitized images of witchcraft that have come to populate mainstream culture.[128] The representation of physical pain here harks back to the depiction of psychological pain in 'Lemonade'. Yet, as recent scholars on pain have observed, drawing a clear distinction between physical and mental pain is not tenable: 'My burn depresses me; my sadness weighs down my body.'[129] Bourke elaborates on this idea in the following passage, which also calls into question the binaries that structure gendered experience:

> The assumption that there is a clear distinction between the mind (characterized as disembodied, rational, computational, and male) and the body (caricatured as pre-social, emotional, impetuous, and female) has been attacked from all sides. Feminists have led the assault on the representation of the mind as some kind of superior, active, unique entity, which 'feeds' information to a passive, universal, and inferior physiology.[130]

In a way reminiscent of Haraway's cyborg politics Bourke, with reference to the work of David B. Morris,[131] seeks to overturn 'the myth of two pains': that is, the idea of 'emotional versus bodily pain'.[132] Instead, she argues, the two are intertwined, even though the sufferer might choose to focus on one aspect of the 'pain-event' over another.

Yet pain, in all its manifestations, can signal something of 'a blank', as poet Emily Dickinson (d.1886) recognized:

> Pain has an element of blank;
> It cannot recollect
> When it began or if there were
> A day when it was not.[133]

This was reinforced when Deborah Padfield spoke to pain sufferers about their experience of severe pain. These sufferers tended to talk about their experience in terms of 'blackness, emptiness and aloneness'.[134] They experienced a sense of their world collapsing or their sense of self disintegrating, highlighting the debilitating and socially isolating feelings associated with pain.

A further difficulty arises when one seeks to communicate the 'pain-event' to another person; for as Elaine Scarry has noted, 'To have pain is to have certainty; to hear about pain is to have doubt.'[135] Visualizing pain therefore holds an important social and political function. Images of pain, Padfield argues '[allow] us to look at aspects of ourselves, and experiences, which have hitherto been too painful to acknowledge'.[136] One need only think of Frida Kahlo's repetitive self-portraits of pain and suffering to see the work that externalizing pain in representation can do.[137] These were not dream worlds, Kahlo was creating. '"I never painted dreams," she said; "I painted my own reality"'.[138] That reality included pain but also mortality, represented by the skull on her forehead: a harbinger of death (Thinking of Death, 1943).

Representing pain in imaginative ways is a key means of moving away from 'pain's contractive potential'.[139] Words, images and films can transform a person's experience of pain into something that is communicable: 'If pain can be trapped within an external representation such as a photograph, instead of within the body, then it passes from a private hidden individual space into a visible, public and collective space.'[140] The viewer cannot experience the pain in the same way as the person depicted in the image, but they can nevertheless be startled, awoken or moved by the image. This idea is encapsulated in Barthes' notion of punctum: 'A photograph's punctum is that accident which pricks me (but also bruises me, is poignant to me).'[141] In this way, it is possible to move from 'pain's contractive potential' to 'a story about the expansive nature of human sentience' through art and imagination.[142] As such, 'this most radically private of experiences begins to enter the realm of public discourse' – the 'extreme privacy' of pain 'begins to be sharable'.[143] CocoRosie provide cultural forms as stand-ins for pain's outward invisibility. This speaks to 'the dream of communicating experience' that underpins Haraway's politics.[144]

Yet it is also the dream of not repeating the past. In 'Lemonade', the psychological suffering of ageing and revisiting childhood memories is rendered visible – whether theorized through Butler, Lacan or Freud. By contrast in 'We Are on Fire' a collective past is reawakened and we are reminded of the suffering endured by those accused of witchcraft, lest we forget. For as Zika argues, 'The history of the European witch-hunt … is a history from which we can and need to learn.'[145] Fire was the 'earthly punishment' which reminded witches of the 'eternal flames of hell' they would endure in life after death.[146] The Casady sisters keep our memory alive by rendering visible the pain experienced on account of punishment and persecution by neighbours, communities, church and state. In this way we are reminded of 'how much people can suffer from other people's descriptions of themselves'.[147]

Eden is polluted

In spite of the suffering endured by victims of witch-hunts, the figure of the witch has been reclaimed by feminists precisely because she exists 'in the interstices of traditional categories'.[148] It makes sense therefore that Coco Rosie would turn to this trope since liminality is central to the way they destabilize social categories. They merge feminine and masculine, they ventriloquize the bodies of young and old, they defy gravity, they are burnt at the stake, they transmogrify from water to solid and back once more. As Bianca explains: 'It's like embracing that middle space, not a grey mediocre middle space, but something risky and provocative. I wouldn't want to be creating something I felt completely sure of, or that the world could easily swallow.'[149] In their videos they 'disassemble' and 'reassemble' elements of the male and female face, incorporating furry elements to signal hair rejected for its alliance with the 'wrong' side of the male/female, human/animal binaries.[150] The blurring of human/animal boundaries is further achieved through the introduction of animal sounds alongside human voices.

Bianca's comments chime with Haraway's vision of cyborg politics

> 'as the struggle for language and the struggle against perfect communication, against the one code that translates all meaning perfectly, the central dogma of phallogocentrism. That is why cyborg politics insist on noise and advocates pollution, rejoicing in the illegitimate fusions of animal and machine. These

> are the couplings which make Man and Woman so problematic, subverting the structure of desire, the force imagined to generate language and gender, and so subverting the structure and modes of reproduction of 'Western' identity, of nature and culture, or mirror and eye, slave and master, body and mind.'[151]

The idea of 'pollution' of course again harks back to Mary Douglas and the mass of cultural attempts to 'purify' the female face by removing facial hair and making up the face to normative standards.

CocoRosie make reference to the work of mistranslating (deciphering at odds) in the following statement:

> If there's a poetic to our work, a weird continuity of deliberate mistranslation, it's what comes off the streets and is purified and reduced by the dollar stores .... I've been reborn by the idea of artificial paradise, an urban hustle. We are finding our way to God through the dollar stores of this world.[152]

This version of Eden is quite different from Sierra's nostalgic, pastoral view of childhood in 'Lemonade'. As Haraway notes, 'We cannot go back ideologically or materially.'[153] This chimes with Butler's comment regarding our inability to locate the original law from which our gendered selves are cited into being. We cannot reconstruct male and female faces from scratch or recover any 'natural' self. In order to say something meaningful, in order to make a progressive change, one must use the means available, reworking and remaking what we already have. Rather than reaching back nostalgically to some 'natural', pre-discursive sense of self, which cannot be found, if we follow Butler and Haraway, the sisters build their 'artificial utopia' through the discourses and material culture available in the present, including the 'artifice' of cosmetics, consumer culture and the dark areas of human history and the self. The sisters' 'urban hustle' seems to suggest a fraud or swindle, just as make-up has been seen as a magic practice or trickery. Yet 'The Cyborg would not recognize the Garden of Eden'; it renders 'thoroughly ambiguous the difference between natural and artificial'.[154] This recalls CocoRosie's earlier comment about facial hair being 'more of an adornment' and the way they see themselves 'turning more and more to nature and becoming more a part of the landscape'.[155] The question for CocoRosie seems to concern the 'effect a non-patriarchal discourse would have had or would now have (supposing we could learn to ungender our discourse)'.[156] How would the face emerge from some such discourse?

Conclusion

This chapter has explored the music videos of CocoRosie, with particular emphasis on the surrealist strategies used to make up the female face. The musicians turn to cosmetics and adornment, not to smooth over imperfections or approximate a feminine ideal but rather to bring abjections back into being. Traces of animality, facial hair, wrinkles, black bile and blackened teeth are introduced as disruptive elements on the sisters' classically beautiful faces. By presenting selves that are not theirs they permit the skeletons of repressed or outlawed subjects to shine through: selves made visible through make-up, adornment and intertextuality. These theatrical subjects bear the burden of contradiction and inconsistency, transcending the binary codes that haunt the human subject and allowing for the expression of pain, humanity and loss in the present. Pain is the thread that ties these images together, ranging from suffering in the founding of (gendered) identity, with the abjections this entails, to the witch's pain at the fire.

The metaphor of the cyborg allows for the idea of disassembling familiar binaries and reassembling them in the wrong order. This provides a vocabulary for explaining CocoRosie's attempt to sidestep the perfect coding of the female face. Yet, as noted above, the sisters' faces risk being 'decoded' perfectly if located in the world of childhood, where the subversive elements cease to be disruptive in a political sense, instead becoming props in the child's dressing-up box – not unlike the critique of postmodern versions of identity. Nevertheless, the very visceral feelings of pain conveyed through their facial expressions, performance and operatic vocals move the imagery away from this interpretation. Such expressive performances embolden and reorient the myths that CocoRosie bring to light, particularly when it comes to the figure of the witch. The sisters succeed in indicting the stigmatizing medieval discourse on the witch-as-hag while simultaneously using it to signal potential for different visions of the face and subject in the future. For if a figure as potent as the witch can be authored through discourse and recede again, then surely there is room for alternative versions of being that transcend the binaries that cut off and delimit human potentiality.

Notes

1 A. Breton, 'Manifesto of Surrealism', in *Manifestoes of Surrealism*, trans. R. Seaver and H. R. Lane, ed. A. Breton (Ann Arbor: University of Michigan Press, 1969 [1924]), 5. Surrealism can be understood to have extended well beyond the art movement itself, with Alison Bancroft suggesting that we see it as 'a way of thinking, of working creatively, a way of understanding and representing the self and the object': A. Bancroft, *Fashion and Psychoanalysis* (London and New York: I. B. Tauris, 2012), 38.

2 D. Haraway, 'A Cyborg Manifesto: Science, Technology and Socialist-Feminism in the Late Twentieth Century', *The Cybercultures Reader*, ed. D. Bell and B. M. Kennedy (London and New York: Routledge, 1991), 291–324, 311.

3 Haraway, 'A Cyborg Manifesto', 312.

4 A. Petridis, 'Tub Thumping', *The Guardian*, 12 November 2004, https://www.theguardian.com/music/2004/nov/12/popandrock1 (accessed 24 February 2017).

5 F. Eberstadt, 'Twisted Sisters', *The New York Times Magazine*, 6 July 2008, http://www.nytimes.com/2008/07/06/magazine/06cocorosie-t.html (accessed 25 February 2017); F. Eberstadt, 'Alien Girls', *The New Yorker*, 26 October 2015, http://www.newyorker.com/magazine/2015/10/26/alien-girls (accessed 25 February 2017).

6 In *Maison de Mon Rêve* (2004) Bianca uses a number of unorthodox objects for percussion, including 'one of those old-fashioned tin toys that moos like a cow when you turn it over': Petridis, 'Tub Thumping'.

7 N. Olah, 'Antony Hegerty: Future Feminism', *DazedDigital* 2013, http://www.dazeddigital.com/artsandculture/article/14121/1/antony-hegarty-future-feminism

(accessed 24 February 2017); The Hole, 'Future Feminism', *The Hole NYC* 2014, http://theholenyc.com/2014/08/15/future-feminism-2/ (accessed 26 February 2017).

8 T. Bendix, 'CocoRosie Explains their Fairytale', *Afterellen*, 2010, http://www.afterellen.com/more/79810-cocorosie-explains-their-fairytale (accessed 1 March 2017).

9 J. Miller, *Fashion and Music* (Oxford and New York: Berg, 2011), 69.

10 J. Parales, 'Review: In "This is Acting," Sia Adhered to Her Victim-to-Victory Formula', *The New York Times*', 27 January 2016, https://www.nytimes.com/2016/01/28/arts/music/review-in-this-is-acting-sia-adheres-to-her-victim-to-victory-formula.html?smid=tw-nytimesarts&smtyp=cur (accessed 24 February 2017).

11 Haraway, 'Cyborg Manifesto', 302.

12 J. Butler, *Gender Trouble* (New York and Oxford: Routledge, 2006 [1990]); J. Butler, *Bodies That Matter* (London and New York: Routledge, 2011 [1993]).

13 Butler, *Gender Trouble*, 24.

14 S. Hall, 'The Spectacle of the "Other"', in *Representation: Cultural Representations and Signifying Practices*, ed. S. Hall (London: Sage, 1997), 223–90.

15 The 'superior' attribute is here presented in bold: a strategy used in Hall, 'Spectacle', 235.

16 Butler, *Bodies that Matter*, 23 and 59.

17 Ibid., 74.

18 Ibid.

19 Ibid., 77.

20 Ibid., 75.

21 A. Smelik, 'A Close Shave: The Taboo on Female Body Hair', *Critical Studies in Fashion and Beauty* 6.2 (2015): 233–51.

22 J. Craik, *The Face of Fashion: Cultural Studies in Fashion* (London: Routledge, 1993).

23 Ibid., 158.

24 Ibid., 160.

25 P. Black, *The Beauty Industry: Gender, Culture, Pleasure* (London: Routledge, 2004); Craik, *Face of Fashion*. Indeed, theatre, and music performance, remain spaces where men are permitted to make-up their faces with relative freedom: J. Miller, 'Heroes and Villains: When Men Wear Make-up' in *Fashion Cultures Revisited*, ed. S. Bruzzi and P. Church Gibson (Oxford and New York: Routledge, 2013).

26 Webb, in this book, 33.

27 Piesse cited in B. Moeran 'The Portrayal of Beauty in Women's Fashion Magazines', *Fashion Theory* 14.4 (2010): 491–510, 494.

28 Moeran 'Portrayal of Beauty', 494.

29 Craik, *Face of Fashion*, 160.

30 Ibid.

31 Moeran, 'Portrayal of Beauty', 502.

32 Ibid., 501.

33 P. Skinner, 'Makeup isn't a "lie" for People Living with Facial Disfigurements', *The Conversation*, 16 February 2017, http://theconversation.com/makeup-isnt-a-lie-for-people-living-with-facial-disfigurements-70768 (accessed 4 April 2017).

34 Moeran, 'Portrayal of Beauty', 496.

35 Butler, *Bodies that Matter*, 77.

36 K. Lesnik-Oberstein, 'The Last Taboo: Women, Body Hair and Feminism', *The Last Taboo: Women and Body Hair*, ed. K. Lesnik-Oberstein (Manchester: Manchester University Press, 2006), 1–17, 2. And see Bonnie Millar's chapter in this book for an example of how medieval writers used hairiness as a sign of the grotesque hag.

37 Ibid.
38 Spencer as cited in Smelik, 'Close Shave', 237.
39 Lesnik-Oberstein, 'Last Taboo', 2; Smelik, 'Close Shave', 244.
40 Lesnik-Oberstein, 'Last Taboo', 2.
41 S. L. Bartky, 'Foucault, Femininity, and the Modernization of Patriarchal Power', in *The Politics of Women's Bodies: Sexuality, Appearance, and Behaviour*, ed. R. Weitz (3rd edn, Oxford: Oxford University Press, 2010), 89.
42 M. Douglas, *Purity and Danger: An Analysis of Concept of Pollution and Taboo* (London: Routledge, 2002 [1966]), 5.
43 Ibid., 50.
44 Lesnik-Oberstein, 'Last Taboo', 11.
45 Smelik, 'Close Shave', 238.
46 Ibid., 239.
47 Ibid., 238.
48 Craik, *Face of Fashion*, 158–64.
49 N. Heinich, 'The Van Gogh Effect', in *The Sociology of Art: A Reader*, ed. J. Tanner (London: Routledge, 2003), 122–31, 123–5.
50 M. A. Lindauer, *Devouring Frida: The Art History and Popular Celebrity of Frida Kahlo* (Hanover and London: University Press of New England, 1999).
51 S. Thornton, *Club Cultures: Music, Media and Subcultural Capital* (Cambridge: Polity, 1996).
52 M. Zamora, *Frida Kahlo: The Brush of Anguish* (London: Art Data, 1990).
53 Smelik, 'Close Shave', 240.
54 Ibid.
55 C. Zika, *The Appearance of Witchcraft: Print and Visual Culture in Sixteenth-Century Europe* (London and New York: Routledge, 2007), 145, 127, 131.
56 A good example of this can be found in Kahlo's *Self-Portrait with Monkey* (1938).
57 In relation to this point, Janice Miller has noted: 'While any mainstream male musician who puts on makeup will find that this remains worthy of popular comment, the idea of a female music performer not wearing makeup remains, it seems, far more controversial': Miller, 'Heroes and Villains', 348.
58 Craik, *Face of Fashion*, 158. Furthermore, 'Flawless skin is the most universally desired human feature, according to zoologist Desmond Morris, and flowing, healthy hair runs close behind': N. Etcoff, *Survival of the Prettiest: The Science of Beauty* (New York: Doubleday, 1999), 95.
59 Keller as quoted in Kirkup, 'Introduction', 4.
60 Lesnik-Oberstein, 'Last Taboo', 3.
61 Both male and female musicians are featured in this documentary, but the term 'eternal child' has quite specific connotations when applied to a woman. Simone de Beauvoir used the term 'eternal child' to critique the 1940s idea of passive, disempowered, non-agentic housewife: S. de Beauvoir, *The Second Sex*, trans. H. M. Parshley (London: New English Library, 1970 [1949]), 329. I would argue that when CocoRosie inhabit the position of eternal child, they critically rework and reconfigure it so that it emerges as empowering (although I lack the space to fully explore this argument here).
62 Bendix, 'CocoRosie Explains'.
63 Zika, *Appearance of Witchcraft*, 11, 13. See, in particular, *The Three Fates* (1513), discussed in ibid., 128.
64 See, for instance, *Circe with Ulysses and his Men Transformed into Animals*, woodcut, 1473 as discussed in Zika, *Appearance of Witchcraft*, 133.

65 M. Warner, *Monuments and Maidens* (London: Vintage, 1996 [1985]), 97; Zika, *Appearance of Witchcraft*, 133.
66 Zika, *The Appearance of Witchcraft*, 133–4.
67 Gaule as cited in R. Briggs, *Witches and Neighbours* (London: Harper Collins, 1996), 20. For further descriptions of the 'hideous hag' see Millar in this book.
68 W. Shakespeare, *Macbeth*, ed. K. Muir (London: Methuen, 1964 [1623]), Act I, Scene iii, lines 32–3.
69 Zika, *The Appearance of Witchcraft*, 1.
70 Ibid., 1.
71 Ibid., Chapters 5 and 6, respectively.
72 Ibid., 2.
73 Other functions included '[supporting] calls for reform of the moral order … [forefronting] the moral lessons of classical literature, [reinforcing] the power of Scriptural precedents, [strengthening] secular authorities in the disciplining of their states, [interpreting] social crimes as signs of divine displeasure, and [helping] incorporate the New World into the cultural frameworks of the old'.: ibid., 4–5.
74 Haraway, 'Cyborg Manifesto', 302.
75 Ibid.
76 Ibid.
77 Zika, *Appearance of Witchcraft*, 7, 40–1.
78 Ibid., 1.
79 Ibid., 59.
80 Ibid., 63.
81 Shakespeare, *Macbeth*, 15. Act I, Scene iii, lines 39–47. Compare the medieval descriptions in Millar's chapter, above.
82 Douglas, 'Purity and Danger', 17.
83 M. Wollstonecraft, *A Vindication of the Rights of Woman* (London: Penguin, 2004 [1792]), 15.
84 Bianca Casady, quoted in Bendix, 'CocoRosie Explains'.
85 Haraway, 'Cyborg Manifesto', 293, 301.
86 Ibid., 316.
87 Ibid., 307.
88 Kirkup, 'Introduction', 5.
89 Childhood in the plural is something explored in P. S. Fass, 'Childhood and Memory', *Journal of the History of Childhood and Youth* 3.2 (2010): 155–64.
90 A vision of childhood put forward by Jean-Jacques Rousseau in J. Rousseau, *Émile* (London and Vermont: Everyman, 1993 [1762]).
91 Romanticism: ibid. Trauma: Fass, 'Childhood and Memory'.
92 Fass, 'Childhood and Memory', 158.
93 Ibid.
94 Ibid.
95 J. Bourke, *The Story of Pain: From Prayer to Painkillers* (Oxford: Oxford University Press, 2014), 17.
96 Quoted in Eberstadt, 'Alien Girls'.
97 M. Laing, 'The *Lula girl* as "Sublime and Childlike": Nostalgic Investments in Contemporary Fashion Magazines', *Critical Studies in Fashion and Beauty* 5.2 (2015): 271–93.
98 Bianca Casady as quoted in Eberstadt, 'Alien Girls'.

99 Breton, 'Manifesto of Surrealism', 16.
100 Bancroft, *Fashion and Psychoanalysis*, 42.
101 J. Lacan, *Écrits: A Selection*, trans. A. Sheridan (London: Routledge, 2001 [1966]), 3–5.
102 Bancroft, *Fashion and Psychoanalysis*, 24.
103 Necromancy, the practice of communicating with the dead, was one of the most prevalent forms of late-medieval magic: Zika, *Appearance of Witchcraft*, 12.
104 Ibid., 24.
105 Ibid., 12.
106 Ibid., 127.
107 Zika, *Appearance of Witchcraft*, 77–8.
108 Ibid., 12, 78, 82, 101.
109 Ibid., 196.
110 In contemporary usage, a grass widow denotes a woman whose husband is away often or for a long period. However, its original usage in the early sixteenth century was for an unmarried woman with a child (the idea being that the couple laid in the grass rather than in the matrimonial bed). Oxford Living Dictionaries, *Grass Widow*, 2017, https://en.oxforddictionaries.com/definition/grass_widow (2017) (accessed 5 April 2017).
111 Zika, *Appearance of Witchcraft*, 81.
112 The words used by Zika to describe Baldung's vision of witchcraft: Zika, *Appearance of Witchcraft*, 13.
113 Butler, *Bodies that Matter*, 107–8.
114 Zika, *Appearance of Witchcraft*, 64.
115 Petherbridge, *Witches*, 41.
116 Zika, *Appearance of Witchcraft*, 87.
117 Ibid., 145, 17, 79, 80, respectively.
118 Ibid., 120.
119 Petherbridge, *Witches*, 57.
120 Butler, *Gender Trouble*, 24.
121 Zika, *Appearance of Witchcraft*, 219.
122 Ibid., 101.
123 Petherbridge, *Witches*, 27.
124 Ovid as quoted in Warner, *Monuments*, 299.
125 Zika, *Appearance of Witchcraft*, 233.
126 Moeran 'Portrayal of Beauty', 493.
127 See Petherbridge, *Witches* for visual examples.
128 Petherbridge, *Witches*, 111.
129 Bourke, *Story of Pain*, 25.
130 Ibid., 24.
131 D. B. Morris, *The Culture of Pain* (Berkeley, Los Angeles and London: University of California Press, 1991).
132 Bourke, *Story of Pain*, 12.
133 Emily Dickinson, 'Pain has an Element of Blank', in *The Complete Poems of Emily Dickinson* (Boston: Little Brown, 1924), Part I, poem 19.
134 D. Padfield, 'In Search of a Visual Language for Pain', in *Perceptions of Pain*, ed. D. Padfield (Stockport: Dewi Lewis Publishing, 2003), 16–27, 17.
135 Scarry, *Body in Pain*, 13, emphasis in original.
136 Padfield, 'In Search', 21.

137 Kahlo is also notable here for her imaginative representations of physical and psychological suffering (Zamora, *Frida Kahlo*): examples include her self-representation as a deer shot with arrows (*The Wounded Deer* 1946), thorns around her neck (*Self-Portrait with Thorn Necklace and Humming-bird* 1940), nails hammered into her flesh (*The Broken Column* 1944) and her internal organs visualized exterior to her body (*Henry Ford Hospital* 1932), and a skull painted on her forehead as a harbinger of her death (*Thinking of Death* 1943).
138 Quoted in Zamora, *Frida Kahlo*, 114.
139 Scarry, *Body in Pain*, 169.
140 Padfield, 'In Search', 20.
141 R. Barthes *Camera Lucida* (London: Vintage, 2000 [1980]), 27.
142 Scarry, *Body in Pain*, 22.
143 Ibid., 6, 170.
144 Haraway, 'Cyborg Manifesto', 315.
145 Zika, *Appearance of Witchcraft*, 9.
146 Ibid., 38.
147 A. Phillips, *Unforbidden Pleasures* (London: Penguin, 2016), 173–4.
148 Petherbridge, *Witches*, 14.
149 Bianca Casady as quoted in Petridis, 'Tub Thumping'.
150 Smelik, 'Close Shave'.
151 Haraway, 'Cyborg Manifesto', 312.
152 Bianca Casady, as quoted in Eberstadt, 'Twisted Sisters'.
153 Haraway, 'Cyborg Manifesto', 301.
154 Ibid., 293.
155 Bianca Casady, as quoted in Bendix 'CocoRosie Explains'.
156 Keller as cited in Kirkup, 'Introduction', 3.

Part Three

Materiality

9

Archaeological Facial Depiction for People from the Past with Facial Differences

Caroline Wilkinson

Introduction

The significance of the face in the acquisition of a physical and psychological self and in social interaction can hardly be overestimated. The face indicates gender, age, ancestry, health, mood, emotion and even perhaps character and personality and, as a 'mirror to the soul', it permeates the very fabric of human experience.[1] The face has the power to cause reference beyond itself and yet to be the central site of signification.[2] Our faces help us understand who we are and where we come from,[3] and the 'unveiling of the soul' that results from the consideration of pure appearance reveals an understanding that we see more than we take in.[4] It is well established that the face is the primary determinant in the perception of attractiveness and that this perception is key to self-esteem and life opportunities. We know that humans have a cognitive bias towards attractive people and 'good-looking' people have higher incomes,[5] are judged to have good personality qualities,[6] are thought to be less likely criminals,[7] receive less severe sentences for criminal convictions,[8] are more likely to be called to a job interview,[9] and have higher mating success.[10] In addition attractiveness has been shown to be related to higher social status,[11] and we make rapid, unreflective judgements on competence, aggression and leadership qualities based on facial appearance.[12] Therefore, facial appearance appears to be crucial to power, status, resilience and reputation. Is it little wonder, then, that we desire to see the faces of people from the past and make similar judgements about their identity, personality and character?

Since the face is fundamental to identity, status and reputation, facial difference (such as malocclusion and deformity) has been shown to reduce opportunity and success. This is, in large part, due to perceptions of lower intelligence,[13] lower attractiveness[14] and higher criminal behaviour,[15] leading to increased social isolation[16] and bullying.[17] Facial appearance was also critical in ancient populations, where it is suggested that facial disfigurement resulted in a diminished social status.[18]

Facial depiction from human remains is now commonplace in archaeological investigations, and allows us to visualize faces from the past, compare ancient faces with contemporary cohorts and make judgements on status, character and lifestyle. Facial depiction from skeletal remains has primarily been applied in forensic investigations, when other approaches are not possible or few clues are available to aid in the identification.[19] The technique is used to rebuild a face onto a skull to recreate the ante-mortem appearance of the individual in order to recognize and identify the decedent.[20] The methods utilized for facial depiction from skeletal remains are well described and tested, and are based on anatomical standards, craniometrics and morphological interpretation.[21] It is widely accepted that the first scientific facial depiction from skeletal remains was attempted by the German anatomist, Wilhelm His, who rebuilt the face of the composer Johann Sebastian Bach in order to authenticate his remains.[22] Thereafter, the demand for facial depiction in archaeological research has increased and interdisciplinary studies have been performed worldwide to rebuild faces from the past.[23] Faces of historical figures have been depicted from skeletal analysis and previous examples have contributed to the changing views of the media and the public in relation to these historical figures. Cases include the face of Richard III,[24] Henri IV,[25] Nefertiti and Tutankhamun.[26] In the case of Richard III, the king was depicted as an ordinary man, without the monstrous deformity or embodiment of evil suggested by Shakespeare, and this depiction informed the academic debate surrounding the king in relation to Tudor propaganda, demonization and historical inaccuracy.[27]

Over the years, there have been some facial reconstructions depicting healed facial trauma or disease, most notably that of Phillip II of Macedon with a healed battle wound across his left eye[28] and Robert the Bruce with leprosy and a healed sword wound on his forehead.[29] The incidence of facial difference in archaeological depictions is not recorded, but the visibility of facial difference in ancient populations must influence our opinions relating to people from the past. Therefore, this chapter describes some depiction cases where facial difference has been demonstrated, and describes the effects of these depictions in the media and public domain.

Healed trauma in archaeological facial depiction

Examples of healed trauma on skulls from archaeological investigations are numerous and well documented. These include ante-mortem wounds attributed to battle,[30] medical treatment,[31] accidents[32] and interpersonal violence.[33] However, there are a much smaller number of individuals with healed wounds (ante-mortem) to the face, where a facial depiction has been produced as part of the archaeological investigation.

Ante-mortem skeletal trauma has a different appearance to peri and post-mortem skeletal trauma, and is usually distinguished by signs of healing to the bone. The physical properties of bone change after death, leading to different fracture characteristics; fresh or living bone is more pliable and more resistant to tensile forces, as it contains living vessels and fibres. Living bone tends to splinter and have irregular edges to fractures, whereas dried bone shatters into small, more regular fragments, as it is harder and more brittle. Furthermore, the concentric and radiating fracture lines, and stellate fractures normal for fresh bone do not occur in dry bone.[34] In this way, anthropologists can distinguish between trauma at or around the time of death (peri-mortem), and trauma to dry bone due to the environment (post-mortem). The time required for significant loss of organic and moisture content of bone depends on the post-mortem environment.[35] However, ante-mortem wounds occur earlier in life and since they prove not to be lethal (or at least not right away), healing processes around the wound will be visible. Where an injury occurred at least a week before death there will be slight evidence of remodelling,[36] and advanced healing is indicated by smooth remodelled bone that often obscures the original injury.[37] It is usually inappropriate for a facial reconstruction to depict the trauma associated with the death of the individual (unless the manner of death is the primary motivation for the depiction), but healed trauma will have affected the facial appearance of the individual during life and this is considered relevant to any facial depiction from skeletal remains.

One example of minor healed trauma can be seen in the depiction of St Nicholas. St Nicholas, also called Nikolaos of Myra, was a fourth-century Christian saint and Greek bishop.[38] He is credited with many miracles and his legendary habit of secret gift giving led to the traditional model of Santa Claus.[39] The human remains of St Nicholas have been housed in the crypt of the Basilica di San Nicola, Bari, where they are contained in a sealed tomb. In the 1950s the crypt needed extensive work and the human remains were removed for the first and only time since they were interred by Pope Urban II in 1089.[40] A special Pontifical Commission, led by Monsignor Enrico Nicodemo, archbishop of Bari, arranged for the examination of the relics to be carried out in the commission's presence. Luigi Martino, professor of human anatomy at the University of Bari, was engaged to perform the examination.[41] Both anatomical and anthropological examinations were performed, including detailed scientific drawings, photographs and x-rays, along with thousands of careful anatomical measurements. These examinations suggested a man over seventy years of age, of average height and slender-to-average build. The skeleton showed evidence that he suffered from severe chronic arthritis of the spine and pelvis. In 2004 a facial depiction from skeletal assessment was produced at the University of Manchester in collaboration with Professor Franco Introna, a pathologist from the University of Bari.[42] The skull was recreated as a digital 3D model using the craniographs and drawings as templates and utilizing the craniometrics and photographs to check the model and apply detail. The resulting skull model was then employed, along with the photographic records, in order to produce a facial depiction following the anatomical and morphological method. The assessment established that the skull exhibited a clear nasal bone fracture, causing a deviation of the nasal bridge to his right and a flattening of the nasal profile. This fracture demonstrated healed bone and was clearly an ante-mortem injury. An updated depiction was produced by Face Lab at Liverpool John Moores University in 2014.[43] Texts relating to St Nicholas describe a man who suffered great persecution and imprisonment for his religion, and he was tortured and beaten during the years of his imprisonment.[44] St Nicholas also destroyed pagan temples and confronted pagan worship, so he was not a stranger himself to violent action. The depiction of the healed trauma to the nose of St Nicholas was a feature noted by the media who, with reference to the historical documentation of his Christian persecution and torture, suggested that interpersonal violence was the most likely cause of his broken nose, and further promoted the view of St Nicholas as a robust and resilient bishop.[45]

Sometimes healed wounds can provide additional evidence of medical treatment and lifestyle in medieval periods. One such investigation related to the human remains of a soldier from the Battle of Towton, 1461.[46] Towton 16 was a middle-aged (45–50 years) man with extensive peri-mortem (at or around the time of death) wounds to the forehead and temporal bones, probably delivered by a poleaxe. However, there was also a well-healed blade injury to the left body of the mandible: the wound was 65 millimetres in length, running along the cheek, from the top of the third molar to the chin. The blade wound had penetrated the internal surface of the lower jaw, removing the apex of the first molar and impacting the internal surface of the other side of the mandible. A gaping hole remained between the left first and second molars where the bone had been removed. All margins of the wound were well healed and exhibited no evidence of infection. There was also a well-healed fracture to the mandible from the right first incisor running inferiorly through the corpus. This injury was probably the result of the blade wound to the left side of the mandible. Towton 16 was reconstructed and depicted with an extensive soft tissue scar and related drooping due to severance of facial nerve branch/es. Knüsel and Boylston state that the wounds seen on the skeletal remains of the Towton soldier suggest some knowledge of the effective treatment of battle injuries, as they do not exhibit any evidence of infection.[47] They claim that treatment on the battlefield was the province of the barber surgeon and that Towton 16 may have had preferential access due to his social standing. They also suggest that treatment consisted of wound cleaning with oil and the application of a salve or plaster with lint and a bandage and that herbs or roots with healing properties may have been applied to the wound. The form of the healed wound was modelled to indicate that it had been treated with the appropriate medicine of the time, leaving a rather rough and heavily scarred face (see Figure 9.1). It is possible that the face of Towton 16 was more significantly affected than represented, and there may have

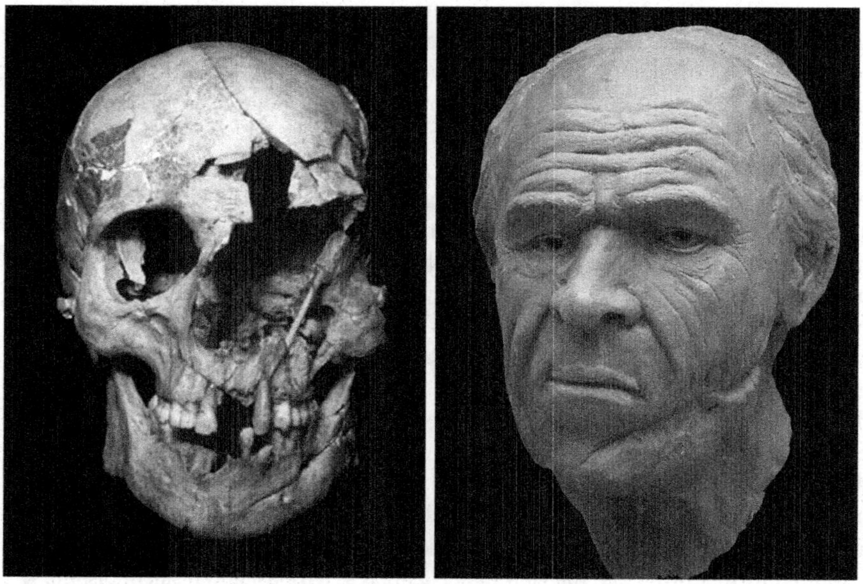

Figure 9.1 The skull (left) and facial depiction (right) of the Towton 16 soldier. (Images courtesy of the University of Manchester).

been more extensive tissue necrosis across the upper face and side of the nose due to the damage to the facial vessels and corresponding muscle wastage caused by the decreased blood supply (ischaemia).

Disease in archaeological facial depiction

Signs of disease to skulls in archaeological investigation are well documented, including leprosy,[48] syphilis,[49] carcinomas,[50] Paget's disease,[51] periodontal disease,[52] infection,[53] scurvy[54] and tuberculous osteitis.[55] There are also some examples where facial depiction has been utilized to illustrate these conditions and pathological conditions depicted on the face can shed light on the social, medical and cultural life of an individual from the past.[56]

The Surgeons' Hall Museum in Edinburgh houses a skull documented as 'Neanderthaloid'.[57] This skull was given to the College by David M. Greig, a former conservator of Surgeons' Hall Museum, who obtained it from a 'medical friend, who had inherited it without knowledge of its origin'. The skull was described as having Neanderthal characteristics, with features such as the low forehead and heavy brow ridge. The back of the skull was also described as protruding at the base. In 2008, part of the mystery of this skull was solved when it was sent away for radiocarbon dating; the results showed that the skull was that of a much more modern human, dating from the mid-seventeenth century, as opposed to being from a Neanderthal.[58] In 2012, the skull was assessed at the University of Dundee and was determined as a young adult male, displaying characteristics associated with otopalatodigital spectrum disorder (OPD). Frontometaphyseal dysplasia (FMD) was determined as the most likely condition within this spectrum disorder.[59] The cardinal manifestations of OPD include a deformed palate, prominent brow ridges, widely spaced and downward slanting eyes, a wide nasal bridge and an underdeveloped lower jaw.[60] In general, these disorders do not affect intellect and may involve hearing loss and other skeletal abnormalities involving the fingers and/or toes (digits). Additional FMD manifestations seen on this skull included thickening of the cranium, under-pneumatization of the mastoid processes (small neck attachments) and small, missing and misaligned teeth.[61] The mandible was absent from this specimen in the Surgeons' Hall Museum, but in FMD the mandible is described as micrognathic (underdeveloped) with anterior constriction (pointed chin).[62] The temporomandibular joints (where the lower jaw articulates with the cranium) were flat and deformed, suggesting that there was limited mandibular movement and the missing, absent and misaligned teeth further suggested that this individual would not be able to chew food effectively. If this person suffered from FMD, then it is likely that he also had joint deformities called contractures that restrict the movement of certain joints, bowed limbs, an abnormal curvature of the spine (scoliosis), and abnormalities of the fingers and hands.[63] In addition to skeletal abnormalities, individuals with FMD may have obstruction of the ducts between the kidneys and bladder, heart defects or problems with breathing.[64] This individual had therefore reached adulthood without the ability to chew food and with limited movement and mobility. This suggested that this individual must have had someone

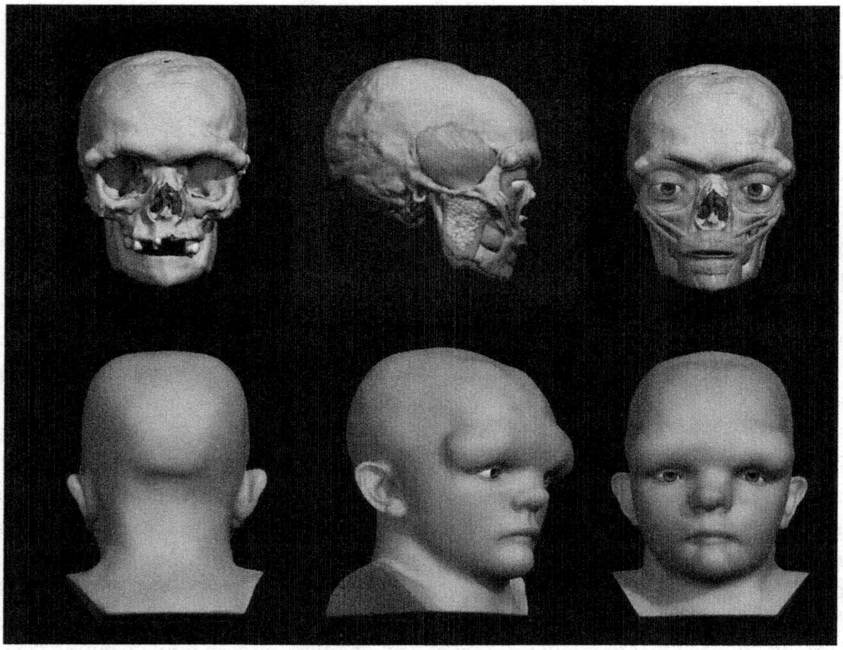

Figure 9.2 The facial depiction of the skull from the Surgeons' Hall Museum, Edinburgh, 2012. Courtesy of A. Davison.

who provided soft food/liquids and tended to his personal care. Davison[65] depicted the living face of this young man using digital facial reconstruction (see Figure 9.2), estimating the mandible using computerised tomography (CT) scans from individuals with FMD, modelling anatomical structures and following anthropometrical standards. In addition to evaluation of his facial appearance and diagnosis of his likely condition, the interpretation and analysis of this skull provided a rare insight into the care of children from the seventeenth century.

Some osteological assessments of craniofacial disease in archaeological investigations have been controversial, and this controversy may demonstrate our discomfort with the interpretation of facial disfigurement in ancient populations. One such example is the chariot burial excavation in 2001 funded by English Heritage in Wetwang, East Yorkshire.[66] The excavation uncovered the grave of an Iron Age chief, who was buried alongside her richly decorated chariot in the fourth century BCE.[67] The aristocratic Celt was female, 35 to 45 years of age, 5 feet 9 inches tall and she may have had a facial disfigurement. She was laid in a crouched position, head to the south, on a mat or blanket at the southern end of the grave pit.[68] An iron mirror was rested against her legs, joints of pig were placed on her upper body and the dismantled chariot was arranged around her.[69] The chariot was particularly ornate and was decorated with coral from the Mediterranean, and the grave also contained remains of a harness for two horses, including two bronze bits with iron cores and five bronze rein rings. It is thought that this chariot may have been the Iron Age equivalent of a state coach, used to take the queen to her final resting place. It may have been placed in the grave to help her pass to the afterlife.[70] There has been very little serious research on gender in Iron Age societies in UK contexts, but this burial raised major questions about the role and status of women in that period.[71]

The skull of the Iron Age chief was fragmented and when the skull was reassembled it appeared distorted, with the right side longer than the left. It is not uncommon for skulls in certain burial conditions to exhibit plastic deformation when subjected to external forces,[72] and after death, bone resilience progressively diminishes.[73] In fact, bones are able to undergo post-mortem distortion of the entire original geometry while the integrity remains relatively intact and several such cases have been described in the literature.[74] One notable example is Ötzi, the Iceman, discovered in the Alps in 1991, who exhibited localized post-mortem distortions of the facial skeleton and the basicranium.[75] Lynnerup states that 'under these conditions, etiological attribution of pathology and trauma lack certainty'.[76] Jurda and colleagues described post-mortem distortion patterns to crania and found that the shape change associated with the in situ orientation agreed with the compression of the cranial shape in the direction of the vertical, that is, the earth's gravity.[77] However, they also found that the original morphology of the cranium affected the final resting position recorded on site. For example, the crania with broad and short braincases and flat occipital regions were more inclined to persist in their supine burial positions since larger contact surface provided them with stability, while the long-headed specimens with the protruded occipital region were unstable in their original positions and slid onto their side.

The Wetwang skull was excavated lying on its left side and there was a bilateral difference in the length of the facial skeleton. However, the rear of the cranium did not appear distorted and the overall pattern did not appear consistent with vertical compression from its right side down to the left. While bone can be distorted in shape by post-mortem pressure, there is no evidence that bone can be stretched in the pattern seen for the Wetwang skull, and the pathologist determined that the most likely cause of the asymmetry was a soft tissue growth on the right side of the face leading to related bone remodelling (see Figure 9.3). Dr Robert Stoddart, from the University of Manchester, suggested a haemangioma as the most likely diagnosis.[78]

Past interpretations of Iron Age grave goods have linked mirrors with women of high status,[79] but it is possible that this was an association to her facial disfigurement. It cannot be ignored that mirrors are utilized to reflect appearance and by the end of her life her facial appearance would have been different and noticeable. Since there was academic disagreement at the time of the excavation as to whether the skull exhibited post-mortem or ante-mortem deformation, the condition was not sensationalized with extreme texture or condition, and if this woman had a haemangioma then it is possible that it was more severe than shown in the depiction. It may be that the mirror is a statement of her facial difference and if this interpretation is correct then this provides an interesting insight into the status of people with facial disfigurement in ancient populations, and disagrees with previous theories linking disfigurement with loss of status and exclusion.[80] In this case her high status could have trumped her facial disfigurement, or it may be that ancient populations were more tolerant and empathetic to facial disfigurement than previously thought.

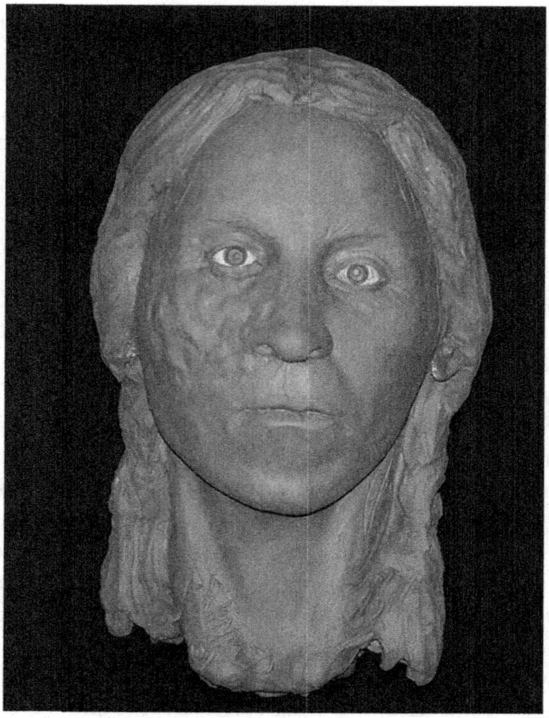

Figure 9.3 The facial depiction of the Wetwang Chariot Queen. Image courtesy of the University of Manchester.

Deliberate body modification in archaeological facial depiction

Signs of deliberate craniofacial modification are well documented in archaeological investigations, including trepanation,[81] dental filing[82] and cranial deformation.[83] It is well established that some cultural practices can leave their mark on the skeleton as altered morphology. Cranial deformation is the result of pressure influence on bone modelling during childhood and can be accidental or intentional. Accidental modification may be the result of the long-term influence of cradle devices, and intentional modification may be due to the application of bindings or boards resulting in the elongation of the vault.[84] The bioarchaeological record of South America is characterized by the high frequency of individuals with a variety of cranial deformations concentrated in three areas (North-West, Central-West and South) of this subcontinent, and this practice is thought to have been aesthetic and related to high status.[85]

One case of an unusual artefact from Elgin Museum in Scotland demonstrates this phenomenon. Elgin Museum houses a large glass dome containing a mummified human body in a foetal position, and this was donated to the museum in 1846 by James

Fletcher Esq of Rosshaugh, Ross-shire, who was born in Elgin. Along with his brother, James founded a trading house in Liverpool, and became involved in the importing of alpaca wool from the Andes of South America. The wool of the alpaca was used to produce a lustrous cloth, which in the 1830s was fashionable and in great demand. During a visit to Elgin in 1846, James donated a number of items from Peru to the Elgin Museum; a case of twelve Peruvian birds, two dressed figures, six specimens of gold and various other items. The glass dome also contains a memorandum notice stating:

> This mummy was taken from a cave in Maretasse, an Islet of Lake Titicaca, adjoining the small peninsula of Capachica, about 10 leagues from the city of Puno, the capital of the department. There was a rope, formed of some vegetable substance, tied around the back and knees, which caused the loss of skin on these parts. The few ignorant and superstitious inhabitants of the Islet regarded the mummy with a species of veneration and its loss was immediately discovered and deeply lamented. The crops were not so abundant that season as they were in some previous years, and this circumstance was attributed to the loss of the Gentil, a term equivalent to Pagan, dead body. Being convinced amongst themselves that the act must have been perpetrated by some person not an inhabitant of the Islet, their suspicions, which turned out to be well-founded, fell upon an Englishman, at the time in Capachica, and to save himself from the fury of the Indians he was obliged to leave that part of the country. The head from its peculiar formation is very singular. The local traditions led one to suppose that this mummy had been interred at a period antecedent to the Reign of the Incas. It is a body of a female, of the computed stature of five foot four inches. The extremities are doubled up so to occupy the smallest possible space. The head is of extraordinary height and of the sugar-loaf form, which seems to have been imparted to it by pressing on the occiput during infancy. From Professor Owen's description of the skulls of the South American tribes this individual would appear to have been of the Carib race. The complete preservation of the subject has probably been accomplished by baking amidst the slow combustion of herbs and substances possessing antiseptic properties.

The mummy was well preserved with some destruction of the soft tissues and evidence of fabric binding marks across the body. This, along with the foetal position, suggests that this was an unwrapped bundle mummy. The teeth and skeleton were in excellent condition and the mummy was mounted onto a wooden pole pushed up through the pelvis. Experts from the University of Dundee, in collaboration with the Scottish Universities Environmental Research Centre, carried out a full analysis of the mummy in 2012, using CT imaging, carbon dating, anthropology and facial depiction.[86] The skull of the mummified girl was sugar-loaf shaped, and cranial modification is known to have been a common practice in Peru, typically suggesting high status in the Aymara tribes. The head would be deformed during infancy using head boards or bindings and such an annular modification is produced when bands are wrapped around the forehead and the back of the skull to force the bone to grow upright. The skull also exhibited shovel-shaped incisors, round orbits, a tented nasal root and a flat mid-face; all characteristics associated with native South American populations. It was a common Aymara custom to inter the dead in a seated or squatted position, with the head pressed down between the knees and arms placed crosswise over the breast, and the foetal position originated as a desire to facilitate rebirth. A common belief is that their first ancestors had risen either from certain fountains and lakes (e.g. Lake Titicaca), or from caves or clefts in certain rocks of extraordinary size, and each person, after death, became a God, who protected the tribe. Each God was given its own Chullpa or shrine. It seems likely therefore that this mummy was an important young woman from the Aymara culture. The facial depiction by Tobias Houlton showed the cranial deformation and a facial appearance similar to contemporary Peruvian faces.

Conclusion

Facial depiction from the interpretation of skeletal remains can play an important part in multidisciplinary archaeological investigations and provides an alternative way to reconstruct stories, lives and people from the past. Each investigation will include numerous agendas, and a facial depiction can counter or justify strongly held beliefs about a historical figure and/or indicate aspects of lifestyle and community in ancient populations. The depiction of people from the past with facial difference can provide insight into medical treatment, status, social cohesion and cultural practice.

In this way the minor trauma exhibited on the skull of St Nicholas led to the perpetuation of the story of a persecuted Christian saint living in a violent time, and yet his broken nose could also have been the result of a clumsy or drunken accident or childhood injury. This contrasts with the facial depiction of Richard III, where his 'unremarkable' face enabled the creation of a different view of the king at odds with the deformed Shakespearean monster influenced by and part of Tudor propaganda. Some examples of the depiction of pathological conditions on the face have added weight to theories relating to medical treatment, cultural traditions and caring communities by demonstrating extensive healed battle wounds in medieval soldiers, deliberate head deformation fashions and the longevity of people with disabling and disfiguring congenital conditions. However, occasionally the facial depiction does not fit with accepted archaeological theory and this can lead to the re-evaluation of accepted norms or the marginalization of the depiction. One such example is the Iron Age Chariot Queen case, where her controversial facial appearance may suggest tolerance, empathy and acceptance of facial disfigurement in ancient populations, especially when the individual is of high status and important to the community. In general the comparison between ancient and contemporary populations continues to inform our perceptions, interpretations and interrogations of facial difference.

Notes

1 A. Synnott, A. 'Truth and Goodness, Mirrors and Masks, Part 1: A Sociology of Beauty and the Face', *The British Journal of Sociology* 40.4 (1989): 607–36.

2 J. T. Siegel, 'George Simmel Reappears: "The Aesthetic Significance of the Face"', *Diacritics* 29.2 (1999): 100–13.
3 N. Rumsey, 'Psychological Aspects of Face Transplantation: Read the Small Print Carefully', *American Journal of Bioethics* 4.3 (2004): 22–5.
4 Siegel, 'George Simmel'.
5 J. Kaczorowski, 'The Good, the Average and the Ugly: Socio-Economic Dimensions of Physical Attractiveness'. MA Thesis, Dept of Sociology, McGill University, Montreal, 1989.
6 D. S. Berry, 'Attractive Faces are not all Created Equal: Joint Effects of Facial Babyishness and Attractiveness on Social Perception', *Personality and Social Psychology Bulletin* 17.5 (1991): 523–31; D. Bar-Tal and L. Saxe, 'Perceptions of Similarly and Dissimilarly Attractive Couples and Individuals', *Journal of Personality and Social Psychology* 33.6 (1976): 772.
7 O. H. MacLin and M. K. MacLin, 'The Effect of Criminality on Face Attractiveness, Typicality, Memorability and Recognition', *North American Journal of Psychology* 6.1 (2004): 145–54.
8 K. L. Wuensch, W. A. Castellow and C. H. Moore, 'Effects of Defendant Attractiveness and Type of Crime on Juridic Judgement', *Journal of Social Behaviour and Personality* 6.4 (1991): 713.
9 F. L. Bóo, M. A. Rossi and S. S. Urzúa, 'The Labor Market Return to an Attractive Face: Evidence from a Field Experiment', *Economics Letters* 118.1 (2013): 170–2.
10 G. Rhodes, L. W. Simmons and M. Peters, 'Attractiveness and Sexual Behavior: Does Attractiveness Enhance Mating Success?' *Evolution and Human Behavior* 26.2 (2005): 186–201.
11 C. Anderson, O P. John, D. Keltner and A. M. Kring, 'Who Attains Social Status? Effects of Personality and Physical Attractiveness in Social Groups', *Journal of Personality and Social Psychology* 81.1 (2001): 116.
12 A. Todorov, A. N. Mandisodza, A. Goren and C. C. Hall, respectively, 'Inferences of Competence from Faces Predict Election Outcomes', *Science* 308 (5728): 1623–6; J. M. Carré and C. M. McCormick, 'In Your Face: Facial Metrics Predict Aggressive Behaviour in the Laboratory and in Varsity and Professional Hockey Players', *Proceedings of the Royal Society of London B: Biological Sciences* 275 (1651) (2008): 2651–6; N. O. Rule and N. Ambady, 'The Face of Success: Inferences from Chief Executive Officers' Appearance Predict Company Profits', *Psychological Science* 19 (2008): 109–11. See also, however, Chapters 4 and 7 in the present book, which expose some of the underlying assumptions of such studies, and propose more nuanced approaches.
13 L. C. Richman, 'The Effects of Facial Disfigurement on Teachers' Perception of Ability in Cleft Palate Children', *Cleft Palate Journal* 15 (1978): 155–60; A. Jesani, A. T. DiBase, M. T. Cobourne and T. Newton, 'Perceived Changes by Peer Group of Social Impact Associated with Combined Orthodontic-Surgical Correction of Class III Malocclusion', *Journal of Dentistry* 42.9 (2014): 1135–42.
14 W. J. S. Kerr and J. M. O'Donnell, 'Panel Perception of Facial Attractiveness', *British Journal of Orthodontics* 17.4 (1990) 299–304; Jesani et al., 'Perceived Changes'.
15 R. Bull, 'The Psychological Significance of Facial Deformity', in *Love and Attraction: An International Conference* 1 (Oxford and Toronto: Pergamon Press, 1979), 21–5.
16 N. Rumsey, R. Bull and D. Gahagan, 'The Effect of Facial Disfigurement on the Proxemic Behavior of the General Public', *Journal of Applied Psychology* 12.2 (1982): 137–50; V. Houston and R. Bull, 'Do People Avoid Sitting Next to Someone who is Facially Disfigured?' *European Journal of Social Psychology* 24.2 (1994): 279–84.

17 N. Rumsey and D. Harcourt, 'Body Image and Disfigurement: Issues and Interventions', *Body Image* 1.1 (2004): 83–97.
18 E. Craig and G. Craig, 'The Diagnosis and Context of Facial Deformity from an Anglo-Saxon Cemetery at Spofforth, North Yorkshire', *International Journal of Osteoarchaeology* 23.6 (2013): 631–9.
19 C. M. Wilkinson, *Forensic Facial Reconstruction* (Cambridge: Cambridge University Press, 2004).
20 Wilkinson, *Forensic Facial Reconstruction*.
21 C. Wilkinson, C. Rynn, H. Peters, M. Taister, C. H. Kau and S. Richmond, 'A Blind Accuracy Assessment of Computer-Modelled Forensic Facial Reconstruction using Computer Tomography Data from Live Subjects', *Forensic Science, Medicine and Pathology* 2.3 (2006): 179–87; C. Wilkinson, 'Facial Reconstruction: Anatomical Art or Artistic Anatomy?' *Journal of Anatomy* 216.2 (2010): 235–50; W. J. Lee, C. M. Wilkinson and H. S. Hwang, 'An Accuracy Assessment of Forensic Computerized Facial Reconstruction Employing Cone-Beam Computed tomography from Live Subjects', *Journal of Forensic Sciences* 57.2 (2012): 318–27.
22 W. His, *Johann Sebastian Bach: Forschungen uber dessen Brabstatte, Bebeine und Antlitz* (Leipzig: F. C. W. Vogel, 1895).
23 F. Cesarani, M. C. Martina, R. Gilletto, R. Boano, A. M. D. Roveri, V. Capussotto, A. Giuliano, M. Celia and G. Gandini, 'Facial Reconstruction of a Wrapped Egyption Mummy using MDCT', *American Journal of Roentgenology* 183 (2004): 755–8; M. Gregersen, J. Boldsen, H. Björn, L. W. Boel and P. Fromholt, 'Examination and Identification of a Danish 17th-Century Nobleman, Laurids Ebbesen: A Multi-Disciplinary Study', *Forensic Science and Medical Pathology* 2 (2006): 51–8; C. Wilkinson, 'The Facial Reconstruction of Ancient Egyptians', in *Egyptian Mummies and Modern Science*, ed. R. David (Cambridge: Cambridge University Press, 2008), 162–80; M. J. Papagrigorakis, P. N. Synodinos, A. Antoniadis, E. Maravelakis, P. Toulas, O. Nillson and E. Baziotopoulou-Valavani, 'Facial Reconstruction of an 11-Year-Old Female Resident of 430BC Athens', *The Angle Orthodontist* 81 (2011): 171–9.
24 C. Wilkinson, 'The Man Himself: The Face of Richard III', *The Ricardian Bulletin* (September 2013): 50–5.
25 H. Samuel, 'Face of France's Good King Henri IV Reconstructed 400 Years After his Death', *The Telegraph* (14 February 2013). http://www.telegraph.co.uk/news/worldnews/europe/9870454/Face-of-Frances-Good-King-Henri-IV-reconstructed-400-years-after-his-death.html (accessed 31 March 2017).
26 T. Friend, 'Could this be the Profile of a Queen?' *Race and History Forum, USA Today*, 15 August 2003, http://www.raceandhistory.com/cgi-bin/forum/webbbs_config.pl?md=read;id=1368B; B. Handwerk, 'King Tut's New Face'. *National Geographic News*, 11 May 2005, http://news/nationalgeographic.com/news/2005/05/photogalleries/tut_mummy/photo4.html (accessed 31 March 2017).
27 S. Pappas, 'Did Richard III really have a Friendly Face?' *Live Science: History*, 8 February 2013, http://www.livescience.com/26959-richard-iii-friendly-face.html (accessed 31 March 2017).
28 J. H. Musgrave, R. A. H. Neave and A. J. N. W. Prag, 'The Skull from Tomb II at Vergina: King Philip II of Macedon', *Journal of Hellenic Studies* 104 (1984): 60–78.
29 I. B. Macleod and B. Hill, *Heads and Tales: Reconstructing Faces* (Edinburgh: National Museums of Scotland, 2001).
30 V. Fiorato, A. Boylston and C. Knüsel, *Blood Red Roses: The Archaeology of a Mass Grave from the Battle of Towton, AD1461* (2nd edn, Oxford: Oxbow Books, 2007);

T. Anderson and I. Hodgkins, 'Healed Cranial Weapon Injury from Medieval Coventry, England', *Neurosurgery* 50.4 (2002): 870–3; A. Kjellström, 'A Sixteenth-Century Warrior Grave from Uppsala, Sweden: The Battle of Good Friday', *International Journal of Osteoarchaeology* 15.1 (2005): 23–50; F. Kanz and K. Grossschmidt, 'Head Injuries of Roman Gladiators', *Forensic Science International* 160.2 (2006): 207–16.

31 A. Nerlich, O. Peschel, A. Zink and F. W. Rösing, 'The Pathology of Trepanation: Differential Diagnosis, Healing and Dry Bone Appearance in Modern Cases', *Trepanation History, Discovery, Theory* 17 (2003): 43–51; N. D. Powers, 'Cranial Trauma and Treatment: A Case Study from the Medieval Cemetery of St Mary Spital, London', *International Journal of Osteoarchaeology* 15.1 (2005): 1–14.

32 V. G. Standen and B. T. Arriaza, 'Trauma in the Preceramic Coastal Populations of Northern Chile: Violence or Occupational Hazards?' *American Journal of Physical Anthropology* 112.2 (2000): 239–49; M. A. Judd, 'One Accident Too Many', *British Museum Studies in Ancient Egypt and Sudan* 3 (2002): 42–54.

33 R. J. Schulting and M. Wysocki, 'In this Chambered Tumulus were found Cleft Skulls: An Assessment of the Evidence for Cranial Trauma in the British Neolithic', *Proceedings of the Prehistoric Society* 71 (2005): 107–38; L. Fibiger, T. Ahlström, P. Bennike and R. J. Schulting, 'Patterns of Violence-Related Skull Trauma in Neolithic Southern Scandinavia', *American Journal of Physical Anthropology* 150.2 (2013): 190–202.

34 W. R. Maples, 'Trauma Analysis by the Forensic Anthropologist', in *Forensic Osteology: Advances in the Identification of Human Remains*, ed. K. J. Reichs (Springfield, IL: Charles C. Thomas, 1986), 218–28; N. J. Sauer, 'The Timing of Injuries and Manner of Death: Distinguishing among Antemortem, Perimortem and Postmortem Trauma', in *Forensic Osteology*, ed. Reichs, 321–32; M. H. Kaufman, D. Whitaker and J. McTavish, 'Differential Diagnosis of Holes in the Calvarium: Application of Modern Clinical Data to Paleopathology', *Journal of Archaeological Science* 24 (1997): 193–218.

35 E. R. Fitzgerald, 'Postmortem Transition in the Dynamic Mechanical Properties of Bone', *Medical Physics* 4 (1977): 49–53; Maples, 'Trauma Analysis'.

36 Sauer, 'Timing of Injuries'.

37 Kanz and Grossschmidt, 'Head Injuries'.

38 C. W. Jones, *St Nicholas of Myra, Bari and Manhattan: Biography of a Legend* (Chicago: University of Chicago Press, 1988).

39 Ibid.

40 St Nicholas Center (2002–2007) Anatomical Examination of the Bari Relics. http://www.stnicholascenter.org/pages/anatomical-examination/ (accessed 31 March 2017).

41 L. Martino, 'Ricognizione anatomica a studo antropometrico delle reliquie ossee di San Nicola di Bari', *Bollettino di San Nicola, Special Issue*, April to December 1957, http://www.stnicholascenter.org/pages/anatomical-examination/ (accessed 31 March 2017).

42 P. Curtis, 'Researchers Find the Real Face of Father Christmas'. *The Guardian*, 13 December 2004, https://wwwtheguardian.com/education/2004/dec/13/highereducation.uk (accessed 31 March 2017).

43 I. Hughes, 'Is this the Real Face of Santa Claus? Presenting St Nicholas, the Inspiration for Father Christmas', *The Mirror*, 5 December 2015, http://www.mirror.co.uk/news/uk-news/real-face-santa-claus-presenting-4755042 (accessed 31 March 2017).

44 Jones, *St Nicholas*.

45 B. Handwerk, 'St. Nicholas to Santa The Surprising Origins of Mr. Claus', *National Geographic*, 20 December 2013, http://news.nationalgeographic.com/

news/2013/12/131219-santa-claus-origin-history-christmas-facts-st-nicholas/ (accessed 31 March 2017).

46 C. Wilkinson and R. Neave, 'The Reconstruction of a Face Showing a Healed Wound', *Journal of Archaeological Science* 30.10 (2003): 1343–8.

47 C. Knüsel and A. Boylston, 'How has the Towton Project Contributed to our Knowledge of Medieval and Later Warfare?', in *Blood Red Roses*, ed. Fiorato, Boylston and Knüsel, 169–88.

48 D. L. Weiss and V. Møller-Christensen, 'Leprosy, Echinococcosis and Amulets: A Study of a Medieval Danish Inhumation', *Medical History* 15.3 (1971): 260–7.

49 T. Anderson, C. Arcini, S. Anda, Å. Tangerud and G. Robertsen, 'Suspected Endemic Syphilis (Treponarid) in Sixteenth-Century Norway', *Medical History* 30.3 (1986): 341–50.

50 L. Kilgore, R. Jurmain and D. Van Gerven, 'Palaeoepidemiological Patterns of Trauma in a Medieval Nubian Skeletal Population', *International Journal of Osteoarchaeology* 7.2 (1997): 103–14; P. Hacking, 'A Pituitary Tumour in a Medieval Skull', *International Journal of Osteoarchaeology* 5.4 (1995): 390–3; T. Anderson, 'A Medieval Example of Meningiomatous Hyperostosis', *British Journal of Neurosurgery* 5.5 (1991): 499–504.

51 J. L. Price, 'The Radiology of Excavated Saxon and Medieval Human Remains from Winchester', *Clinical Radiology* 26 (1975): 363–70; J. Byock, 'The Skull and Bones in Egil's Saga: A Viking, a Grave, and Paget's Disease', *Viator* 24 (1993): 23–50.

52 S. Sagne and G. Olsson, 'Studies of the Periodontal Status of a Medieval Population', *Dentomaxillofacial Radiology* 6.1 (1977): 46–52.

53 M. Djurić-Srejić and C. Roberts, 'Palaeopathological Evidence of Infectious Disease in Skeletal Populations from Later Medieval Serbia', *International Journal of Osteoarchaeology* 11.5 (2001): 311–20.

54 D. J. Ortner and M. F. Ericksen, 'Bone Changes in the Human Skull Probably Resulting from Scurvy in Infancy and Childhood', *International Journal of Osteoarchaeology* 7.3 (1997): 212–20.

55 J. D. Schuster, T. A. Rakusan, T. Chonmaitree and Q. T. Box, 'Tuberculous Osteitis of the Skull Mimicking Histiocytosis X', *Journal of Pediatrics* 105.2 (1984): 269–71.

56 C. Needham, C. Wilkinson and C. J. Knüsel, 'Reconstructing Visual Manifestations of Disease from Archaeological Human Remains', *Journal of Audiovisual Media in Medicine* 26.3 (2003): 103–7.

57 Surgeons Hall Museum, 'The Enigma of the Neanderthaloid Skull (GC 9413)', 8 October 2014, https://surgeonshallmuseum.wordpress.com/2014/10/08/the-enigma-of-the-neanderthaloid-skull-gc-9413/ (accessed 31 March 2017).

58 Ibid.

59 M. Roscoe, 'Differential Diagnosis of an Unknown 17th-Century Pathological Skull' (Honours Dissertation, University of Dundee, 2012); A. Davison, '3D Facial Depiction of a Specimen Displaying Frontometaphyseal Dysplasia' (MSc Forensic Art thesis, University of Dundee, 2012).

60 Robertson, S. 'Otopalatodigital Spectrum Disorders', *Gene Reviews* 2005–2013, https://www.ncbi.nlm.nih.gov/books/NBK1393/ (accessed 31 March 2017).

61 A. Verloes et al., 'Fronto-otopalatodigital Osteodysplasia: Clinical Evidence for a Single Entity Encompassing Melnick-Needles Syndrome, Otopalatodigital Syndrome Types 1 and 2, and Frontometaphyseal Dysplasia', *American Journal of Medical Genetics* 90 (2000): 407–22; S. Robertson, 'Otopalatodigital Syndrome Spectrum Disorders: Otopalatodigital Syndrome Types 1 and 2, Frontometaphyseal Dysplasia and Melnick-Needles Syndrome', *European Journal of Human Genetics* 15 (2007): 3–9.

62 R. J. Gorlin and M. M. Cohen Jnr, 'Frontometaphyseal Dysplasia: A New Syndrome', *American Journal of Diseases of Children* 118 (1969): 487–94; D. M. Danks et al., 'Frontometaphyseal Dysplasia: A Progressive Disease of Bone and Connective Tissue', *American Journal of Diseases of Children* 123 (1972): 254–8.
63 Robertson 'Otopalatodigital Spectrum Disorders' (2005).
64 Ibid.
65 Davison, '3D Facial Depiction'.
66 J. D. Hill, 'Wetwang Chariot Burial', *Current Archaeology* 15.178 (2002): 410–2.
67 D. Derbyshire, 'Chariot Queen's Grave Unearthed', *The Telegraph*, 7 April 2001, http://www.telegraph.co.uk/news/uknews/1310654/Chariot-queens-grave-unearthed.html (accessed 31 March 2017).
68 Hill, 'Wetwang'.
69 Ibid.
70 Derbyshire, 'Chariot Queen's Grave'.
71 Hill, 'Wetwang'.
72 M. Jurda, P. Urbanová and M. Králík, 'The Post-Mortem Pressure Distortion of Human Crania uncovered in an Early Medieval Pohansko (Czech Republic) Graveyard', *International Journal of Osteology* 25.4 (2015): 539–49.
73 T. A. J. Crist, A. Washburn, H. Park, I. Hood and M. A. Hickey, 'Cranial Bone Displacement as a Taphonomic Process in Potential Child Abuse Cases', in *Forensic Taphonomy: The Postmortem Fate of Human Remains*, ed. W. Haglund and M. Sorg (Boca Raton, Boston, London, New York and Washington DC: CRC Press, 1997), 319–36.
74 G. W. Weber and F. L. Bookstein, *Virtual Anthropology: A Guide to a New Interdisciplinary Field* (Vienna: Springer, 2011); J. L. Thompson and B. Illerhaus, 'A New Reconstruction of the Le Moustier 1 Skull and Investigation of Internal Structures using 3-D-mCT Data', *Journal of Human Evolution* 35.6 (1998): 647–65; M. S. Ponce de León and C. P. E. Zollikofer, 'New Evidence from Le Moustier 1: Computer-Assisted Reconstruction and Morphometry of the Skull', *The Anatomical Record* 254.4 (1999): 474–89.
75 H. Wilfing, H. Seidler, D. Nedden, G. Weber, H. Preuschoft, W. Platzer, R. Knapp, G. Haouser and W. A. Murphy, 'Cranial Deformation of the Neolithic Man from the Hauslabjoch', *Collegium Anthropologicum* 18.2 (1994): 269–82.
76 N. Lynnerup, 'Medical Imaging of Mummies and Bog Bodies, a Mini-Review', *Gerontology* 56 (2010): 441–8.
77 Jurda et al., 'Post-Mortem Pressure'.
78 *Meet the Ancestors*, series 5, episode 1: The Chariot Queen, 27 November 2007, http://www.bbc.co.uk/programmes/b008fj5s (accessed 31 March 2017).
79 C. Fox, *Pattern and Purpose: A Survey of Early Celtic Art in Britain* (Cardiff: National Museum of Wales, 1958); J. Joy, *Iron Age Mirrors: A Biographical Approach* (Oxford: Archaeopress, 2010).
80 Craig and Craig, 'Diagnosis and Context'.
81 J. W. Verano, 'Trepanation in Prehistoric South America: Geographic and Temporal Trends over 2000 Years', *Trepanation History, Discovery, Theory* 17 (2003): 223–36; V. A. Andrushko and J. W. Verano, 'Prehistoric Trepanation in the Cuzco Region of Peru: A View into an Ancient Andean Practice', *American Journal of Physical Anthropology* 137.1 (2008): 4–13; W. T. Clower and S. Finger, 'Discovering Trepanation: The Contribution of Paul Broca', *Neurosurgery* 49.6 (2001): 1417–26.
82 Mower, J., 'Deliberate Ante-Mortem Dental Modification and its Implications in Archaeology, Ethnography and Anthropology', *Papers from the Institute of*

Archaeology 10 (1999): 37–53; B. C. Finucane, K. Manning and M. Touré, 'Prehistoric Dental Modification in West Africa: Early Evidence from Karkarichinkat Nord, Mali', *International Journal of Osteoarchaeology* 18.6 (2008): 632–40.

83 E. J. Dingwall, *Artificial Cranial Deformation* (London: John Bale, Sons and Danielson, 1931).

84 D. Brothwell, 'Possible Evidence of a Cultural Practice Affecting Head Growth in Some Late Pleistocene East Asian and Australasian Populations', *Journal of Archaeological Science* 2 (1975): 75–7.

85 S. I. Perez, 'Artificial Cranial Deformation in South America: A Geometric Morphometrics Approximation', *Journal of Archaeological Science* 34.10 (2007): 1649–58; C. Torres-Rouff, 'Shaping Identity: Cranial Vault Modification in the Pre-Columbian Andes' (Unpublished PhD thesis, University of California, 2003).

86 'Mummy Returns to Elgin', *The Northern Scot*, 25 September 2015, http://www.northern-scot.co.uk/Features/People-and-Places/Mummy-returns-to-Elgin-25092012.htm (accessed 31 March 2017).

10

'Trotule (Trotula) Puts Many Things on to Decorate and Embellish the Face but I Intend Solely to Remove Infection': L'Abbé Poutrel and his *Chirurgerie* c.1300 in Context

Theresa Tyers

Cultural mores not only determine, at different times in different cultures, who or what is desirable, but also attempt to manipulate our primitive and intuitive sense about what we, as a society, assume to be beautiful and visually acceptable. It is with this in mind that this chapter examines a surgical and medical text that aims to cure a range of facial problems that, at the time the work was composed, would have been considered forms of disfigurement. It examines the role of the text in authorizing a practice – scarification – which would, itself, have caused noticeable scarring and briefly looks at the ways in which this aim to correct facial differences to procure a complexion that reflected society's idea of beauty also caused tension in the mind of its clerical author.

On later-medieval surgical practice

Modern scholars have noted that the work of surgeons and physicians in the later-medieval period often overlapped, and their respective arenas of work were continually disputed. Surgeons who also practised as physicians reserved authority over both the practical and theoretical arts arguing that their art involved both medicine and surgical work such as bloodletting. To add further to these deliberations Pope Innocent III's 1215 conciliar legislation introduced changes to canon law, in which the decree, *Sententiam sanguinis*, prohibited clerics from shedding blood. The result of this change was that ordained clerics were no longer able to carry out the range of activities which included surgical procedures: 'Nor may a subdeacon, deacon or priest practice the art of surgery insofar as it involves burning and cutting.'[1] At this same time vernacular texts on surgery began to appear which reveal how medical treatment was given, or prescribed, alongside the use of a range of therapeutic treatments, including both pharmacological and medicinal (herbal) elements.[2] The resulting hybrid texts, such as the *Chirurgerie* examined here in MS Vatican Library, Reg. lat. 1211, provided its commissioner and readers with an easily consulted manual, and in this case, for household use.[3] This hybrid text, therefore, not only includes the surgical procedures of incisions and excisions, followed by suturing, and the use of cauterization, but also a wide range of medicinal preparations intended to deal with other surface problems on the skin such as tumours, abscesses and skin (dis)colouration. The use of potions, ointments, powders and medicinal plasters and even purgations belonged, according to the author of the text, to the *l'uevre de cyurgie* (the work of 'surgery'). This statement strongly suggests that the changes which resulted in this vernacular hybrid form of a 'surgical work' allowed ordained men to continue to offer medical advice on a wide range of medical conditions, including facial treatments, to their patients.

By the thirteenth century, some university-trained physicians were claiming that to be granted the title of *medicus* a man should not only be a learned physician but also a surgeon, as the practice of bloodletting and purgation were a fundamental part of carrying out their work. This battle to regulate surgical tasks would last for centuries and would not be resolved, in England at least, until the beginning of the sixteenth century when the Royal College of Physicians was founded. This was the result of a move to try to eliminate bad practice and to ensure that physicians had much more than the guild status with which surgeons and apothecaries were then regulated. In the thirteenth and fourteenth centuries, the move to restrict clerics from any manual procedures that entailed the spilling of blood and damage to the skin – for example by cauterization – and to regulate physicians, surgeons and apothecaries (or spicers), was in its infancy. However, with still only a small number of university-trained physicians among the population, healthcare provision was sought from an eclectic range of practitioners.[4]

The text discussed in this chapter is dated to around 1300, and from internal evidence we know that Gui de Châtillon III, Count of St Pol (1254–1317) in what is now the Pas-de-Calais in northern France, instructed his clerk, a physician and monk, Jehan de Prouville, to translate a Latin text, attributed to 'Abbé Poutrel', for his own and his friends' use.[5] It is precisely the type of hybrid text that reflects the changes to both clerical medical practice and the need for a wide-ranging group of therapies. These therapies deal with not only facial problems but also the internal workings of the body that were often considered to be the cause of the problems. A section on battle wounds and trauma surgery in Gui's text is followed by advice for a full range of medical treatments which are largely based on herbal, and to a much lesser extent pharmacological medicine.[6] These recipes and remedies include instructions for making potions, simple plasters, the use of medicinal gums and other treatments, along with dietary and fertility advice. Much of this practical advice was intended to treat a range of diseases which would have been visible on the face and neck. These include leprosy, 'measles', apostemes, *noli me tangere,*[7] (facial ulcers) and gout, in its many forms, and, as with many other texts of this period it also includes methods of whitening the face and camouflaging blemishes (see below, Appendix, for examples).[8]

Incentives for text production

Montserrat Cabré and other scholars have demonstrated how difficult it is to define the term 'beauty' as it was understood in the Middle Ages,[9] and the widespread dissemination of cosmetic advice in medical texts in this period also raises questions of whether the knowledge disseminated in these manuscripts was solely intended for women. Modern research is demonstrating that the onlooker's perception of women who wear cosmetics can be influenced by their gender; this is particularly relevant in view of the attitudes of male, medieval moralists and commentators on women's behaviour.[10] Furthermore, Laurence Moulinier-Brogi has analysed the contents of Latin surgical texts and found that, along with routine surgical advice, they also generally include material on *ornatus*.[11] Its inclusion, even in Latin medical texts, hints at the possibility that men, too were seeking out cosmetic advice, such as this, for themselves.[12] Moulinier-Brogi also sought to define what the phrase *soins de beauté* (beauty care) implied to the creators and readers of Latin texts of this period. For example, the semantic range of *le soin – care* – carries with it not only the sense of 'caring for something or someone', implicit in the relationship between a surgeon or physician and their patient, but also infers a sense of worry (*souci*).[13]

While these nuances may, therefore, reflect an empathic response by both the practitioner to his patient and society's drive to ensure that the widely publicized norm was met, they may also be seen as the incentive behind the production and popularity of these texts. In other words, the writers' incentives were driven by their own sense of the need to conform to cultural and social norms, a norm that many of the medieval readers of these texts would also aspire to achieve.[14] Berthold de Ratisbonne in the thirteenth century, and Jacques de la Marche in the fourteenth, also provide evidence of male interest in fashion trends as they both criticized men for what must have been a long-standing fashionable trend of tying and plaiting their hair,[15] while advice for colouring the hair black, brown, gold or as red as a fox are also found throughout remedy collections. Recipes and remedies such as these, therefore, suggest that the advice to be found in texts such as Gui's was of interest to men and women alike.

This later-medieval cultural emphasis on efforts to deal with differences in skin tone, or removing blemishes in these texts inevitably draws attention to the contemporary concern with appearance. Medical texts and recipe and remedy collections (such as that of the anonymous author of the thirteenth-century Anglo-Norman *Ornatus* discussed below), along with a range of literary works, reveal that overwhelmingly a white face with highlights of rosy pink was desired. In the words of theologian Bishop Robert Grosseteste (1128–1253): 'Light, as the principle of colour, is the beauty and ornament of all that is visible.'[16] The ideal of beauty here is centred on the philosophical ideals of light, proportion and clarity, a search for which is still evident today in the production of Western cosmetic products. Facial treatments included in the later-medieval *chirurgeries* can therefore easily be seen as equivalent to modern-day 'cosmetics' which, both then and now, include ingredients to lighten the skin tone and introduce a healthy rosy colour (in white skin) that would indeed shine. While medieval moralists, therefore, criticized the use of cosmetics, medical writers and contemporary surgeons willingly continued to offer such advice. Thus, when examined against the writings of other surgeons, the claim made in Gui's version of l'abbé Poutrel's *Chirurgerie* that he 'only intends to treat infection' reveals a tension in his writings between this aim and the contents of the text itself, some of which deal with precisely the issue of embellishment that he claims to ignore.

Wound care: Surgery in context

The opening lines of the first section of the medical remedies and recipes in the book commissioned by Gui de Châtillon (which I shall refer to hereafter as 'Gui's book') deals with physical trauma and reveals the conscious response made by a society attempting to deal with battle wounds, at the time when the dispute between surgeons and physicians was particularly active. The surgeon Lanfranc of Milan (c. 1250–1306), a student of William of Saliceto (1212–77), refers to this ongoing debate in his small teaching manual on bloodletting.

> Oh Lord, why is there such a distinction made today between the physician and the surgeon? Perhaps because physicians abandoned manual operation to laymen? Or because, some say they disdain to work with their hands? Or, as I believe, they do not understand the method of operation which is necessary.[17]

Gui's book on *cyurgie* also provides its reader with a number of general rules, referred to as 'Commandments', which were to be followed by anyone attempting to treat a patient. Among these he stresses that no physician should attempt to 'heal a man whose skull has been broken, or whose entrails are injured, except for (no other reason) than (certain) death'.[18] With this advice dealt with, he then enumerates another set of rules which outline the basic elementary teaching for dealing with wounds and the body's four humours. The detailed instructions listed for treating a range of wounds include the preliminary act of washing the wound with wine (now known to be an antiseptic agent),[19] followed up by methods of dealing with the proud flesh left by a scar caused by inadequate knitting of the flesh around the area using corrosive powders. The text also stresses the correct way of making incisions in the body in that they should not be made lengthways, but should always be carried out across the skin to avoid any unnecessary scarring. However, he points out that in the case of incisions made on the face another rule must be applied:

> L'autre riule est que nule incisions ne doit estre faite en lonc ne en travers fors que en test pardesous les sorcius et desus les oreilles et desus le hatereil. En toutes les autres parties du cors doit on incisions faire dou lonc.[20]
>
> No horizontal or lengthwise incisions should be made on the head except just below the eyebrows, above the ears, and above the nape of the neck. In all other parts of the body the incisions must be made lengthwise.

The care for the facial scar is also found in the writings of eminent surgeons, for example, Henri de Mondeville (1260–1320), Guy de Chauliac (1300–1368) and the physicians Taddeo Alderotti (1215–1295) and Arnau de Villanova (1235–1311).[21] The writings of these eminent authorities on surgical practice also testify to a continuing fascination with finding methods of maintaining, or perfecting, the face and the body's image.

For the Italian surgeon William of Saliceto (d. 1277), for example, it is clear that surgical scars were arousing an aesthetic concern in this period:

> Some people sew it up as skinners sew up skins, and this suture gives the prettier scar. It is also done with [individual] knots and stitches, and here the threads are wrapped around twice the first time and once the second, so that the knot will stay tight, leaving a certain distance between one stitch and the next.[22]

The surgical work of Henri de Mondeville (1260–1316), surgeon to Philip the Fair and his son Louis X of France, also warns its readers about the correct way to deal with incisions. In his advice to remove encysted growths, which would have included those on the face and neck, and repair the wounds caused by the excisions, he stresses that if the procedure is not carried out with due care and attention, the results may be an aesthetically unpleasant scar:

> An ugly scar remains there, due not only to the broad cut but to the skin and flesh that originally covered the growth, because since they had been stretched to cover and surround the growth, and nothing of them is removed with the growth, many wrinkles are left in the scar … . Others cut off the growth together with the skin and flesh, and if it is a big one, this makes the ugliest of scars.[23]

On the whole, translations of simple medical texts into the vernacular were aimed at a broader audience than their Latin counterparts, and their readers were, although outside of the educated medical milieu, still eager to learn for themselves the art of *chirurgerie* and medicine. The proliferation of manuscripts in this period provides the evidence for an audience eager to have a range of cosmetic advice: to improve the complexion by removing blemishes, to remove unwanted and unsightly facial hair through the use of depilatories, along with methods of colouring the face. Gui's text on surgery and medicine, despite its earlier affiliation with l'abbé Poutrel's version of Frugard's text, is now little more than a collection of recipes and remedies (*receptaria*) which were designed to answer to the immediate needs of families and households, and its contents reveal contemporary concerns with facial appearance.[24]

Facial remedies: Beyond the battle scar

Although Gui's book starts by listing a range of surgical treatments that someone would find useful when dealing with battle wounds, the major part of the text (fol. 14–71) is dedicated to dealing with a range of skin conditions, along with other simple advice that provides a household with medical knowledge. This text, examined in the context of an example of written household medicine, provides the opportunity to consider in what ways it allowed a high-status family, who were caught up in turbulent times, to provide their own medical care. The knowledge it contains enabled the treatment of skin conditions such as suppurating abscesses and pustules, contagious erysipelas and impetigo, zoonotic anthrax (here referred to as *maus d'espagne*: the Spanish illness), and disfiguring leprosy, all of which affected facial appearance. This care also included advice on cosmetics, understood here as the application of a range of medicinal preparations that was intended to treat infection and also preparations that altered the appearance of the skin and which camouflaged both the outward signs of disease and its surgical treatment. The surgical and cosmetic advice made available to Gui's family members and their entourage, through Poutrel's original text, and other similar texts, reveals the later-medieval search for empirical methods of altering and improving facial appearance.

The compiler of Gui's treatise is unequivocal in stating that his rationale for creating his work is solely to treat infection. He furthermore overtly criticizes another well-known medical practitioner, this time a woman who, unlike him uses 'many things to adorn the face'. The practitioner to whom he alludes was the well-known Trotula, who at the time was believed by writers to be the author of a group of texts which, on the whole, dealt with women's medicine and were eventually brought together to form a corpus of writing on women's conditions which took its purported author's name.[25] The popularity of these texts coincided with the move to exclude women from medical practice; as such the disparaging remarks made in Gui's translated work suggest an awareness of this move.[26] Nevertheless, by distancing himself from such advice, I suggest that our writer was not only claiming authority above that of a female practitioner, but was hoping that his audience would value his work as a learned physician rather than that of a mere surgeon and a compiler of cosmetic advice.

Yet the text includes a range of treatments that should be considered to be both therapeutic and cosmetic. The relevant section of Gui's text that I want to explore in more detail here highlights purifying the skin to eliminate unwanted marks which were considered to be blemishes and worthy of scar-*producing* therapeutics. It begins with dealing with a common manifestation of humoural imbalance which was thought to have been the cause of a condition often referred to as *apostumes*, a widely cited umbrella term for a range of skin problems.

> After some remedies to treat apostumes in general and especially those that manifest themselves outwardly and which belong to the art of 'cyurgie' we will speak of some infections which appear from outside of our bodies, firstly: outwardly there are filthy marks produced from outside [of the body] and in this case they are called freckles [spots]. They are made sometimes of an outside cause, such as the heat of the sun or of the air or from hot blood which does not dissipate and more importantly is not removed. So if it is not taken away [by bloodletting] or if only a small amount is removed then the face is black or marks [stains] are caused [to appear] on the skin.[27]

The writer is here separating out the role that bloodletting played in removing an excess of blood believed to cause the disfigurement of stains, or dirty marks, or even freckles on the skin, which was expected to be pristine white with a healthy rosy glow. He then turns to causes which lie within the body and therefore would, usually, fall under the remit for treatment by a physician and not a surgeon. However, this distinction of external and internal medicine as reflected in vernacular medical writings suggests that later-medieval medical writers also found great difficulty in maintaining the theoretically defined boundaries. Thus when combining both surgery and medicine their repertoire and range of expertise was, lucratively, expanded. The following extract demonstrates not only this blurring of the boundaries of expertise but also how interrelated the idea of inward disorder was to the external appearance of unsightly and, therefore, socially unacceptable marks:

> Freckles or spots are caused sometimes from within because nature throws the filth outside. These *infections* (my emphasis) occur in man and are caused by the various humours ... be informed that: cupping under the chin is of great value and [the use of] scarification or 'routures'.[28]

Such treatments figure in some of the Latin surgeries discussed above, but ones that would result in visible and long-lasting scarring were viewed with ambivalence. Henri de Mondeville, writing slightly later, was also concerned with providing treatments to correct an array of skin conditions which would detract from the prevailing cultural idea of beauty and yet while he supported the correction of a wide range of therapeutics he, unlike the author of Gui's treatise, specifically advised that these treatments should *not* be used for *lentilles* (freckles or spots) or for *pannus* (?).

A number of the ingredients for remedies for the complexion which are cited in Gui's work are also recorded in other recipe and remedy collections. However, there is one intriguing remedy that follows on immediately from the author's statement in Gui's treatise that his aim is to remove infection. This calls for a compound treatment that is almost identical to that in another vernacular popular thirteenth-century work, Aldobrandino de Siena's *Livre de physick*. The oldest surviving copy of this text is written in Picard, the same dialect that is used by Jehan de Prouville for Gui's treatise. This popular work was addressed to a woman, Beatrix of Savoy, and was intended to help its readers maintain a healthy body, eat well and care for both their bodies and their psychological health. Like many other texts it aims to treat minor imbalances caused by malfunctioning bodily humours. Although it targets its readers with such simple advice it also recommends bloodletting and cupping as prophylactic treatments. The result is, therefore, a handy manual of health advice that remained popular right through to the late fifteenth century. A comparison of the remedy that follows on from the advice for scarification to 'clean the face' reveals that both Gui's text and the *Livre de physicke* contain a very similar remedy; nevertheless, their authors' stated intentions are contradictory, raising once again the question of why Jehan de Prouville's treatise for Gui, unlike the *Trotule* he references, specifically claimed that he only intended to treat infection.

Gui's hybrid *Book of Medicine* MS Ref. Lat. 1211	Translation of Gui's cosmetic recipe	Aldobrandino's *Livre de physicke*
Trotule met mout de choses a ourner la face, ***mais je entenc seulementoster l'infection*.**	Trotule puts many things on the face to adorn it **but I intend solely to remove infection**.	*Le maladie oster n'est mie no ententions, mais de faire bele coleur et removoir les malvaises qui vient par les mauvaises okisons, si com nous aprenderons.* … (**The aim is not to remove infection but to provide a beautiful colour** and to remove unwanted imperfections that are brought about by ill causes (*okisons*))
Pren entale et dentale, bourrache, sarcacolle, marbre blanc, corax blanc, de cascun yvelment, ferine de rouges cerises, semence de chanvre, feves frasees, o viel oint, destempre toutes ces coses en eve de limons et de feves et i met eve rose et .i. peu de musc, s'il est riches, secie en ombre et fai ent tronchons qui peuent ester gardé par mult d'ans et quant il sera mestiers, destemprent .i. en ta paume ou en .i. vaissel net et oing au soir la face de çou et au matin le leve o couleure de bren ou d'eve u feves aient longement temper che taut seur toutes chose lentilles et netoie la fache.[29]	Take *entale* and *dentale*, borage, sarcocol (a resinous gum), white marble, white coral, of each one an equal amount, red cherry powder, flax seed, ground beans, with old grease mix all of these things together with the juice of lemons and beans and (then) put in some rosewater and a small of amount of musk, if one is rich, dry this in the shade and then make this into small pieces (slices) which can be kept for many years and when there is a need for it moisten one of these pieces in the palm of your hand, or in a clean vessel, and in the evening anoint the face and, then, in the morning remove it with a filtered bran or oat water water might for a long time moderate that surely all things (such as) freckles and spot and clean the face.	… *prendés ental dental, borrache, sarcocol, marbre blanc, coral blanc, cristal blanc, ferine de feves, gis, et goume dragant, et en faites pore et destemprés à l'iauve que noumé nous avons et en faites petis torcis; cist torcis sot piles scachees à maniere de lupins et les faites sechier en l'umbre, et quant user les voles, si les destemprés à .i. pau d'iauves rose ou d'iave de flors de feves, et oigniés le visage, et de ces meismes iauves vous laves au matin.*[30]

While there are minor differences, these recipes are, to all intents and purposes, the same. The slight differences may be caused by the availability of ingredients and the season, for example the bean flowers that Aldobrandino's remedy calls for would have a short season although the medicinal water made from them could be stored away for later use. Both remedies would produce a long-lasting solid powder which would effectively cover any facial blemishes and the compiler of Gui's text, perhaps aware of his wider audience, even suggests that a 'small amount of musk' can be added if one is rich. I suggest that remedies such as these highlight the discrepancy between what is claimed to be the intention of the compiler of Gui's text, that is his stated aim to treat infection, and the reality of his task, which was to provide his patrons, members of Gui's family and entourage, the means to conform to the cultural mores of the perception of beauty of his time.

The advice to treat freckles or spots on the face with invasive and scar-producing treatments which *netoient la fache* (clean the face) is then followed up in Gui's book with gentle treatment which, the author suggests, the reader can carry out for themselves. This involves the simple use of honey and oat straw or oats, along with two other plants with traditional anti-inflammatory, antimicrobial and antiseptic use, which should be made into a facial wash to treat the skin.[31]

> After that make your cure, purified honey (*la pourre*), sweet bay with skimmed honey and anoint the face with this, but before this make in any way whatsoever steam baths with a decoction of wormwood and oat straw or oats, and in the morning wash the face with the strained bran water.[32]

It is immediately after this remedy that the compiler of Gui's book draws the reader's attention to the practices transmitted by the texts attributed to the figure of the female practitioner Trotula.

To evaluate whether the motivation for the inclusion of cosmetic remedies such as these in Gui's text are culturally motivated, while at the same time being disparaged, I am using another vernacular text. This is found in one of the earliest and most important compilations of vernacular medical works which also includes a range of cosmetic advice. Referred to as the *Physique rimee*, the thirteenth-century writer draws attention to his concern in setting out his expertise in helping women and girls who, despite their shapely form and noble birth, were still in need of advice to improve their appearance. He explains that he has provided this remedy *par corteysie* (out of consideration) for those females who, although of a beautiful and shapely form, nevertheless fear the shame and distress caused by any man who upon seeing them, not only takes pity, but adds to their burden by not (even) coveting them:

> A lotion for women out of consideration
> A lady and young woman of good birth
> Who fears shame and hurt,
> Such that man must take pity (on her)
> So that he does not even covet her.

To alleviate their distress he gives them a remedy to use and finishes this off with the reassurance that 'if you do this every day, both at the morning and the evening, a white face you will have, clear and healthy'.[33] In other words this author's concern for a young woman's feelings reflects a deeply held contemporary concern for women's well-being and peace of mind which, through their faulty complexions, and despite their shapely forms and noble birth, required the attention of male medical writers.

At the time this text was translated the aim, therefore, in Western Christendom, appears to have been to achieve a white and rosy complexion which was completely free from blemishes or stains.[34] The inclusion of cosmetic advice in Gui's text is not at all unusual as practical vernacular medical texts often include a wide range of cosmetic treatments for the skin and hair, including depilatories. For example, an author of another thirteenth-century, and in this case blatantly cosmetic, collection of remedies and recipes, the *L'ornement des dames*, addresses his work directly to women and spins a tale of how hard he has worked to gather his advice together. He writes quite confidently that he has composed this work specifically for women and that his advice will allow them not only to retain their beauty but also improve it. He also adds, praying to God the heavenly King that he hopes his readers will recognize these endeavours, which he has carried out on their behalf, and in return he hopes that they may always love him.[35] The incentive for authors to provide cosmetic advice, however, is not expressed as explicitly in other texts.

From minor blemish to major disfigurement

Poutrel's remedy for freckles addressed minor blemishes which as we have seen were freighted with associations of filth. His remedy, however, potentially introduced scarring where previously there had been none. At the same time, moralists were quick to condemn the vanity of women who engaged in such treatments. For example, Jacques de Vitry (d.1240) recounts the cautionary tale of a young woman who wanted to remove a black mark from her face and paid a physician who gave her a potion (*un jus*). According to Vitry this potion proved not only to be extremely effective, but also damaging, as he warns that the results were that it not only removed the unwanted mark but also her skin.[36]

Poutrel's misunderstanding of *Trotule*'s intentions, however, might derive from a practice that was carried out in the same geographical area as the origins of much of the knowledge which is now embedded in Gui's late-thirteenth-century treatise. In the *De curis mulierum* (On Treatments for Women), one of the texts in the *Trotula* compendium referred above that Monica Green suggests may have been written by the female medical practitioner, Trota, the following skin remedy is included:

> 167 An ointment that the Salernitan (the women of Salerno) made that is very good for sunburn and fissures of any kind. … This is the ointment with which the Salernitan women anoint themselves for rivulets and floods (of tears?) made in mourning the dead; also it is good for suppressing pustules in lepers and as a cleanser.[37]

In this instance the remedy has been cited to help a woman conceal the self-inflicted scarring which had taken place during an act of ritual mourning, a practice which long survived in southern Europe,[38] but is also noted to be of use in dealing with the facial disfigurement caused by skin changes due to leprosy. Its contents would have softened and eased the puckering of scars and raised flesh. This was not 'embellishment' but therapy.

Conclusion

The practice of *routures* or scarification which the compiler of Gui's treatise recommends for dealing with freckles, or stains on the face, would have resulted in noticeable scarring to treat what today would be seen as a natural response to sun exposure. Admittedly, identifying what was actually meant by a number of medical terms, including *conchiures* or *pannus* in medieval medical texts, is fraught with difficulties. However, by looking at the range of meanings implied by the word *conchiures* they all suggest the opposite of cleanliness, purity and clarity and point towards something that is dirty or undesirable. I noted above that the concept of illness and health was tightly bound to a holistic vision of both bodily and spiritual cleanliness. The scars left on the face or neck by the process of scarification or cutting would have been the outward sign of the treatments which had been carried out in an attempt to cleanse the body of its evil humours and, as such, they would have been subject to considerable efforts to conceal them: whether by the application of cosmetics or by covering the neck.

Modern scholars have noted that the work of surgeons and physicians in the later-medieval period often overlapped, and their respective arenas of work were continually disputed. This chapter has focused on one of a number of early vernacular translations of a practical medical text which were generally aimed at a broader audience than their Latin counterparts, but who were, although outside of the educated medical milieu, still eager to learn for themselves the art of medicine and well-being. The author of Gui's version of l'abbé Poutrel's book of *Chirurgie* is unequivocal in stating his own incentive for creating his work as this advice, he states, is solely to treat infection and yet his remedies do far more. In attempting to achieve the perfect complexion to reflect the cultural mores of his era he advocates treatments that would result in scarring for a simple condition that was essentially incurable. Perhaps this is why, ultimately, he feels the need to distance himself explicitly – if inaccurately – from texts that promise only 'beauty' and 'embellishment'.

Appendix I

This remedy immediately follows on from the treatment cited in this chapter, page 184.

Gui de Châtillon's text (translated from *La Chirurgie*, ed. Södergård, 48)

Or you (can) mix vernis (juniper resin or varnish) with bean flour and lemon water and then add rosewater and with this anoint the face in the evening and at rising in the morning, as we have said, and if they are on the face or on the forehead , anoint this with your saliva before breakfast and take endive and hart's tongue fern cook these in water and then foment the face and then take laurel berries and trule and put in red honey and then make accordingly like an ointment and anoint the face in the evening and in rising in the morning.

Against this and blisters (vessies) take oil rosat, the juice of wild celery (smallage), egg yolk, litharge (lead monoxide yellow in colour), corn starch, gum Arabic, frankincense, mix these together and anoint the face in the evening and then cover it with a wimple, remove it in the morning as said above and then fumigate the face with a decoction of oats and after take corn starch and bean flour and mix these with vinegar and anoint the face often and then remove it and the juice of red dock boiled with pork fat will remove stains from the face if one washes it often, and if they are caused by retention of the flowers then bleed from the sophone ie saphenous vein under the heel of the foot on the inside this is of much value in cleansing the face of lentilles and of blisters, or equally of smallpox (variole); if they come from choler take l'erbe de moleté and barley straw or oat straw cook these in water in any vessel which is well stoppered and then steam the patient with their eyes closed so much that s/he sweats and when he has sweated enough, recoif an ointment of pumpkin with rosewater and anoint the face with this. After this cover it with a wimple and leave it all night and take it off in the morning, then wash the face with bran water or with water of beans and then take powdered rice and some tragacanth (resin), and anis and make small pills with rose water and when here is need then take some and then anoint the face and cover do this in the evening and in the morning on rising and that you must also do soon after; take tragacanth, gum Arabic, starch, mastic and make a powder with rosewater and then mix one with your saliva and then anoint the face and then remove it as before.

Know that you cannot cure natural lentilles but one is able to make them paler, if they are made by the force of nature pushing forth its superfluous humours. I will give you this rule that which you must nothing put before and that is that the body will be cleansed, and this is the retention of the flowers, for this be bled from the saphenous vein underneath the ankle on the inside; and then put in place these remedies: if it is because of a cold cause then anoint with French soap, improved with nitre (sodium carbonate) and mix with vinegar, but also carry out an immersion in water with a decoction of wormwood. If they are from a hot cause, mix almonds and bean flour with vinegar or with pomegranate wine, or macianer (marciaton ? – a green ointment Rg.Frug), and anoint, but first make a subfumigation with water of a decoction of oats or take a steam bath or cotton dyed with the blood of a cockerel or a pigeon and put this lukewarm on the face, this is much valued for this (condition). Hare blood hot, or of a goat, put below and left until it is ciee is of value in either cold or hot causes. Triacles mixed with fennel juice is of value for the stains that occur after childbirth if put below. Against the bruises from injuries of blows or of falls, take bay berries and make them into a power and small balls with honey and make this into a plaster and then lay it on warm. This is of value against lentilles and marks which arise from a cold cause. Breaking the skin 'routoires' with cantarides put beneath the chin is of value in this. Ventouse on the fontanelle of the neck and bleed from the end of the nose is of value in this.

Notes

1 Finbarr McAuley, 'Canon Law and the End of the Ordeal', *Oxford Journal of Legal Studies* 26.3 (2006): 473–513.

2 Tony Hunt, *The Medieval Surgery* (Woodbridge: The Boydell Press, 1992).

3 The advice discussed in this chapter is from MS Vatican Library, Reg. lat. 1211, edited in *La chirurgie de l'abbé Poutrel: texte picard de 1300 environ*, ed. Östen Södergård (Stockholm: Almqvist and Wiksell International, 1980). All citations refer to this edition.

4 Healing could be sought by men and women of all social ranks, and the healing process could encompass both bodily and spiritual concerns: Faye Getz, *Medicine in the Middle Ages* (Princeton: Princeton University Press, 1988). For the larger of theme of who had access to texts in the vernacular, see Monica Green, *Making Women's Medicine Masculine: The Rise of Male Authority in Pre-Modern Gynaecology* (Oxford: Oxford University Press, 2008) and Theresa Tyers, '"In the Merry Month of May": Instructions for Ensuring Fertility in MS British Library, Lansdowne 380', *Social History of Medicine* 29.2 (2016): 267–89. For the use of another household manual that includes recipes and remedies such as those found in MS Reg. Lat. 1211, see L. Herbert Mcavoy, Patricia Skinner and Theresa Tyers, 'Family, Feud and Fertility at Manorbier Castle, Pembs., 1200–1400', *Transactions of the Honourable Society of Cymmrodorion* 22 (2016): 8–25.

5 *Dieus, nostre Sires, Jhesu Cris, qui sovrains mires est de cors et d'ame, il saut et gart le boin et le gentil contre de saint Pol, Gui de Casteillon, a qui commandment et requeste li siens clers, Jehans de Prouvile, phisiciens, explane ceste ciurgie, dite de l'abé Poutrel, traite en roumant du latin, a lequel cose faire Dieus se sgrasce nous otroit*: *La chirurgie*, ed. Södergård, 13. The identity of Poutrel remains a mystery.

6 Tony Hunt has identified the basic source text as the northern Italian Roger Frugard's *Chirurgia*: Tony Hunt, *Anglo-Norman Medicine*, i (Oxford: D.S. Brewer, 1994), 3–145 at 8.

7 There is an extensive list of treatments in the manuscript, many of which use corrosive substances such as realgar to deal with the festering sores and ulcers of this disease. For the history of the description of the disease by medical authorities, including Roger Frugardi, see *Systemic Lupus Erythematosus*, ed. Caroline Gordon and David Isenberg (Oxford: Oxford University Press, 2016).

8 See also Luke Demaitre, *Medieval Medicine: The Art of Healing, from Head to Toe* (Santa Barbara: Praeger, 2013).

9 Montserrat Cabré, 'Beautiful Bodies', in *A Cultural History of the Human Body in the Middle Ages*, ed. Linda Kalof and William Bynum (London and New York: Berg, 2010), 127–48.

10 Viktoria R. Mileva et al., 'Sex Differences in the Perceived Dominance and Prestige of Women With and Without Cosmetics', *Perception* 45.10 (2016): 1166–83.

11 Laurence Moulinier-Brogi, '*Esthétique et soins du corps dans les traités médicaux latins à la fin du Moyen Âge*' *Médiévales*, 46 (2004).

12 Ibid., 27.

13 Ibid., 55.

14 For the ways in which society developed and cultural norms were established, see David Crouch, *The Birth of Nobility: Constructing Aristocracy in England and France 900–1300* (Harlow: Pearson, 2005).

15 Moulinier-Brogi, 'Esthétique', 57.

16 Sarah-Grace Heller, 'Light as Glamour: The Luminescent Ideal of Beauty in the *Roman de la Rose*', *Speculum* 76.4 (2001): 934.

17 Quoted in translation in *Medieval Medicine: A Reader*, ed. Faith Wallis (Toronto: University of Toronto Press, 2010), 281.

18 *Le premiers commandemens de cirugie est tells: que tu ne pregnes nul home en cure qui ait le ties brisiet u les entrailles blecies fors que mort*: *La chirurgie*, ed. Södergård, 13.

19 See M. Daglia, 'Antibacterial Activity of Red and White Wine against Oral Streptococci', *Journal of Agricultural Food Chemistry* 55.13 (2007): 5038–42.

20 *La chirurgie*, ed. Södergård, 15.

21 Neither Mondeville nor Chauliac are cited in Poutrel's book on surgery, however, suggesting it predates these authors.

22 Guglielmo da Saliceto, *Chirurgia*, II. 3: '*Nam vulnus a quibusdam suitur ut pelliparii suunt pelles, et hec sutura pulchrioris est cicatrizationis. Fit etiam cum nodis et nectatione, et in hac sutura revolvitur filium in prima nectatione bis et in secunda semel ut nodus firmior remaneat et dimittatur inter nectum et nectum distantia aliqua*': quoted in Michael McVaugh, *The Rational Surgery of the Middle Ages* (Firenze: Edizione del Galluzzo, 2006), 215–29.

23 Henri de Mondeville, *Chirurgia*, III.1: '*Quia cicatrix remanet ibi turpis tam propter incisionem secundum latum quam propter cutem et carnem, quae erant super apostema, quae propter comphrensionem et defensionem apostematis distendebantur et quia de eis nihil amovebant cum apostemate, ideo ibi in cicatrice remanent multae rugae … Alii totam excrescentiam cum cute et carne amputant et isti, si sit magna, turpissimam faciant cicatricem*': Quoted in McVaugh, *Rational Surgery*, 124. Additional contemporary comments on the ugly effects of medieval scars include Alaisina Yselda, 'Na Carenza al bel cors avienenz', in *Songs of the Women Troubadours*, ed. Matilda Tomaryn Bruckner, Laurie Shepard and Sarah White (New York: Garland, 2000), 96–7, ll. 17–20; see also Paolo Santoni-Rugiu and Philip J. Sykes, *A History of Plastic Surgery* (Berlin: Springer, 2007), 55; Marie-Christine Pouchelle, *The Body and Surgery in the Middle Ages* (New Brunswick: Rutgers University Press, 1990), 167, 227.

24 See Claude de Tovar, 'À propos de la Chirurgie de l'abbé Poutrel', *Romania* 103.410–11 (1982): 345–62.

25 Monica Green *The* Trotula: *A Medieval Compendium of Women's Medicine* (Philadelphia: University of Pennsylvania Press, 2001).

26 The question of the rise of male authority in women's healthcare and their exclusion from formal training has been discussed by Green, *Making Women's Medicine Masculine*.

27 *Aprés aucuns remedes contre les apostumes en general et en especial qui aperent par defors et qui apartienent a l'uevre de cyurgie, nous dirons d'aucunes infections aparans par defors en nos cors et au prenier: defors a unes conchiures engenrees illuec et eles sunt dites lentilles. Eles sunt faites a la fie de cause dehors, si comme de caleur du soleil or de l'air ou de sanc boullant qui dissout plus qu'il ne degaste. S'il la detrait et il ne desout peit quant il degaste, la face est noire ou taches sunt engendrees u cuir, lentilles sunt faites a le fie de cause dedens, car nature jete ses cunkiures par defors*: *La chirurgie*, ed. Södergård, 47.

28 *Ces infections sunt faites es homes et eles sunt faites de diverses humeurs … que saces tu: venteuses sous le menton valent mult o scarification ou fait routures*: *La chirurgie*, ed. Södergård, 47.

29 *La chirurgie*, ed. Södergård, 48.

30 *Le Régime du Corps de Maître Aldebrandin de Sienne: Texte Français du xiiie siècle*, ed. Louis Landouzy et Roger Pépin (Paris: Librairie Ancienne Honoré Champion, 1911), 98–9.

31 Results of a number of trials suggest that oatmeal is also effective on a variety of dermatologic inflammatory diseases: E. S. Kurtz and W. Wallo, 'Colloidal Oatmeal: History, Chemistry and Clinical Properties', *Journal of Drugs Dermatology* 6.2 (2007): 167–70.

32 *Après çou, si fai ta cure, melle la pourrre, baie laurier o miel escumé et oing la face de çou, faites toutes voies estuved avant la decoction de l'aloisne et de paille d'orge ou d'avaine, au matin leve la face o couleure de bren*: *La chirurgie*, ed. Östen Södergård, 47.

33 Tony Hunt, *Popular Medicine in Thirteenth-Century England* (Cambridge: D.S. Brewer, 1990), 216. This manuscript also contains a comprehensive copy of Roger Frugard's text which is unlike that of Gui's treatise as set out in MS Reg. lat. 1211.

34 Modern research into the psychology and of perception of colour repeats this finding. Alex Jones et al., 'Coloration in Different Areas of Facial Skin is a Cue to Health: The Role of Cheek Redness and Periorbital Luminance in Health Perception', *Body Image* 17 (2016): 57–66.

35 *Pour ceo vus fas jeo cest livre/Que tres bien seez deliver/Vus memes en bauté guarder/Et vos acunit el amende r … Maintenant, je prie Dieu, le roi celeste, que ce livre puisse me valoir les remerciements des dames, grâce à quoi je sois toujours aimé d'elles*: *L'Ornement des dames (Ornatus Mulierum) Texte Anglo-Norman du XIIIe siècle*, ed. Pierre Ruelle (Brussels: Presses Universitaires de Bruxelles, 1967), 34–5.

36 Laurence Moulinier-Brogi, '*Esthétique et soins du corps dans les traités médicaux latins a la fin du Moyen Âge*', *Médiévales* 46 (2004): 57.

37 Green, *Trotula*, 135. 'Lily root, white lead, mastic and frankincense – of each half a dram, one dram of camphor, one ounce of animal grease [and] rose water as needed. … And so with this ointment the patient ought to anoint herself in the evening in front of the fire so that in the morning they might preserve themselves in daylight hours from the above mentioned conditions. The [ointment] raises the skin and colours it beautifully.'

38 'Those two gentle, reserved butterflies with their black and white ribbons were suddenly transformed into furies. They tore their veils, pulled their clothing out of place, scratched their faces until blood came and began to dance with long steps around the room, beating their heads and singing on one high note the death story': Carlo Levi, *Christ Stopped At Eboli: The Story of a Year* (New York: Farrar, Strauss and Company, 1947), 226.

11

Disrupting Our Sense of the Past: Medical Photographs that Push Interpreters to the Limits of Historical Analysis

Jason Bate

Rethinking the photographic face

From their outside appearance, the photograph albums held in the archive of the Royal Army Medical Corps look like domestic Edwardian photograph albums. On the inside back cover of each album, in the bottom left corner, is stamped 'photo department Harrods Ltd'. The covers do not prepare the viewer for the medical records that are displayed inside, which are small, closely cropped headshots, of soldiers with devastating facial injuries, documenting before-and-after surgery. These albums are housed in the Archives, Manuscripts and Rare Materials section of the Wellcome Library and form part of the Royal Army Medical Corps Muniment Collection. This collection was transferred by the RAMC Historical Museum to the Wellcome Institute in 1986–7. As a result, the collection moved from being kept closed within a military-medical context and became accessible to the public.

The photographs in the RAMC albums were produced to document new cases of interest that were not known in civilian surgical practice; examples of extensive destruction of bone, muscle and skin that required complex stages in plastic and prosthetic treatment. Photography's value as an analytical tool for recording (and classifying) cases provided important evidence for understanding the complexity of these injuries as they healed after surgery.[1] Enabling facial injuries to be catalogued and placed in relation to each other, these photographs provided a way of systematizing, associating cases with particular methods of treatment and anticipated result.

By systematically recording a case's condition at key stages in reconstruction, a photograph registered enough data for surgeons to assess treatment or to plan out their own sessions. Before-and-after photographs came to define clinical practice in the hospital, showing how much skin, muscle or bone needed to be taken off or attached in theatre. By evaluating photographs of patients' results, surgeons could gauge the best time to introduce dental splints, how long to wait in-between operations, the different healing times of skin and bone, to evaluate the practicality of borrowing skin from other parts of the body, and the healing properties of new tissue.

The RAMC albums proved essential resources for setting out rules and guidance for others to learn from. A letter accompanying the RAMC albums in the Wellcome Library provides evidence of the photographs being used as teaching aids for many years.[2] Surgeons used these albums as instruction manuals to guide readers through facial reconstructive procedures and predicted outcomes. The albums display the photographs in sequential order so that courses of treatment and recovery times could be conceptualized and followed, highlighting the scientific value that was attached to these surgical images. The albums consolidated the value of the photographs as reference sources and made their comparative information accessible to individuals or groups of surgeons. The small size of the albums allowed them to be handheld or rested on the lap of individual readers, or shared within a larger group. Surgeons at meetings, consultations, in medical libraries or the operating theatre, could therefore exchange the visual information contained within these albums.

Some of the First World War facial injury photographs record the nature of a wound, while others record the outcome of a surgical technique. The way the camera frames each case's face through a clinical frontal or side view suggests an analytical reading of the patients' wounds; men become abstract categories, wounds are detached from bodies and turned into specimens (Figure 11.1).

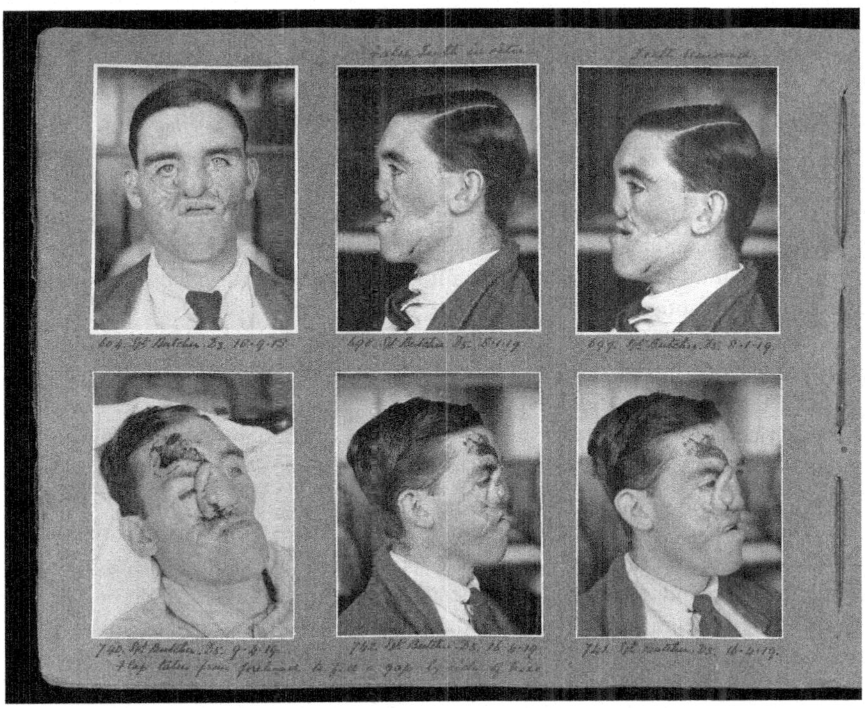

Figure 11.1 Sergeant George W. Butcher, 1918–19, photographs by Dr Albert Norman, 1915–19. RAMC album 2, RAMC 760. Courtesy of Wellcome Library, London.

In these images, some viewers will see specimens, while others will see men. This juxtaposition gives an indication of the conflicting elements embedded in the photographs.[3] The archive quickly creates problems of context and viewing because one's reading of the photographs keeps shifting between an objective medical study of the patients' injuries and surgical reconstruction, and a more subjective human response that seeks to find out about the injured men and their experiences.[4]

Newspapers explain the context of British military hospitals and of facially injured patients' recuperation once they returned home for treatment. These sources show the social impact of the wounded servicemen's recoveries and reveal how supportive local communities were. However, because of the politics between the two organizations that ran military hospitals, the RAMC and the British Red Cross Society, the surgical photographs are embedded in a contradictory story about the wounded soldiers' rehabilitation that shifts between public and private spaces of healing.[5]

Reflecting on the meanings that photographs afford as historical evidence, this chapter focuses on how material confrontations with facial disfigurement shapes historians' interpretations – and therefore affects what is taken to be 'the past'. What kinds of histories can be written if we start from the photographic image as the basis of a lived experience of facial disfigurement rather than a representation? Can photographs get under the skin? Are photographs a face-centred medium, meaningless unless they include the subject's face? Is the viewer's emotional experience of the photographs a product of a specific social–historical context? In this chapter I argue that the historical meaning of a photograph can exceed the meaning that was intended within the historical context in which it was produced. As evidence, I focus on my experience of viewing photographs of men with facial disfigurement (a reading which is intensely emotional for me) to show that present-day viewers may not share the 'medical gaze' of the surgical photographers and surgeons. First, I draw on Michel Foucault's concept of the medical gaze to help us understand how photography transformed patients into cases within medical practice by distancing trauma and objectifying the body. Then, I take up Frank Ankersmit to ultimately argue that at this juncture of intended meaning and experienced meaning there is an excess which takes the form of untameable emotion.

This chapter will also draw on the work of Ewa Domanska, who argues for an approach that brings the consideration of the material back into the centre of historical understanding and interpretation, for it is the material that shapes existences and experiences.[6] Focusing on 'mattering' as opposed to 'signifying', Domanska has pointed out the historiographical significance of the presence of the past, not in terms of representation, or signifying discourse, but in focusing those discourses through 'the materiality and thingness' of the trace rather than on its 'textuality and content'.[7] Domanska says we might wish to reconsider the (ontological) status of relics of the past and their function in mediating relations between people and things, for reconceptualizing the study of the past.[8] She considers the spatial dimension of presence – a presence that is in front of us, a presence of things, in order to challenge our relationship with the past, and says it is 'the future rather than the past, and the presentification of the past rather than its absence, that provide a reference point for reflection' (on what the past, and history, is).[9] Key for my argument here is Domanska's suggestion that we ask 'how things matter', because 'mattering' leads us to the concerns of those being studied and allows a space for the subjective. By reflecting upon the sensual and material qualities of the object, and by extension facial disfigurement, we are able to unpick the more subtle connections with cultural lives and values that are objectified through medical history. Significantly, her contention is that 'instruments' – here, the camera, printing papers, albums, captioning and the archive – mediate experiences and articulate desires to the extent that they 'co-constitute the reality studied by scholars'.[10] Central to my argument therefore is the role of materiality, both of photography and the disfigured face, as hermeneutic devices that shape the ways that I gain access to reality and where things become readable.

To examine how the historical subjects of photography and those engaging with photographs today are bound into this set of socially meaningful experiences, and towards how photographs as affective objects socially enfold historical experience, this chapter expands on Elizabeth Edwards' concerns with the ways in which social relations and the materiality of things are mutually constituted.[11] Exploring the potential of photographs as historical evidence as a means to investigate the forms through which historical imagination might be made possible and experienced, Edwards argues that photographs have the ability to perform a sense of the dynamic presence of the past.[12] This is because photographs exist in the present and therefore within 'the historical context of the viewer'. 'Things' such as photographs emerge as powerful and active players that extend or replace embodied experiences and connect with spiritual ones, such as historical imagination.[13]

Frank Ankersmit has argued that it is through emotional expression, engendered by our physical, intellectual and affective connections with the time before now, that the past becomes present. The past is here through our encounter with it in some sort of memorial or realizable form that is highly personalized and empathic; it is the person experiencing the past either through personal recollection or coming into contact with a remnant of the past.[14] For Ankersmit, what connects the past and the present is the human mind, we reflect on what happened in the past within the context of our own personal experience of it. This experience is demonstrated through a sensory knowledge of artefacts and objects, which imprints a 'felt' meaning on us. This imprinting occurs as we subjectively experience the object through our engagement with it. Ankersmit's analysis of our experience of the past is formed via his notion of the continuing 'presence' of the past in our perpetual here and now. 'While there is no direct experience of the past because it is gone, touching and sensing its artefacts allows us some kind of sublime contact.'[15] The archival photographs simultaneously prompt the viewer to read them through both our analytical and sensuous capacities. On the one hand, we read the photographs cognitively through medical language. On the other hand, we read them through visceral sympathy and repulsion. They are simultaneously close because they have a presence but are also distant and of the past. These images of horrific facial injuries make the past more present in particular ways.

The impact that these photographs have on me needs to be understood within the problematic perceptual and cultural engagement with human faces. To engage with the face as the focus of the photograph, it is useful to include Giorgio Agamben's idea of the face as an abyss. Agamben looks for resistance in faciality, as an interstice, a threshold through which the potentiality of de-identification is discernible. He argues that only human beings want to take possession of an opening; to seize hold of their own appearance and of their own being manifest. Language is this appropriation, and so appearance is a problem for human beings because it becomes the location of a struggle for truth. The face's revelation is revelation of language itself. 'It is *only* opening, *only* communicability … The face is at once the irreparable being-exposed of humans and the very opening in which they hide and stay hidden.'[16] When I look into the eyes of the patients and their eyes look back at me – shamelessly – they exhibit their own emptiness as if there was another abyssal eye behind it that knows this emptiness and uses it as an impenetrable hiding place.[17] In the surgical photographs, the disfigured soldiers look into the camera, thereby exhibiting an awareness of being exposed to the gaze, to other people's seeing. This is especially acute for me as a viewer when viewing photographs which block out the eyes of the subjects because I can look at these images without feeling the awkward discomfort of my gaze being returned and questioned.[18]

There is a politics here because we seek the opening and try to take possession of our own exposition; we care about it.[19] We are interested in mirrors, in the image as image. As human beings, we want to recognize ourselves. The soldiers' facial disfigurements thus let the shapeless and bottomless abyssal background that threatens them emerge. 'As soon as the face realizes that communicability is all that it is and hence that it has nothing to express – thus withdrawing silently behind itself, inside its own mute identity – it turns into a grimace, which is, what one calls character.'[20] Agamben speaks here of a 'character' in relation to its manifestation in the face, in the ability to present the face beyond any representational schemes.[21] The soldiers' facial disfigurements prevent us from taking possession of them, or understanding them through language, because their visage is absent, destroyed and unobtainable.

The face is the truth presented beyond representations, beyond the limitations constitutive of the human being 'as construct'. This is also, of course, a presentation that runs contrary to a 'will to total self-possession', something that a vulnerable presentation of the self would ultimately seek to destabilize. Facial disfigurement therefore not only prevents the wounded veterans from regaining or obtaining total self-possession, but it also denies us possession of these men as individuals. The face is a heterogeneous 'presence', posited solely through its externality, the 'being outside' of the multiple underlying desires that constitute the self.[22] Yet the face is also an externality that demonstrates a certain indifference to the internal/external boundary in the first place:

> My face is my *outside:* a point of indifference with respect to all of my properties, with respect to what is properly one's own and what is common, to what is internal and what is external. In the face, I exist with all of my properties; but this happens without any of these properties essentially identifying me or belonging to me. The face is the threshold of de-propriation and of de-identification of all manners and of all qualities – a threshold in which only the latter become purely communicable. And only where I find a face do I encounter an exteriority and does an *outside* happen to me.[23]

The surgical photographs confront us with our very being as human, as social creatures, and bring to the fore our worst fears about corporeal boundaries between interior and exterior and our vulnerabilities about the face as a route to the intimate.

Faces are among the most important objects seen, indispensable for relations of any kind for they (and people's reactions to them) define who we are.[24] It is the human face, or rather, what Elkins perceives a face to be that triggers such a strong response. A face is a particular thing; it has a nose, and a mouth, ears, two eyes, cheeks, a chin and hair.[25] However, when a face disfigures, when ripped open, the skin torn apart, its interior exposed, and when features become inflamed, contorted, dysfunctional and unrecognizable, the effect is an erasure of the face.

When looking at portraits we have a desire to seek an account of the soul of the sitter that in some way will be transparent in the present 'and will provide a means by which the boundary between being and representation might be possibly breached'.[26] However, facial disfigurement has the potential to erase the character of the person underneath in the eyes of the spectator.

Looking afresh at the indifference to the corporeal, Margrit Shildrick reassesses differently excluded others, those who are monstrous, and those who are physically disabled or whose bodies radically disrupt morphological expectations.[27] Drawing on current feminist thought to embodied difference as the grounds for a specific reappraisal of the conventional paradigms of ontology, epistemology and ethics, Shildrick suggests that the body that is recovered in its difference, remains highly normative.[28] She proposes a new form of ethics that answers more fully to the multiplicity of embodied difference, and that undoes the singular category of the monster. Shildrick embraces instead the ambiguity and unpredictability of an openness towards the monstrous other. It is a move that acknowledges both the vulnerability to the other, and the vulnerability of the self.[29] The question of value here is disrupted, suspended in the face of an encounter that cannot be known in advance. Despite persistent desire in medical discourse, which seeks to categorize and explain monstrosity (disfigurement) through the pathology of abnormal corporeality, there is another more disruptive intuition that the monstrous cannot be confined in the place of the other.

It is the anonymity of these faces and subsequently their potential spaces for the inscription of meaning and the projection of fantasy that makes them fit so awkwardly into most available discourses.[30] We have an inherent wish to understand a face as someone and to empathize with the face as a person. We have an instinctive wish to do this, but when the face is injured it interrupts our attempts to know and understand the person and the soul of self, exposing 'the larger cultural question of how we can know the other, how we can understand the subject'.[31] I want to know about the soul of the person in the photograph. I experience a strong desire to personalize the men that the photographs have objectified as medical discourse. I find that medical discourse is impersonal and lacks any acknowledgement of, or interest in, human feelings. Gaining access to, and providing a way of knowing about, the individual or their private selves become out of reach to me. Considered within the traditional conventions of representing the face, medical discourse blocks the codes through which I can understand a person's identity and private life.

The limits of medical discourse

Michel Foucault presents a way of putting the photographs into the past, and provides a framework to understand when, how and with what consequences the photographs entered into medical discourse. Within medical discourse, language dominates and photographs are subordinate. Foucault described how doctors see their subjects, and how bodies become legible within a textual coding: 'The transformation of symptom into sign and the passage from patient to disease and from the individual to the conceptual'.[32] The clinic brings into play the fundamental relation between the doctor's perception and the point at which the body is read and interpreted within medical language.[33] Patients were defined by their symptoms and photography helped to codify these into visual signs.

According to Foucault, 'What is perceptible is only what the conceptual space defines in advance.'[34] Surgeons had to accommodate vision with language in order to make their seeing consistent with their discursive practice. It is the integration of the visible and legible, spatial and verbal that allows the doctor to perceive the patient's body. Once the surgeon describes the injury's visible characteristics in medical terms and any changes in its condition, then, the injury becomes legible and conceptually coherent. Medical discourse therefore makes knowledge of the visible possible and allows certain details to appear. The surgeon's vision only detects what the discourse says is significant. This suggests that the surgical photographs embed within a discursive terrain that theorized the body and transformed it into a text.

The RAMC photographs show patients whose courses of treatment divided into textual codes of reference. They are part of a process of description that sought to understand the changes in a patient's condition. The photographs contain evidence of the symptoms of facial injuries, before-and-after treatment, and these observations would have been invaluable to surgeons for analysing their characteristics. Important factors such as knowing how long infected wounds wept or remained swollen prior to treatment; understanding how different parts of the face responded to grafting techniques or comparing healing processes between cases would have provided essential data for surgeons to evaluate.

Articles published in medical journals in 1915 seem to be demanding visual evidence.[35] Existing literature on facial injuries was limited. Surgeons who were treating these wounds saw a need to record these new injuries so that they could be recognized as cases. Once these injuries could be recognized the surgeons could know them. In the *British Dental Journal* in 1916, an article by Surgeon-Major Leon Frey, director of the central service of prosthetics at Val-de-Grâce, Paris, called for dental surgeons to identify and document war injuries of the face and jaw as opposed to cases already known to the profession.[36] The Val-de-Grâce hospital was one of the first specialized maxillofacial centres and made efforts to facilitate facially injured veterans back into civilian life as well as surgical reconstruction.[37] For the purposes of dental literature, the cases with loss of bone and skin were to be thoroughly recorded, while the better-known wounds, those normally found in civilian practice, were not to be recorded and discussed as a priority because they had already been identified within the discipline and in text books on surgery.

In another article in the *British Dental Journal* in 1916, Kazanjian, an American dental surgeon from Harvard University treating facial injuries in France under the British Medical Services, pointed out the importance of systematically recording these new cases. He argued that photographs should be taken of the patient as soon as possible:

> A good photograph procures, in such cases, sufficient data on the state of the case without the need for a plaster cast model which would be uncomfortable for the patient, and is often more accurate than plaster models.[38]

Photography could deliver types of result that other kinds of images could not manage.[39] For example, Lewin Payne uses a photograph to illustrate a class of case that had presented particular difficulties in his wards. As a solution, he devised a lingual clamp to prevent haemorrhage:

> In fractures of the mandible, bleeding from the lingual artery is rather common and at times serious, and it is rendered difficult to arrest because the parts are tender and the patient cannot open his mouth wide.[40]

The photograph contains enough data to identify the type of jaw injury at hand, the design of the dental splint devised, and the position and means of fitting the appliance. At the Royal Society of Medicine meeting held in February 1916, Dr Hayes, chief dental surgeon at the American Ambulance at Neuilly, France, pointed out that

> many questions involving the treatment of these wounds are undergoing renewed and general study, much of this work is new and the best means and methods have yet to be determined, careful statistics have yet to be compiled … so proper classification of these cases in reference to the line of treatment to be followed is required.[41]

It was important that each surgeon documented his observations and findings and circulated the results so that guidelines could be set in place for the treatment of these injuries. In the *British Dental Journal* in 1916, Charles Bubb described the dental appliances he and the surgeon Percival Cole had jointly designed, based on his experiences as honorary dental surgeon to the King George V Hospital and assistant to Cole:

> As a result of my experience I feel justified in recording some points in technique in the hope that help may thereby be afforded to others engaged in the treatment of similar cases.[42]

Bubb declared that owing to Cole's skills as both dentist and surgeon, and because of the pooling of knowledge and the thorough records that were being produced in their wards, their efforts had achieved a high level of success. Even though this article is not supported by visual evidence, it is of interest because it contextualizes the unique characteristics of the photographs within the RAMC albums, which illustrate a large number of the dental splints designed by Bubb.

In the president's address at a meeting held by the Royal Society of Medicine in October 1917, J. H. Badcock said:

> The very mass of clinical material garnered by the cruel hand of war has afforded an unequalled opportunity for the study and advancement of medical science, and facilities for research and practice have arisen such as would never have been available in peacetime.[43]

However, Badcock warned that still more needed to be done and urged every worker in the field to make it his business to record his observations and experiences for the benefit of others. He hoped practitioners would consider it their duty to publish reports from time to time, half yearly being not too often for a subject so new and so rapidly developing. He argued that if such a system of records were generally adopted it would allow the development and advancement of knowledge and methods, 'both for the sake of our soldiers today and the benefit of posterity'.[44] Badcock pointed out that comparison of these injuries was crucial to advancing knowledge because it stimulated observation. Such records allowed surgeons to check results and draw conclusions from them, and 'form a mine of wealth for the independent investigator'.[45]

A contribution to this meeting by G. Northcroft, from the First London General Hospital's jaw department, declared that the importance of the work of radiographers, photographers and those making plaster casts could not be too strongly emphasized.[46] Northcroft encouraged doctors to publish full and accurate records of the results so that this valuable information could be disseminated more widely. He said, 'Photographic records where any plastic work is involved are well worth the time and trouble … . It is astonishing how quickly one forgets the original appearance of a wound after seeing the patient for two or three months progressively improving.'[47] Photographs not only functioned as records of the reconstructive techniques but also provided evidence of the practicality of the plastic work being carried out through the visible improvements made by the patient. Northcroft relies heavily on his photographic evidence to structure his argument and provides a rich visual analysis to support his discussion. He talks about each image individually and not only explains their injuries but also contextualizes them, providing a fuller understanding of the wound and a justification for each specific treatment. Each photograph in this publication had the eyes of patients blocked out. This could have been a visual device to maintain a degree of anonymity in the patient, or it could have been a way of pointing out what to look at by drawing the viewer's attention to the injury. This begs the question, why photograph the whole person or face at all? It was important to include each patient's whole face, and often his head and shoulders, in order to document general health as well as facial transformation.

Northcroft says because of the curious nature of jaw work, 'whenever one tries to lay down a fundamental principle it always seems to be upset by the next half dozen cases one sees'.[48] Because dental splints had to be specifically designed according to each patient's particular condition, the most accurate way of describing these appliances was visually. These splints were complex designs and made from a number of component parts. Some parts needed to be adjustable or interchangeable, depending on whether it served as an interior or exterior support. While a reader may not have been able to gauge the exact size or measurement of a dental splint from its illustration, its ingenious shape and fitting to the patient could be conveyed so that other surgeons could use it as a template or guide.

Photography opened up surgical procedures for debate and helped surgeons to observe, test and learn about the face's healing properties from the visible results on display. This knowledge could then be written into a body of work on the subject and integrated into medical literature through publications. Even in black and white, photographs showed a wealth of details about cases and their treatment because they documented the recovery and healing times of patients and provided important information on the unpredictable effects and outcomes of surgery. They represented procedural texts that offered instructions on how to perform specific reconstructive techniques. Photography was a more practical means of visually documenting facial surgery compared with wax models, casts and drawings. Although wax and plaster-casting techniques were used to produce three-dimensional models of the face so that surgeons could study the contours and shape of the face from all angles and observe its structure at close range, these techniques were time-consuming procedures. Drawings were also time-consuming. Photographs were potentially more valuable than sketches because their fast exposure times made it possible to follow the rapid changes of an injury more easily. The pastel drawings produced by Henry Tonks and Daryl Lindsay only record patients prior to treatment or just after all surgery was complete. Drawings were not used to observe lengthy course of treatment but as a means of comparing the severity of the injury on admission with the success of the overall treatment. While drawings were helpful for summarizing courses of reconstruction, as well as providing colour records that conveyed the subtleties in skin inflammation, wound infection, bruising or blood supply, photography proved to be a better and more efficient means of recording and monitoring the changing conditions of cases during treatment because it was easier, quicker, cleaner and less invasive. Photographs were not necessarily better at illustrating facial injuries than drawings, but the visual evidence they contained and their ability to be reproduced and disseminated more widely also made them valuable for communicating with colleagues.

The RAMC and British Red Cross never showed the surgical photographs in society, to non-medical viewers, or used them as propaganda presumably because they were unfit for public consumption.[49] Newspapers did not write many articles on facial injury patients and never published the surgical images. Those that did report on this type of injury were aware of the sensitive nature of this class of case. The surgical photographs were unsuitable for the public because they were explicit and exposed the hideous nature of war wounds. These images showed wounded servicemen in a weak and vulnerable state. To circulate such records to a public audience would have exploited the men's privacy and created anxiety in the public because of the shock of such traumatic wounds and a shared sympathy towards the devastating injuries and suffering inflicted on the average soldier.[50] The surgical photographs highlight a contrast between war propaganda's romanticized view and the visual evidence of the war's catastrophic bodily consequences. Carefully orchestrated group photographs of facially injured servicemen were used in Britain as propaganda tools to raise money from the public to support wounded veterans, as long as disfigurements were bandaged and concealed. In France, by contrast, photographs of facially disfigured individuals and groups were circulated to the public not only to raise funds to support rehabilitation but also to draw attention to anti-war sentiment.[51]

Civilians were called upon to make donations to help maintain the running of British military hospitals and to support the rehabilitation of injured servicemen (not just those with facial injuries) across the country. Newspapers published weekly appeals to the public for financial help. In addition to these appeals, newspapers published regular updates during the war on the amounts that were being raised and how the money was being spent to meet the sick and wounded men's needs while in hospital. By April 1915, the Red Cross Compassionate Fund had raised over £1 million from public donations.[52] Group photographs were taken and circulated to keep the public up to date on the progress of the patients and their medical care, and to foster a dialogue between the Red Cross and the local community. Throughout the rest of the war the public continued to send in money, and these donations finally reached over £11 million by September 1918.[53] The total wartime income was £22 million of which the majority came via *The Times* appeal.[54] Local and national newspaper appeals across the country played a highly significant role in influencing public opinion and in advertising the activities of wartime charities, and reveal public support as both humanitarian and patriotic.[55] Nevertheless, the group photographs create a contradictory story about disfigured soldiers' rehabilitation that shifts between public and private spaces of healing.

Newspaper publishers could not show the visible effects of these injuries on the men but told of the human story. An article written in 1921 describes just how shocking such images were to the non-medical viewer and reveals something of the ethical role that medicine took regarding the sensitivity of patients with facial disfigurements. Commenting on a visit to the facial injury wards at Sidcup and the attached museum, the journalist Robert Blatchford describes his thoughts on seeing the surgical photographs and says he will never talk about these photographs, not ever, though he may dream of them, and would not even talk about the operating theatres, or the instruments, or the gifted and untiring doctors or the wounds:

> One can imagine it all with awe and sympathy. When a jagged fragment of shell strikes the side of a brave face, or when shrapnel tears a mouth or cheek … one can picture the result.[56]

The journalist concluded that

> we often say of our brave dead that they made the supreme sacrifice but these mutilated heroes at Frognal have paid a higher price than life for their unselfish valour.[57]

There is a mixture of absence and presence in the newspaper articles. The wounded men, and the surgical photographs, are out of reach from public view. There was 'an unofficial censorship of facially disfigured veterans in the British press and propaganda' because the wounded face 'presents the trauma of mechanized warfare as a potentially contaminating, and shameful, loss of identity and humanity'.[58] Britain did not represent the horrors of facial mutilation visually, outside the professional contexts of clinical medicine, and this is characterized as a visual anxiety and aversion to such injuries.[59] Contrasting the surgical photographs with the newspapers, Britain's war propaganda was in direct conflict with the restriction of images of patients' injuries and with what was being portrayed about wounded servicemen and their recoveries. Newspapers and official photographs focused on the efficiency of the military hospitals and not the horrors of the maimed.

The most detailed first-hand account of facial injury patients is in a book written by Ward Muir in 1918. His book describes his experiences as a medical orderly at the Third London General Hospital. Muir talks about the ordeal of working in the facial ward and what he encounters there disturbs him. He says, 'I never felt any embarrassment in confronting a patient, however deplorable his state, until I came in contact with certain wounds of the face.'[60] His feeling towards these men is not an objective medical reasoning but one based on an emotional encounter manifested through embarrassment and discomfort at the visible appearance of these patients. Muir also talks directly about the before-and-after photographs; these, he says, tell the whole story:

> Showing the macabre and brutish physiognomies being metamorphosed into sane and reasonable ones, they cannot register the cheeriness which has been brought about the brains which those facades hide. Some hint at it is directly conveyed by certain of the 'after' portraits.[61]

The surgical photographs suggest personal narratives of the patients' reconstructions and hospital experiences and this helps to acknowledge them as individuals rather than cases. The sequence of before-and-after images of each patient represents a story of individual repair and renewal through visible proof of recovery. Although the pre- and post-operative photographs represent the patients as texts, this tells a story of improvement because they show wounded soldiers being given the best medical care available in an attempt to rebuild their faces and physical health. The progressive surgery recorded by these photographs is therefore a positive one because it shows patients patched up and rebuilt so that they could rejoin society again.

These intimate spaces of the patients' hospital experiences personalize the men and symbolically re-historicize their marginalized stories into a message that suggests the family and the home. Articles in newspapers presented patients' recovery settings as being as homelike and domestic as possible. This provides an alternative story of their experiences and gives a personal voice to the photographs. The newspapers link the surgical photographs to narratives outside of medical discourse. However, despite the support for servicemen in rehabilitation they were very isolated and the longer they were in hospital the more difficult their situation became and the harder their struggle back to normal life.[62]

Group photographs taken in military hospitals suggest that the facial injury patients did, at some points or to some extent, occupy a domestic space resembling that of a family. These photographs create intimate domestic spaces. By photographing groups of patients with similar injuries and celebrating their camaraderie through their shared war experiences, these images construct a 'hospital family' by foregrounding the men's close relationships and personal sacrifices. Such group photographs reveal that facial injuries were being organized in a particular way for public consumption. The nurses in the photographs have a presence of female 'family members'. Female presence played a significant part in the social rehabilitation of facially disfigured men.[63] These familial relationships helped men to cope with their disfigurement and participated in their preparation for reintegrating back into the family once they left the hospital and returned to society.

The familial becomes a space to heal the traumas of disfigurement. The maternal care provided by female staff reveals the importance of emotional support for rehumanizing facially disfigured soldiers.[64] By censoring facial injuries to make them less shocking, the group photographs constructed familial meanings about the patients and their rehabilitation to appeal to national ideals of family values. To identify with the wider community, group photographs sought to reinforce key familial characters and themes and connect with public sympathy. Soldiers with facial injuries needed to reintegrate into their family and social environment, through relationships with their families, wives and children, to reassert themselves as subjects rather than objects and to regain their humanity.

Newspaper appeals for public funds help to explain the challenges that long periods of rehabilitation had on patients' psychological well-being. In the *Evening Standard* in June 1918, the Red Cross made an appeal to the public to help alleviate patients recovering from facial injuries by giving donations:

> During long days of treatment the majority cannot go into public places, cannot receive friends and visitors, and have to fight acute depression. It is to alleviate this and to provide recreation and amusements only that this special national appeal is being made. The alleviation of suffering is a national matter. It is the business of everyone who cannot more directly aid the nation in the war.[65]

Another article calling for public financial help in the same newspaper described a walk through the blue-curtained wards of the Queen's Hospital as a mixture of emotions, of affection and pain.

> For a fleeting instant one saw the dreadful abyss into which the wounded servicemen had fallen. Not every one of the sailors and soldiers who have been severely wounded in the face or jaw at Frognal suffer from acute depression, but most of them do.[66]

Similar appeals for public support took place in France. Press publications in both countries called for public action, especially for donations to relief funds.[67] There is a very different story emerging here, one that exposes cracks in the smooth surface of propaganda as the personal suffering of facially injured patients is allowed to seep through in order to directly appeal to public sympathy. Another article reads:

> Depression of the most acute kind is one of the greatest enemies for the surgeons. Many suffered from depression and refused to return to their homes, families, friends, until they were convinced everything possible had been done for them. Outsiders did not hear much about facial wounds, for all severe cases were taken directly to Frognal.[68]

The newspapers continued to run stories on the recovering patients at Frognal long after the war had ended. These stories of patient recovery were part of a national appeal to keep the convalescent hospitals open after the war. However, newspapers show that in the early 1920s many patients had been in hospital for several years. As an article in July 1921 says, some men had been in hospital for two or three years: 'When they first went out again they were hypersensitive and insisted on screened vehicles and curtained boxes,' and could not 'function well outside anymore'.[69] This article is significant because it reveals that the men perceived themselves as 'others' and detached from life outside of the hospital, its title also uses an interesting language of 'veiling' and revealing these men's disfigurements from society.

The photographs show us the surface of these tensions and also hint at the depths of what was involved in the process of dismembering and re-membering. The press publications expose the tensions involved in the war veterans' isolation, their fear, shame or humiliation of difference. Articles published from 1918 reveal a change in tone and suggest a shift in public sentiment; they started to forget patients and sent fewer funds to the voluntary-aid organizations. Newspaper articles served as a means to remind the public that there were patients still under treatment that needed their help, but support for these men began to wane. While rehabilitation in the military hospitals had acknowledged the importance of returning wounded soldiers to civilian life as soon as possible, this reintegration had relied heavily on the support of the public to make this transition. Due to limited funds being donated from the public after the war ended and thousands of patients still in need of several more years of surgical treatment before they were fit enough to rejoin society, reintegration into civilian life was proving an expensive task for the government, war charities and the public. The longer these patients remained in hospital, the more difficult it was for them to adjust back into public life or return to a normal existence.

By paying attention to the different levels on which the surgical photographs operate it has been possible to map out their multilayered meanings. These images demonstrate medical–scientific investigation, whereby they enabled surgeons to identify, recognize and accurately assess a range of facial injuries and surgical procedures and allowed them to think about their practice and methods of treatment. Another narrative that emerges in the before-and-after photographs is the recreational activities and occupational therapies for patients as part of the healing process. The surgical photographs intersect with official group photographs that were used to promote the day-to-day running of this military hospital to the public, and became absorbed into readings of national sentiment and political propaganda. Through the provision of activities, recreation and workshops within the daily routines of patients undergoing long periods of convalescence, military hospitals introduced therapy exercises to aid effective healing, relieve trauma and provide work-related retraining to help reintegrate them back into the workplace and community.[70]

I turned to Foucault's work on discourse theory and clinical ways of seeing to gain a better understanding of how these images functioned. By employing a Foucauldian methodology to the analysis of these photographs, the aim was to establish how the images operated within a specialized language, such as medical discourse, and how surgeons originally read them. However, a focus on photographs as 'medical discourse' does not adequately account for the affective nature of these photographs.

A non-medical viewer or any viewer as distant in time as we are from the First World War does not 'see' in terms of medical structures. The more I consider the authority of the medical gaze which structured these photographs and the issues of trained or untrained seeing that they raise, the more I become conscious that such a reading does not actually address my viewing encounter, the experience of being upset. These images are coded objects of medical photography that fix suffering into medical discourse, but they also operate simultaneously to reveal extraordinary stories of human beings. They are both fraught with trauma and proof of the huge progress made in plastic surgery.

Shildrick's theorization of embodiment provides a useful framework for explaining the personal affect caused by these photographs. Shildrick reflects on the figures of difference that haunt the Western imaginary, which she terms as the monstrous – those physically disabled or whose bodies radically disrupt morphological expectations – in order to uncover and rethink a relation with the standards of normality that proves to be uncontainable and unknowable. The notion of vulnerability is firmly linked to encounters with disabled others because their very presence signals a threat.[71] Facial difference compromises notions of self and identity and profoundly affects social relations.[72] This encounter defines my own boundaries of normality. My immediate connection to the images, my sense of seeing men, rather than 'cases' or 'conditions', is not contained within the limits of medical 'discourse', nor, in fact by any contemporaneous discourse with them. The photographs can clearly mean many other things that are quite independent of medical discourse. This study of one specific group of photographs leads me to question the identity and function of a 'historical' photograph as implied, to suspect that any photograph could detach itself from the position it may have within an inevitably written, historical account of events, and initiate a very different meaning.

Sensing photographs and materiality

Forced by open access, the political need to be open and transparent, and digitization, policy dictates that medical archives now make their collections accessible to the public in order to promote the understanding of medicine and its role in society, past and present. As part of their mission to widen public access, archives are committed to handling personal and sensitive information in these research collections lawfully, ethically and responsibly. A number of archive collections fall within the scope of the Data Protection Act 1998 as they contain personal data, particularly case studies and patient records, in which information about specific individuals can be readily retrieved. Such material contains sensitive personal data relating to the physical or mental health of individuals. And so, protected by patient confidentiality laws, they are generally closed or restricted for a finite period based on an assumed lifespan of each individual of hundred years. To demonstrate their commitment to the ethical handling of all personal data, archives may sometimes restrict the availability of sensitive data even after the death of an individual because of the impact on close family members connected to the deceased person referred to in the material.

As non-public bodies, medical archives are not subjected to the Freedom of Information Act 2000 nor do they acquire materials that fall within its scope. Access is considered on a case-by-case basis subject to clear explanation of what information is being sought and the purpose to which it will be put. To meet the varied and changing demands of its audiences, and provide an appropriate gateway to their resources as well as physical access to the material, archives are ensuring that collections of long-term historical value are placed in the most appropriate publicly accessible repository. A particular expectation for research in recent years has been for wider public engagement, to expose special collections to a broader lay audience, and so more records of potential long-term value are being created in digital format. Archives have embraced digitization because of the principles that underpin the accessibility of their collections.

Today, surviving relatives of patients are accessing medical archives and engaging with the surgical photographs to reconstruct their ancestors' war experiences. For family members, the facially injured soldiers seen in these photographs can be reclaimed as ancestors through their presence in the archive (Figure 11.2).

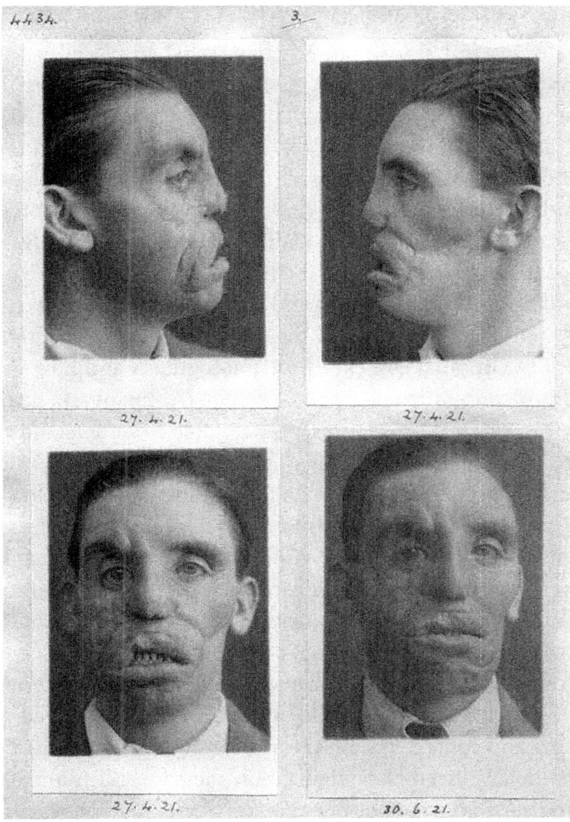

Figure 11.2 George William Butcher, 1920–21, photographs by Sidney Wallbridge. Gillies Archive, British Association of Plastic, Reconstructive and Aesthetic Surgeons, from the Archive of the Royal College of Surgeons of England, London.

The absence of their ancestor is marked by their photographic presence. The historical discontinuity of the photographic document is overwhelmed by the presence of the ancestral image in the present.[73] The ability to see an ancestor's war injuries within the image is central to the photograph's power in the present. Family members become overwhelmed at the photographs and what they see, and to be able to look back in history to see the face of their grandfather or great-grandfather.[74] These archive engagements with the past in the present situate the disfigured soldiers within their relevant kin network today, giving the photograph social meaning. This is the reconceptualization of the archival object with familial networks. This points to an understanding of historical photographs that sees them as the basis for lived experiences of such a real and emotional presence of the past. Such encounters highlight the interwoven nature of historical continuity and discontinuity. Family relatives engaging with the archival photographs today exemplify a dimension of the photograph as an object that materializes the presence of ancestors, and is an integral part of the way the past is experienced and understood in the present. These photographs are historical, but they are history right now.

Photographs are difficult to pin down and can have different meanings at different times. They are porous and absorb other meanings when placed with other objects or when used or viewed in different ways. Photographs are spaces through which different meanings pass through and they take on new readings when their contexts of viewing shift. Joanna Sassoon shows that although image content remains, a great deal of information can be lost in the process of translation. As Sassoon argues, the origin and importance of photographic meaning, and how the photographic object is perceived in relation to other forms of material, are revealed through the styles and content of the documentation relating to the material. 'This can have a profound effect on the understanding of the photographs and the nature of the research questions that can be asked of the material.'[75] Through the tactile viewing encounter, we can learn a great deal about the original context of the photographs and how they functioned as cognitive tools for conceptualizing reconstructive treatment and instructing others on how to carry out surgical procedures, and are able to understand them better as historical and visual resources and their ongoing and shifting meanings. In digital form these objects are reduced to content, although there are ways to knowing about their materiality, and they are reshaped into the archive's own set of descriptions and arrangement that invariably loses vital information on provenance and history of ownership. However, we can understand more about what these photographs mean today and the ways we experience and gain knowledge about them and their histories through our sensory and embodied engagement with them.

The image's content stays the same because the process of translating a photograph from a material to a digital form appears to be neutral and unmediated, therefore their sources of meaning and context remain the same.[76] Yet the digital medium allows for zooming to facial details. Even after digitization they are still traumatic because the translation of their shocking content remains. However, Sassoon warns that while an image's content remains the same, its meaning can become less stable when digitized because of its changing tangibility, for it 'encourages a shift from thinking about the complexity of the material object to viewing the visual surface of an image'.[77] This suggests that digitization shifts the viewer's attention to subject content rather than the materiality of the photographs as objects, which has a profound effect on the viewer's interpretation of the photographs and an understanding of their social biography.

The tactile viewing encounter enhances our understanding of the integrity of the photographs as historical sources by allowing us to see how they have been used at different times and how they have acquired value as documents. Seeing the photographs in their case files or compiled into albums reveals how they were viewed by surgeons, how the images have been reused and positioned into different contexts and how these shifts have impacted on the way the images are read (Figure 11.3). The materiality of the photographs makes our encounter richer by providing an insight into what photographs are capable of doing; we see them as material documents that play active roles in history.[78] Through knowing the various ways that a photograph has been used and the shifting contexts it has been viewed in, it is possible to identify the intimate spaces set up between the image and the viewer and comprehend how various users have contextualized the photograph and constructed its meaning.

Is the digital version of the image therefore as effective in communicating beyond the limits of medical discourse as the paper one? Does the paper quality work in digital form? Today the facial injury photographs have been digitized in order to reach out to new audiences. The effect of digitizing on photographs is material because the digitizing process does not change the physical state of a photograph from the material to the pixel. It is not a transformation of state but a complex cultural process of translation, 'a change between forms of representation'.[79]

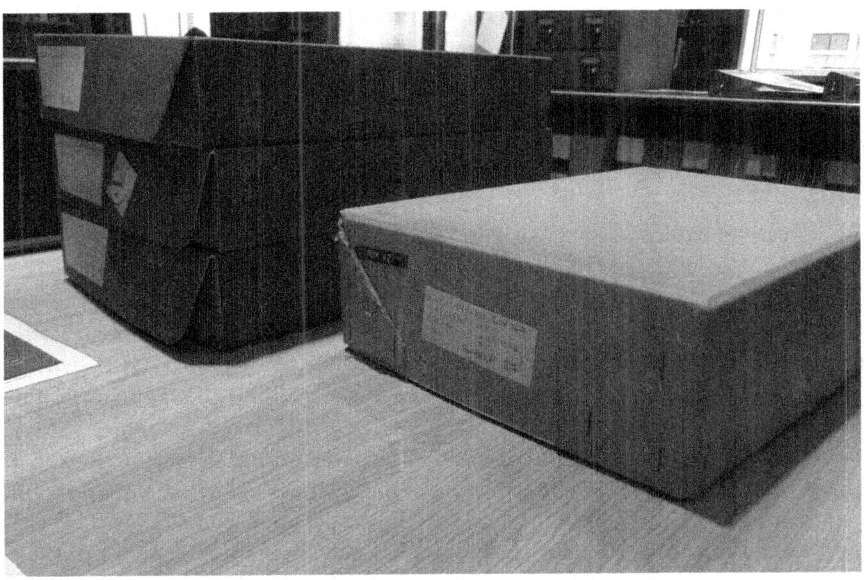

Figure 11.3 RAMC Albums in their Storage Box, RAMC 760. Courtesy of Wellcome Library, London.

Our encounter with physical images is interrupted by a response to the emotional weight of the photographs, 'a punctual viewing at the limits of meaning'.[80] When photographs are experienced, such an encounter opens up the space for historical imagination and the complex networks of relationships and associations that constitute the making of historical narratives.[81] The surgical photographs are sensory objects that become meaningful through our engagement. It is not simply because of a photograph's infinite recodability that its meaning becomes open to question, but because the material presence of a photograph as an artefact impresses itself on our understanding of the past through our viewing of them. The explosion of potential meanings across the sensory and emotional range cannot be contained within the archive.

Our realization that reading the photographic artefact is only one physical interaction with them during their existence, and reflecting on their making, uses, preservation and exchanges, all make us think about, and understand, the relations that have constituted the histories of this archive material.[82] The photographs demand an embodied engagement and experience. They entangle us with the lives of others – of the subjects, the users and the viewers.[83] Archival material includes the possibility of such affects because they are active in the relations between people and the weight of the past in the present, helping us to understand the dialectical nature of the relationship between past and present and our own positioning within this.[84] A photograph is both an image *and* a physical object that exists in time and space and thus in social and cultural experience.[85] A photograph's characteristics cannot be reduced to a set of meanings or ideologies that take the images as its pretext because they occupy and move through different spaces, and at the heart of the materiality of a photograph is the act of viewing the object, for it 'implies specific relations with an embodied viewer and determines responses to photographs'.[86] The demand of objects on the embodied viewer creates a 'torque of materiality that can never be assimilated to a disembodied "linguistic-philosophical closure", "culture", or "history"'.[87]

Cutting through time

The affective nature of the surgical photographs triggered a psychic and phenomenological stress on corporeality that forced me to contemplate the horrors of facial disfigurement. These photographs are traumatic and make me reflect on my feelings towards the patients and the nature of embodiment itself. In seeking confirmation of my own secure subjecthood in what I am not, what I see mirrored in the facially disfigured soldiers are the leaks and flows, the vulnerabilities in my own embodied being.[88] The disfigured faces refuse to stay in place and disrupt internal and external order, overturning the distinctions that set out the limits of the human subject. Family relatives of the patients respond to the surgical photographs through an intense personal connection they feel towards an ancestor, particularly if they recall knowing them in person and their struggles to readjust to and cope with their disfigurement. Such a viewer connects with the photograph through personal memories and close familial bonds. These photographs tell us about human exploitation and the inhumanity of industrial warfare. They force us to think about the extent of war through the physical effects and social impairments that such conditions inflicted on soldiers' bodies.

Ulrich Baer helps to build on the issues raised by the problems of context by addressing how historical photographs can shift in meaning. From the position of literary theory, Baer raises a poignant question about whether the viewer who contemplates traumatic photographs can dissociate the suffering depicted from its political and historical context. The fact that Baer's project uses a theory of literature to think about history is valuable here because it not only makes reference to writing as *the* medium of historiography but also emphasizes the underlying problems, tensions and incompatibilities between the visual and the textual, and in particular, the unsettling relations between picture and context.

Baer helps to understand our position as viewers, and the unconventional reading, meaning and the unpredictable effect experienced by engaging with photographs that force the viewer to consider experiences that resist integration into historical contexts. As Baer says:

> Some images bypass painstaking attempts at contextualization and deliver, straight up and apparently across the gulf of time between the viewer and photographically mummified past, a potent illusion of the real.[89]

Although we assume that the past would be safely apart from our grounding in the present, Baer suggests that photographs unsettle this perceived distance. There is nothing new in Baer's claim that history makes the past real, or tries to, but it cannot do this of course because it is history and not the past. However, Baer's work is similar to my reading because he also argues that we cannot keep traumatic photographs in place. I use Baer's argument to explain how the surgical photographs cut through time and bring their traumatic effects more directly into the present, transcending their historical frame by connecting us with a staged surgical display of horrific First World War facial injuries.

In his explanation of how feelings towards a photograph generate, Baer argues that traumatic photographs in particular have the ability to provoke and stimulate affect. He says that we cannot assimilate such photographs into a continuous narrative or linear accounts of the past because they reveal experiences that expose as a construction the idea that history is ever flowing and pre-programmed to produce an ongoing and linear narrative.[90] Blocking the conception of the past as an unstoppable movement forward, the surgical photographs compel us to think about horrific facial injuries and refuse to integrate into the idea of the past as a forward sweeping movement.[91]

Shocking photographs cross time and a brutal reality can affect the viewer in the present because of the impact of the horror. Photographs of trauma 'interrupt the otherwise seemingly unstoppable flow of thought in the viewer by triggering a sudden experience of overwhelming mental transport, elevating the photographs into representations of suffering that may facilitate transcendence over conventional thought processes'.[92] Such a response arises before thought and understanding.[93] My reaction to the surgical photographs exposes history as a convention – an extremely powerful one, but a convention nonetheless. Engaging with the surgical photographs privileges the moment rather than the story. For a moment, the concept of linear history cannot adequately address my encounter. These photographs make me feel, see and experience the affective dimension to any narrative here and now. This links directly to Ankersmit's notion of historical experience. He argues that because historical experience ultimately aims at a recovery of the past that transcends the barriers between past and present, the past has a presence.[94] The presence of past entities performs through our sensory engagement with the photographic object. The past becomes present through the sensory knowledge that these objects imprint on us.[95] This is precisely the tension that informs and defines my viewing encounter. I come to realize that historical knowledge shapes directly through my emotions. The photograph's past becomes a tangible present experience.[96] Baer argues that traumatic photographs have the potential to resist integration into larger historical contexts for many viewers and not only individuals because they impose ethical demands on us and can open us up to experiences of embodiment on a collective level. We feel a shared sense of moral shock.

Baer has developed his discussion of how photographs compel viewers to think about time from Vilém Flusser's work on photography and conception of history. Flusser understands photographs to cut through time. It is possible to apply Flusser's ideas directly to the viewer's role because he explores the ways that 'photographs dam up the flow of history'.[97] Building on an understanding of history as a form of consciousness inextricably linked to writing, Flusser proposes that photographs actually interrupt the stream of history because they are not part of a linear reading or sense of time flowing, but projections that can be reanimated in the present and future, 'jamming historical happenings'.[98] This emphasis on 'jamming' does not imply 'blocking' but 'improvising' and 'creating' history. Flusser's argument applies directly to my reading of the medical photographs. The surgical photographs contain such traumatic references that they unsettle my sense of history as chronological progression, my sense of the photographs being safely in the past. My response to the photographs is a tangible encounter in the present, they affect me now and make their recorded past into a contemporary concept because it is active, it has a presence. My immediate response to the photographs occurs in the absence of meaning-making discourses, of explanatory narratives.

Conclusion: Projections and possibilities

Flusser's work on photography and conception of time helps to explain how the surgical images bear on the idea of history. Flusser does not suggest that photography is another way of organizing history. For him, it is a means of obliterating the whole idea of putting events in linear, logical or chronological order. The surgical photographs seem in some sense to offer one example of how a photograph could do this. Flusser's work suggests that photographs break through history and even have the potential to resist history altogether. While I agree with Flusser that photographs can disrupt historical understanding, I still believe that photographs embed us in history despite the lack of exactness and stability found in written documents and evidential records to contain them. Flusser does not help to read history in a different way but problematizes it as a philosophical concept. The photographs become windows through which we can see how history is constructed and comprehend how we constitute historical narratives. Whereas Flusser believes that for some viewers, imagination defends itself against a linear conception of the world, we come to realize that the photographs cause us to reflect on and interpret history in a specific way.

Flusser's claim that the way we read photographs is not about a connection to the past, but as a projection, is very useful in our particular context. According to Flusser, 'From the perspective of formal consciousness, photographs are information intentionally produced from a swarm of isolated possibilities … . Photographs are computed possibilities (models, projections, on to the environment).'[99] Photographs are not so much windows into the past but representations that translate their specific moments into infinitely repeatable projections. Rather than taking us back to a specific moment, the image says something about how the world *could* be, or, could have been. We do not go back to the specific events that shaped the surgical photographs' production, but, rather, they cause us to think about the experiences of the wounded soldiers and to imagine what it could have been like and what might have happened. The sight of horrific facial injuries and the fragments that the images showed of the soldiers' rehabilitation allow us to think about their war experiences and reflect on the lasting effects that their injuries might have had on their lives.

Flusser was consistent about the claim that photographs say something about how the world could be rather than taking us back to a specific past moment. I, or anyone else, could agree that the surgical photographs do not really, usually, or necessarily, take us back to the moment of their making. These surgical photographs are pictures of how things could be and they say something about possibilities. They project very grim possibilities. I am confronted with 'excessive' possibilities that are too horrible to contemplate. The impact, the putative threat to my own well-being that facial disfigurement presents between open and closed boundaries of the body springs from my unavoidable realization that as an embodied subject I too am fragile. The facially disfigured soldiers are marked out by open boundaries of the face and the materiality of the(ir) flesh reflects back my unity and completion as a viewing subject.

Although the medical archives steer the photographs into specific narratives, the viewer structures their own narratives from them, as well. The surgical photographs make the viewer think in a certain way about their history. These photographs can both shape the viewer's understanding of history and change our view of it. Surgeons made these photographs work, but we cannot box them up with surgical talk now. These photographs may be 'framed', made to serve specific medical purposes within chronological accounts of the past. They will also always retain the potential to exceed that frame, to cut across time and meet the viewer here and now.

Notes

1 Martin Kemp, '"A Perfect and Faithful Record": Mind and Body in Medical Photography Before 1900', in *Beauty of Another Order: Photography in Science*, ed. Ann Thomas (New Haven: Yale University Press, 1997), 127.

2 This short letter was written by Brigadier James Howden Robertson, a consulting dental surgeon to the army from the Royal Army Dental Corps. In the letter, dated 6 August 1969, he thanked Colonel John George Eric Vachell, from Hove, for the gift of the two albums which he proposed to retain in the Department of Dental Science in the Royal Army Medical College, Millbank. The albums were used as reference sources for training officers before being donated to the Wellcome Library in the 1980s. RAMC Muniment Collection, Wellcome Library, London, RAMC 760.

3 Elizabeth Edwards, *Raw Histories: Photographs, Anthropology and Museums* (Oxford: Berg, 2001).

4 Susan Sontag, *Regarding the Pain of Others* (London: Penguin Books, 2003).

5 For a recent treatment, see Suzannah Biernoff, *Portraits of Violence: War and the Aesthetics of Disfigurement* (Ann Arbor: University of Michigan Press, 2017).

6 Ewa Domanska, 'Material Presence of the Past', *History and Theory* 45.3 (2006): 337–48.

7 Domanska, 'Material Presence of the Past', 337.

8 Ibid., 337.

9 Ibid., 344.

10 Ibid., 341.

11 Elizabeth Edwards, 'Photography and the Material Performance of the Past', *History and Theory* 48.4 (2009): 130–50.

12 Edwards, 'Photography and the Material Performance of the Past', 131–2.

13 Ibid., 135.

14 Frank Ankersmit, *Sublime Historical Experience* (Stanford: Stanford University Press, 2005).

15 Ankersmit, *Sublime Historical Experience*, 79, quoted in Alun Munslow, *A History of History* (London: Routledge, 2012), 162.

16 Giorgio Agamben, *Means Without End: Notes on Politics*, trans. Vincenzo Binetti and Cesare Casarino (Minneapolis and London: University of Minnesota Press, 2000), 91.

17 Agamben, *Means Without End*, 92.

18 On this reciprocal relationship, see the chapter by Jane Frances in the present book.

19 Agamben, *Means without End*, 92.

20 Ibid., 97.

21 Colby Dickinson, *Agamben and Theology* (London: Bloomsbury, 2011), 160.

22 Dickinson, *Agamben and Theology*.

23 Agamben, *Means Without End*, 98.

24 James Elkins, *The Object Stares Back: On the Nature of Seeing* (New York: Harcourt Inc, 1996), 160.

25 Elkins, *The Object Stares Back*, 161.

26 Joanna Lowry, 'Portraits, Still Video Portraits and the Account of the Soul', in *Stillness and Time: Photography and the Moving Image*, ed. David Green and Joanna Lowry (Brighton: Photoworks, 2006), 65.

27 Margrit Shildrick, *Embodying the Monster: Encounters with the Vulnerable Self* (London: Sage Publications, 2002).

28 Shildrick, *Embodying the Monster*, 1.

29 Ibid., 2.

30 Lowry, 'Portraits, Still Video Portraits and the Account of the Soul', 65.

31 Ibid., 75.

32 Michel Foucault, *The Birth of the Clinic: An Archaeology of Medical Perception* (New York: Vintage Books, 1973), 114.

33 Foucault, *The Birth of the Clinic*, 95.
34 Ibid., 113.
35 Anon., 'Dental Surgery', *The Lancet* (26 December 1914), 1482; F. Newland-Pedley, 'The Dental Needs of the Army', *The Lancet* (7 August 1915), 305; Anon., 'Wounds of Jaw and Face', *The British Medical Journal* (25 March 1916): 457–8; Anon., 'Gunshot Injuries of the Jaws', *The Lancet* (24 June 1916): 1266–7.
36 Leon Frey, 'A Survey of Restorative Prosthesis in Fractures of the Jaws and Mutilations of the Face', *British Dental Journal* (War Supplement) 37.2 (1916): 7.
37 Marjorie Gehrhardt, *The Men with Broken Faces: Gueules Cassées of the First World War* (Oxford and New York: Peter Lang, 2015), 80.
38 V. H. Kazanjian, 'Immediate Treatment of Gunshot Fracture of the Jaws', *British Dental Journal* (War Supplement) 37.2 (1916): 22.
39 Kemp, '"Perfect and Faithful Record"', 120.
40 J. Lewin Payne, 'Discussion on War Injuries of the Jaws and Face', *British Dental Journal* (War Supplement) 37.5–6 (1916): 81.
41 George B. Hayes, 'Discussion on War Injuries of the Jaw and Face', *British Dental Journal* (War Supplement) 37.5–6 (1916): 82.
42 Charles H. Bubb, 'Some Technical Details in the Treatment of Jaw Injuries', *British Dental Journal* (War Supplement) 37.9 (1916): 370.
43 J. H. Badcock, 'Section on Odontology. President's Address', *Proceedings of the Royal Society of Medicine* 11.4 (1918): 1.
44 Badcock, 'Section on Odontology', 5.
45 Ibid., 4.
46 G. Northcroft, 'Section on Odontology. A Short Account of a Year's Work at One of the Jaw Injuries Centres of the London Command', *Proceedings of the Royal Society of Medicine* 11.3 (1918): 8.
47 Northcroft, 'Section on Odontology', 7.
48 Ibid., 16.
49 Biernoff, *Portraits of Violence*.
50 This is a similar conclusion that Julie M. Powell has come to in a recent study of personal trauma and the reassertion of masculine virtue in the constructed restorative narratives of shell shock victims in propaganda films from the First World War: 'Shock Troupe: Medical Film and the Performance of "Shell Shock" for the British Nation at War', *Social History of Medicine* 30.2 (2017): 323–45.
51 Gehrhardt, *Men with Broken Faces*, 121.
52 Anon. 'The Times Fund. Great New Hospital for the Wounded, Appeal for Equipment', *The Times* (21 November 1914): 11; Anon. 'The Times Sick and Wounded Fund. Total of £1, 171, 278. The New King George Hospital', *The Times* (13 August 1915): 13.
53 'Today The Times fund on behalf of the British Red Cross Society and the Order of St. John passes the "landmark" of eleven-and-a-half million pounds. The subscriptions acknowledged amount to £16, 623 18s, 7d, and the grand total of the Fund to date is increased to £11, 510, 242 7s, 2d': Anon. 'The Times Fund: Eleven-and-a-Half-Millions', *The Times* (23 September 1918): 12.
54 Peter Grant, *Philanthropy and Voluntary Action in the First World War* (London: Routledge, 2014), 132.
55 Grant, *Philanthropy and Voluntary Action*, 21.
56 This article comes from a scrapbook of newspaper cuttings relating to the Queen's Hospital, Sidcup, dating from 1917–30. Robert Blatchford, 'A Holstery [*sic*] of

Healing: I Visit a Place of Noble Men', *Sunday Chronicle* (June 1918): 46. London Metropolitan Archives, (H02/QM/Y/01/05).

57 Ibid., 46.

58 Suzannah Biernoff, 'Shame, Disgust and the Historiography of War', in *Shame and Sexuality: Psychoanalysis and Visual Culture*, ed. Clare Pajackowska and Ivan Ward (London: Routledge, 2008), 217.

59 Suzannah Biernoff, 'The Rhetoric of Disfigurement in First World War Britain', *Social History of Medicine* 24.3 (2011): 666.

60 Ward Muir, *The Happy Hospital* (London: Simpkin, Marshall, Hamilton, Kent and Co, Ltd., 1918), 143.

61 Muir, *The Happy Hospital*, 153.

62 Anon., 'Men Shattered in the War', *Evening Standard* (June 1918), 50. London Metropolitan Archives, (H02/QM/Y/01/05).

63 Sophie Delaporte, *Les Gueules Cassées: Les Blessés de la Face de la Grande Guerre* (Paris: Noêsis, 1996), 148.

64 Delaporte, *Les Gueules Cassées*, 57–8.

65 Anon., 'Men Shattered'.

66 Anon., 'Public Can Help. Fund for Queen's Hospital Mounting Up', *Evening Standard*, date and page number omitted. London Metropolitan Archives, (H02/QM/Y/01/05).

67 Gehrhardt, *Men with Broken Faces*, 114.

68 Anon., 'A Fund to Fight Depression and Fill Monotonous Days', *Evening Standard* (June 1918), page number omitted. London Metropolitan Archives, (H02/QM/Y/01/05).

69 Anon., 'The Men Behind the Veil', *The African World* (16 July 1921): 428. London Metropolitan Archives, (H02/QM/Y/01/05).

70 British Red Cross Museum and Archives, London, Acc 81/544: *Report by the Joint Committee and the Joint War Finance Committee of the British Red Cross Society and the Order of St John of Jerusalem in England in Voluntary Aid Rendered for the Sick and Wounded at Home and Abroad and to the British Pensioner of War, 1917–1919* (London, 1921), 249.

71 Shildrick, *Embodying the Monster*, 1.

72 Heather Laine Talley, *Saving Face: Disfigurement and the Politics of Appearance* (New York and London: New York University Press, 2014), 12.

73 Christopher Morton, 'The Ancestral Image in the Present Tense', *Photographies* 8.3 (2015): 254.

74 Personal Correspondence with Stephen Dewhurst, great-grandson of George William Butcher.

75 Joanna Sassoon, 'Photographic Materiality in the Age of Digital Reproduction', in *Photographs, Objects, Histories: On the Materiality of Images*, ed. Elizabeth Edwards and Janice Hart (London: Routledge, 2004), 194.

76 Sassoon, 'Photographic Materiality', 188.

77 Ibid., 190.

78 Ibid., 199.

79 Ibid., 188.

80 Elizabeth Edwards, 'Photographs and History: Emotion and Materiality', in *Museum Materialities: Objects, Engagements, Interpretations*, ed. Sandra H. Dudley (London: Routledge, 2010), 26.

81 Edwards, 'Photographs and History', 28.

82 Elizabeth Edwards, 'Entangled Documents: Visualized Histories', in *Susan Meiselas: In History*, ed. Kristen Lubben (New York: International Center of Photography, 2008), 335.
83 Edwards, 'Entangled Documents', 341.
84 Harriet Bradley, 'The Seductions of the Archive: Voices Lost and Found', *History of the Human Sciences* 12.2 (1999): 107.
85 Elizabeth Edwards and Janice Hart, eds, *Photographs, Objects, Histories: On the Materiality of Images* (London: Routledge, 2004), 1.
86 Edwards and Hart, *Photographs, Objects, Histories*, 5.
87 Christopher Pinney, 'Things Happen: Or, From Which Moment Does that Object Come?' in *Materiality*, ed. Daniel Miller (Durham and London: Duke University Press, 2005), 270.
88 Shildrick, *Embodying the Monster*, 4.
89 Ulrich Baer, *Spectral Evidence: The Photography of Trauma* (London: MIT Press, 2002), 2.
90 Baer, *Spectral Evidence*, 1.
91 Ibid., 1.
92 Ibid., 179.
93 John Taylor, *Body Horror: Photojournalism, Catastrophe and War* (New York: New York University Press, 1998), 29.
94 Ankersmit, *Sublime Historical Experience*, 9.
95 Munslow, *A History of History*, 162.
96 Ibid., 162.
97 Vilém Flusser, *Vilém Flusser: Writings*, ed. Andreas Ströhl, trans. Erik Eisel (Minneapolis: University of Minnesota Press, 2002), 128.
98 Flusser, *Vilém Flusser*, 128.
99 Ibid., 129.

Bibliography

Adamantius. *Physiognomy*. Trans. I. Repath. In *Seeing the Face, Seeing the Soul: Polemon's Physiognomy from Classical Antiquity to Medieval Islam*, edited by S. Swain, 487–547. Oxford: Oxford University Press, 2007.

Aeschylus. *Libation Bearers*. Trans. A. H. Sommerstein. Loeb Classical Library 146. Cambridge, MA: Harvard University Press, 2009.

Aëtius. *Sixteen Books on Medicine*. Trans. A. Olivieri. Berlin: Corpus Medicorum Graecorum, 1935.

Agamben, G. *Means Without End: Notes on Politics*. Trans. V. Binetti and C. Casarino. Minneapolis and London: University of Minnesota Press, 2000.

Aldersley-Williams, H., K. Arnold, M. Gordon, N. Kotsopulos, J. Peto and C. Wilkinson, eds. *Identity and Identification*. London: Black Dog Publishing, 2009.

Als, H. and T. B. Brazelton. 'Parenting an Infant with a Birth Defect: The Regulation of Self-Esteem'. *Psychoanalytic Study of the Child* 39 (1984): 561–89.

Anderson, T. 'A Medieval Example of Meningiomatous Hyperostosis'. *British Journal of Neurosurgery* 5.5 (1991): 499–504.

Anderson, T. and I. Hodgkins. 'Healed Cranial Weapon Injury from Medieval Coventry, England'. *Neurosurgery* 50.4 (2002): 870–3.

Anderson, C., O. P. John, D. Keltner and A. M. Kring. 'Who Attains Social Status? Effects of Personality and Physical Attractiveness in Social Groups'. *Journal of Personality and Social Psychology* 81.1 (2001): 116–32.

Anderson, T., C. Arcini, S. Anda, Å. Tangerud and G. Robertsen. 'Suspected Endemic Syphilis (Treponarid) in Sixteenth-Century Norway'. *Medical History* 30.3 (1986): 341–50.

Andrushko, V. A. and J. W. Verano. 'Prehistoric Trepanation in the Cuzco Region of Peru: A View into an Ancient Andean Practice'. *American Journal of Physical Anthropology* 137.1 (2008): 4–13.

Ankersmit, F. *Sublime Historical Experience*. Stanford: Stanford University Press, 2005.

Anon. *A True and Sad Reration [sic] of the Great and Bloudy Murder committed at Ratcliff in Stepney Parish neer the City of London, upon the body of John Hunter, a sea man, who was stabbed to the heart with a long knife, by one Mr. Smith and his wife and a young maid*. London: B.A., 1647.

Anon. *A True Narrative of the Proceedings at the Sessions-house in the Old-Bayly, at a sessions there held on April 25, and 26. 1677*. London: n.p., 1677.

Anon. *Charecters [sic] of some Young Women, belonging to the Changes, who are to be dispos'd of by way of Auction*. London: n.p., 1691.

Anon. 'Dental Surgery'. *The Lancet* 2 (1914): 1482.

Anon. *Five Merry Wives of Lambeth or, the Carpenter Cornuted*. London: R. Burton, 1680.

Anon. 'A Fund to Fight Depression and Fill Monotonous Days'. *Evening Standard* (June 1918): page number omitted. London Metropolitan Archives, (H02/QM/Y/01/05).

Anon. 'Gunshot Injuries of the Jaws'. *The Lancet* 1 (1916): 1266–7.

Anon. Maurizio Cattelan *Il Super Noi* http://www.laboralcentrodearte.org/en/recursos/obras/super-noi-torino-1996.

Anon. 'Men Shattered in the War'. *Evening Standard* (June 1918): 50.

Anon. *Nevvs from Hide-Parke: Or A Very Merry Strange Passage which Happened Betwixt a North-Country Gentleman, and a Very Gaudy Gallant Lady of Pleasure.* London: n.p., 1640–3.

Anon. 'Public Can Help. Fund for Queen's Hospital Mounting Up'. *Evening Standard*, date and page number omitted. London Metropolitan Archives, (H02/QM/Y/01/05).

Anon. 'The Men Behind the Veil'. *The African World* (16 July 1921): 428. London Metropolitan Archives, (H02/QM/Y/01/05).

Anon. 'The Times Fund: Eleven-and-a-Half-Millions'. *The Times* (23 September 1918): 12.

Anon. 'The Times Fund. Great New Hospital for the Wounded, Appeal for Equipment'. *The Times* (21 November 1914): 11.

Anon. 'The Times Sick and Wounded Fund. Total of £1, 171, 278. The New King George Hospital'. *The Times* (13 August 1915): 13.

Anon. 'The World's Most Successful Forensic Artist: How Lois Gibson's Incredible Sketches Have Helped Solve Thousands of Crimes'. Guinness World Records, 2017 http://www.guinnessworldrecords.com/news/2017/1/lois-gibson-facing-crime-with-world-records-458852.

Anon. 'Wounds of Jaw and Face'. *The British Medical Journal* 6.1 (1916): 457–8.

Anselment, R. *The Realms of Apollo: Literature and Healing in Seventeenth-Century England.* Newark: University of Delaware Press, 1995.

Apollodorus. *The Library*. Trans. J. G. Frazer. Loeb Classical Library 121 and122. Cambridge, MA: Harvard University Press, 1921.

Apuleius. *Metamorphosis*. Trans. J. A. Hanson. Loeb Classical Library 44. Cambridge, MA: Harvard University Press, 1996.

Ariès, P. *The Hour of Our Death.* Harmondsworth: Penguin Books, 1981.

Ariès, P. *Images of Man and Death*, Trans. J. Lloyd. Cambridge, MA: Harvard University Press, 1985.

Aristotle. *Categories*. Trans. H. P. Cooke and H. Tredennick. Loeb Classical Library 325. Cambridge, MA: Harvard University Press, 1938.

Aristotle. *Generation of Animals*. Trans. A. L. Peck. Loeb Classical Library 366. Cambridge, MA: Harvard University Press, 1942.

Aristotle. *History of Animals*. Trans. A. L. Peck. Loeb Classical Library 437. Cambridge, MA: Harvard University Press, 1965.

Aristotle. *Problems*. Trans. R. Mayhew. Loeb Classical Library 38. Cambridge, MA: Harvard University Press, 2011.

Athenaeus. *Dinner Sophists*. Trans. S. D. Olson. Loeb Classical Library 327. Cambridge, MA: Harvard University Press, 2010.

Ausonius. *Eclogues*. Trans. H. G. Evelyn-White. Loeb Classical Library 96. Cambridge, MA: Harvard University Press, 1919.

Ausonius. *Twelve Caesars*. Trans. H. G. Evelyn-White. Loeb Classical Library 96. Cambridge, MA: Harvard University Press, 1919.

Badcock, J. H. 'Section on Odontology. President's Address'. *Proceedings of the Royal Society of Medicine* 11.4 (1918): 1–5.

Baer, U. *Spectral Evidence: The Photography of Trauma.* London: MIT Press, 2002.

Bailey, C. 'Revisiting the Station Strangler cases'. IOL (Independent Online), 16 August 2005. http://www.iol.co.za/news/south-africa/revisiting-the-station-strangler-cases-250970 [accessed 15 January 2017].

Baker, N. *Plain Ugly: The Unattractive Body in Early Modern England.* Manchester: Manchester University Press, 2010.

Bar-Tal, D. and L. Saxe. 'Perceptions of Similarly and Dissimilarly Attractive Couples and Individuals'. *Journal of Personality and Social Psychology* 33.6 (1976): 772–81.

Barthes, R. *Camera Lucida: Reflections on Photography.* New York: Hill & Wang, 1981.

Bartley, M., D. Blane and S. Montgomery. 'Health and the Life Course: Why Safety Nets Matter'. *British Medical Journal* 314 (19 April 1997): 1194–6.

Barton, T. *Power and Knowledge: Astrology, Physiognomics, and Medicine under the Roman Empire.* Ann Arbor: University of Michigan Press, 1994.

Bayly, T. *Witty Apophthegms delivered at Several Times, and upon Several Occasions by King James, King Charls, the Marquess of Worcester, Francis Lord Bacon, and Sir Thomas Moor.* London: W. R. for Matthew Smelt, 1669.

BBC iWonder. 'Would Churchill's Depression Disqualify him as PM Today?' *BBC* website, http://www.bbc.co.uk/guides/zqjpsbk.

Beard, M. *Laughter in Ancient Rome: On Joking, Tickling, and Cracking Up.* Berkeley, Los Angeles and London: University of California Press, 2014.

Bennett, J. *Empathic Vision: Affect, Trauma and Contemporary Art. Series: Cultural Memory in the Present.* Stanford: Stanford University Press, 2005.

Bennett, J. 'Art, Affect, and the "Bad Death": Strategies for Communicating the Sense Memory of Loss'. *Signs* 28.1, Gender and Cultural Memory (Autumn 2002): 333–51.

Bennett, J. *Vibrant Matter: A Political Ecology of Things.* Durham: Duke University Press, 2010.

Berrebi, S. *The Shape of Evidence: Contemporary Art and the Document.* Amsterdam: Valiz, 2014.

Berry, D. S. 'Attractive Faces are not all Created Equal: Joint Effects of Facial Babyishness and Attractiveness on Social Perception'. *Personality and Social Psychology Bulletin* 17.5 (1991): 523–31.

Bettella, P. 'The Marked Body as Otherness in Renaissance Italian Culture'. In *A Cultural History of the Human Body in the Renaissance*, edited by L. Kalof and W. Bynum, 149–81. Oxford: Bloomsbury Academic, 2010.

Biernoff, S. 'Medical archives and digital culture'. *Photographies* 5.2 (2012): 179–202.

Biernoff, S. *Portraits of Violence: War and the Aesthetics of Disfigurement.* Ann Arbor: University of Michigan Press, 2017.

Biernoff, S. 'The Rhetoric of Disfigurement in First World War Britain'. *Social History of Medicine* 24.3 (2011): 666–85.

Biernoff, S. 'Shame, Disgust and the Historiography of War'. In *Shame and Sexuality: Psychoanalysis and Visual Culture*, edited by C. Pajackowska and I. Ward, 217–36. London: Routledge, 2008.

Blatchford, R. 'A Holstery of Healing'. *The Clarion* (June 1918), n.p.

Block, R. G. '"Is It Just Me?" Experiences of HIV-Related Stigma'. *Journal of HIV/AIDS and Social Services* 8 (2009): 1–19.

Bóo, F. L., M. A. Rossi and S. S. Urzúa. 'The Labor Market Return to an Attractive Face: Evidence from a Field Experiment'. *Economics Letters* 118.1 (2013): 170–2.

Borland, C. *The Dead Teach the Living: Selected Works 1990–1999.* Zürich, Amsterdam and Porto: Migros museum für geenwartskunst/De Appel/Fundação de Serralves.

Bourke, J. *The Story of Pain: From Prayer to Painkillers.* Oxford: Oxford University Press, 2014.

Bradley, H. 'The Seductions of the Archive: Voices Lost and Found'. *History of the Human Sciences* 12.2 (1999): 107–22.

Brady, C., ed. *A Viceroy's Vindication? Sir Henry Sidney's Memoir of Service in Ireland, 1556–78*. Cork: Cork University Press, 2002.

Bregman, J. *Synesius of Cyrene: Philosopher-Bishop*. Berkeley, Los Angeles and London: University of California Press, 1982.

Breton, A. 'Manifesto of Surrealism'. In *Manifestoes of Surrealism*, edited by A. Breton, trans. R. Seaver and H. R. Lane. Ann Arbor: University of Michigan Press, 1969 [1924].

Briggs, R. *Witches and Neighbours*. London: Harper Collins, 1996.

Broder, H. L., M. Wilson-Genderson and L. Sischo. 'Examination of a Theoretical Model for Oral Health Related Quality of Life Among Youths with Cleft'. *American Journal of Public Health* 104.5 (2014): 865.

Brofenbrenner, U. 'Ecology of the Family as Context for Human Development: Research Perspectives'. *Developmental Psychology* 22.6 (1986): 723–42.

Broomberg, A. and O. Chanarin. *Spirit is a Bone*. London: MACK, 2016.

Brothwell, D. 'Possible Evidence of a Cultural Practice Affecting Head Growth in Some Late Pleistocene East Asian and Australasian Populations'. *Journal of Archaeological Science* 2 (1975): 75–7.

Bruce, V., H. Ness, P. J. B. Hancock, C. Newman and J. Rarity. 'Combining Face Composites Yields improvements in Face Likeness'. *Journal of Applied Psychology* 87.5 (2002): 894–902.

Bubb, C. H. 'Some Technical Details in the Treatment of Jaw Injuries'. *British Dental Journal* (War Supplement) 37.9 (1916): 370–6.

Bull, R. 'The Psychological Significance of Facial Deformity'. In *Love and Attraction: An International Conference*. 1, 21–5. Oxford and Toronto: Pergamon Press, 1979.

Bull, R. and N. Rumsey. *The Social Psychology of Facial Appearance*. New York: Springer-Verlag, 1988.

Bunn, D. 'Sitting for the Postmortem Portrait'. *The Salon* 2 (2012): 1–4.

Butler, J. *Bodies That Matter*. London and New York: Routledge, 1993.

Butler, J. *Frames of War: When is Life Grievable?* London and New York: Verso, 2009.

Butler, Judith. *Gender Trouble*. New York and Oxford: Routledge, 1990.

Bychowski, G. 'The Ego of the Brain Wounded'. *Psychoanalytic Review* 36 (1949): 333–43.

Byock, J. 'The Skull and Bones in Egil's Saga: A Viking, a Grave, and Paget's Disease'. *Viator* 24 (1993): 23–50.

Byrne, R. '"Muse on my mirror": Gawain's Function in Relation to the Supernatural of Arthurian Literature'. *Innervate: Leading Undergraduate Work in English Studies* 3 (2010–11): 57–62.

Calame, J. 'Saving Face: Into the canny valley with the emoticon'. *Cabinet* issue 49 (Death), Spring 2013: 28–31.

Canter, H. V. 'Personal Appearance in the Biography of the Roman Emperors'. *Studies in Philology* 25.3 (1928): 385–99.

Carling, F. *And Yet We Are Human*. London: Chatto and Windus, 1962.

Carney, Jo Eldridge. *Fairy Tale Queens: Representations of Early Modern Queenship*. Basingstoke: Palgrave Macmillan, 2012.

Carrabine, E. 'Just Images: Aesthetics, Ethics and Visual Criminology'. *The British Journal of Criminology*, 52.3 (2012): 463–89.

Carré, J. M. and C. M. McCormick. 'In Your Face: Facial Metrics Predict Aggressive Behaviour in the Laboratory and in Varsity and Professional Hockey Players'. *Proceedings of the Royal Society of London B: Biological Sciences* 275.1651 (2008): 2651–6.

Carroll, R., and Latin America correspondent. 'Childlike Photo-Fit Leads Bolivian Police to Murderer'. *The Guardian*, 24 November 2009, section World news https://www.theguardian.com/world/2009/nov/24/bolivia-photo-fit-murderer [accessed 21 May 2017].

Cèbe, J.-P. 'Catulle et la physiognomie'. *Annales de la Faculté des Lettres d'Aix* 43 (1967): 174–8.

Celsus. *On Medicine*. Trans. W. G. Spencer. Loeb Classical Library 304. Cambridge, MA: Harvard University Press, 1938.

Cesarani, F., M. C. Martina, R. Gilletto, R. Boano, A. M. D. Roveri, V. Capussotto, A. Giuliano, M. Celia and G. Gandini. 'Facial Reconstruction of a Wrapped Egyption Mummy using MDCT'. *American Journal of Roentgenology* 183 (2004): 755–8.

Cicero. *Divisions of Oratory*. Trans. H. Rackham. Loeb Classical Library 349. Cambridge, MA: Harvard University Press, 1942.

Cicero. *On Invention*. Trans. H. M. Hubbell. Loeb Classical Library 386. Cambridge: Harvard University Press, 1949.

Cicero. *On the Laws*. Trans. C. W. Keyes. Cambridge, MA: Harvard University Press, 1928.

Cicero. *On the Orator*. Trans. E. W. Sutton and H. Rackham. Loeb Classical Library 348. Cambridge, MA: Harvard University Press, 1942.

Cicero. *Speech in Defence of Quintus Roscius the Comedian*. Trans. J. H. Freese. Loeb Classical Library 240. Cambridge, MA: Harvard University Press, 1930.

Claes, P., D. Vandermeulen, S. De Greef, G. Willems and P. Suetens. 'Craniofacial Reconstruction Using a Combined Statistical Model of Face Shape and Soft Tissue Depths: Methodology and Validation'. *Forensic Science International* 159S (2006) S147–58. doi:10.1016/j.forsciint.2006.02.035.

Claes, P., D. K. Liberton, K. Daniels, K. M. Rosana, E. E. Quillen, L. N. Pearson, B. McEvoy, M. Bauchet, A. A. Zaidi, W. Yao, H. Tang, G. S. Barsh, D. M. Absher, D. A. Puts, J. Rocha, S. Beleza, R. W. Pereira, G. Baynam, P. Suetens, D. Vandermeulen, J. K. Wagner, J. S. Boster, M. D. and Shriver. 'Modeling 3D Facial Shape from DNA'. *PLoS Genetics* 10.3 (2014): e1004224. doi:10.1371/journal.pgen.1004224.

Clark, J. *Looking at Laughter: Humor, Power and Transgression in Roman Visual Culture, 100 B.C.–A.D. 250*. Berkeley: University of California Press, 2007.

Clement, J. G. and M. K. Marks, eds. *Computer-Graphic Facial Reconstruction*. Cambridge, MA : Elsevier, 2005 [Kindle edition].

Clower, W. T. and S. Finger. 'Discovering Trepanation: The Contribution of Paul Broca'. *Neurosurgery* 49.6 (2001): 1417–26.

Cock, E. '"Lead[ing] 'em by the nose into publick shame and derision": Gaspare Tagliacozzi, Alexander Read and the Lost History of Plastic Surgery, 1600–1800'. *Social History of Medicine*, 28.1 (2015): 1–21.

Coen, S. J. 'The Sense of Defect'. *Journal of the American Psychoanalytical Association* 34 (1986): 47–67.

Cokayne, K. *Experiencing Old Age in Ancient Rome*. London and New York: Routledge, 2003.

Coleman, K. M. 'The Emperor Domitian and Literature'. *ANRW* 2nd series 32.5 (1986): 3087–115.

Connolly, M. '*The Awntyrs off Arthure*: Beyond the Tapestry'. *The Hiyoshi Review of English Studies* 60 (2012–13): 1–21.

Corbeill, A. *Controlling Laughter: Political Humour in the Late Roman Republic*. Princeton: Princeton University Press, 1996.

Corye, J. *The Generous Enemies, or, The Ridiculous Lovers a Comedy*. London: H. Lloyd for James Magnus, 1572.

Costello, D. and D. Willsdon, eds. *The Life and Death of Images: Ethics and Aesthetics*. London: Tate, 2008.

Coward, R. *Female Desire*. London: Paladin Books, 1984.

Craig, E. and G. Craig. 'The Diagnosis and Context of Facial Deformity from an Anglo-Saxon Cemetery at Spofforth, North Yorkshire'. *International Journal of Osteoarchaeology* 23.6 (2013): 631–9.

Craik, J. *The Face of Fashion: Cultural Studies in Fashion*. London: Routledge, 1993.

Cremers-van der Does, E. C. *The Agony of Fashion*. Poole: Blandford Press, 1980.

Cribiore, R. *Gymnastics of the Mind. Greek Education in Hellenistic and Roman Egypt*. Princeton: Princeton University Press, 2001.

Crist, T. A. J., A. Washburn, H. Park, I. Hood and M. A. Hickey. 'Cranial Bone Displacement as a Taphonomic Process in Potential Child Abuse Cases'. In *Forensic Taphonomy: The Postmortem Fate of Human Remains*, edited by W. Haglund and M. Sorg, 319–36. Boca Raton, Boston, London, New York and Washington DC: CRC Press, 1997.

Curtis, P. 'Researchers Find the Real Face of Father Christmas'. *The Guardian* (13 December 2004). https://wwwtheguardian.com/education/2004/dec/13/highereducation.uk [accessed 31 March 2017].

Dalal, F. 'The Social Unconscious and Ideology in Clinical Theory and Practice'. In *The Social Unconscious in Persons, Groups, and Societies Vol.1: Mainly Theory*, edited by E. Hopper and H. Weinberg, 243–63. London: Karnac, 2011.

Danks, D. M., V. Mayne, R. K. Hall and M. C. McKinnon. 'Frontometaphyseal Dysplasia: A Progressive Disease of Bone and Connective Tissue'. *American Journal of Diseases of Children* 123 (1972): 254–8.

Daubert Standard (definition). Legal Information Institute, Cornell University Law School. Available at https://www.law.cornell.edu/wex/daubert_standard [accessed 15 January 2017].

Dauphinée, E. 'The Politics of the Body in Pain: Reading the Ethics of Imagery'. *Security Dialogue* 38.2 (2007): 139–55.

Davis, L. J. *Enforcing Normalcy: Disability, Deafness and the Body*. London and New York: Verso, 1995.

Davison, A. '3D Facial Depiction of a Specimen Displaying Frontometaphyseal Dysplasia'. MSc Forensic Art thesis, University of Dundee, 2012.

Davy, S., T. Gilbert, D. Schofield and M. Evison. 'Forensic Facial Reconstruction Using Computer Modeling Software' Chapter 10 In *Computer-Graphic Facial Reconstruction*, edited by J. G. Clement and M. K. Marks. Cambridge, MA: Elsevier, 2005 [Kindle edition].

Davy-Jow, S. 'The Devil is in the Details: A Synthesis of Psychology of Facial Perception and its Applications in Forensic Facial Reconstruction'. *Science and Justice* 53 (2013): 230–35. doi:10.1016/j.scijus.2013.01.004.

De Beauvoir, S. *The Second Sex*, Trans. H. M. Parshley. London: New English Library, 1949.

De Greef, S., P. Claes, D. Vandermeulen, W. Mollemans, P. Suetens and G. Willems. 'Large-Scale in-vivo Caucasian Facial Soft Tissue Thickness Database for Craniofacial Reconstruction'. *Forensic Science International* 159S (2006): S126–46. doi:10.1016/j.forsciint.2006.02.034.

De Mello, F. *The Government of a Wife, or, Wholsom and Pleasant Advice for Married Men in a Letter to a Friend.* London: Jacob Tonson and R. Knaplock, 1697.

De Pascale, E. *Death and Resurrection in Art* (Guide to Imagery series). Los Angeles: Getty Publications, 2009.

Decker, S., J. Ford, S. Davy-Jow, P. Faraut, W. Neville and D. Hilbelink. 'Who is this Person? A Comparison Study of Current Three-Dimensional Facial Approximation Methods'. *Forensic Science International* 229 (2013): 161.e1–161.e8.

Delaporte, S. *Les Gueules Cassées: Les Blessés de la Face de la Grande Guerre*. Paris: Noêsis, 1996.

Deleuze, G. and F. Guattari. 'Year Zero: Faciality'. In *A Thousand Plateaus – Capitalism and Schizophrenia*, edited by G. Deleuze and F. Guattari, Trans. B. Massumi. Minneapolis and London: University of Minnesota Press, 1987.

Dennis, K. '"She was a Queen and therefore Beautiful": Sidney, his Mother and Queen Elizabeth'. *The Review of English Studies* 169 (1992): 18–39.

Derbyshire, D. 'Chariot Queen's Grave Unearthed'. *The Telegraph* (7 April 2001). http://www.telegraph.co.uk/news/uknews/1310654/Chariot-queens-grave-unearthed.html [accessed 31 March 2017].

Dewey-Hagborg, H. 'Stranger Visions'. http://deweyhagborg.com/projects/stranger-visions.

Dickinson, C. *Agamben and Theology.* London: Bloomsbury, 2011.

Dingwall, E. J. *Artificial Cranial Deformation*. London: John Bales, Sons and Danielson, 1931.

Dio, C. *Roman History*. Trans. E. Cary and H. B. Foster. Loeb Classical Library 176. Cambridge, MA: Harvard University Press, 1925.

Djurić-Srejić, M. and C. Roberts. 'Palaeopathological Evidence of Infectious Disease in Skeletal Populations from Later Medieval Serbia'. *International Journal of Osteoarchaeology* 11.5 (2001): 311–20.

Dolan, F. E. 'Taking the Pencil out of God's Hand: Art, Nature, and the Face-Painting Debate in Early Modern England'. *PMLA* 108.2 (1993): 224–39.

Domanska, E. 'Material Presence of the Past'. *History and Theory* 45.3 (2006): 337–48.

Douglas, M. *Purity and Danger: An Analysis of Concept of Pollution and Taboo*. London: Routledge, 1966.

Douzinas, C. and Nead, L., eds. *Law and the Image: The Authority of Art and the Aesthetics of Law*. Chicago: University of Chicago Press, 1999.

Draycott, J. 'Hair Today, Gone Tomorrow: The Use of Real, False and Artificial Hair as Votive Offerings'. In *Bodies of Evidence: The Anatomical Votive Past, Present and Future*, edited by J. Draycott and E.-J. Graham, 77–94. London: Routledge, 2017.

Draycott, J. 'The Lived Experience of Disability in Antiquity: A Case Study from Roman Egypt'. *Greece and Rome* 62.2 (2015): 189–205.

Draycott, J. 'Prosthetic Hair in Ancient Rome?' In *Prostheses in Antiquity*, edited by J. Draycott. London: Routledge, forthcoming.

Dubovsky, S. L. and S. E. Groban. 'Congenital Absence of Sensation'. *Psychoanalytic Study of the Child* 30 (1975): 49–73.

D'Urfey, T. *New Poems, Consisting of Satyrs, Elegies, and Odes Together with a Choice Collection of the Newest Court Songs Set to Musick.* London: J. Bullord and A. Roper, 1690.

Durston, C. 'Hewson, John, appointed Lord Hewson under the protectorate (*fl.* 1630–60)'. In *Oxford Dictionary of National Biography*, online edition, edited by D. Cannadine. Oxford: Oxford University Press, 2004. http://www.oxforddnb.com/view/article/13157.

Dyer, J. G. and T. M. McGuiness. 'Resilience: Analysis of the Concept'. *Archives of Psychiatric Nursing* 5 (1996): 276–82.

Earle, E. M. 'The Psychological Effects of Mutilating Surgery in Children and Adolescents'. *Psychoanalytic Study of the Child* 34 (1979): 527–46.

Eberstadt, F. 'Alien Girls'. *The New Yorker* 26 October 2015. http://www.newyorker.com/magazine/2015/10/26/alien-girls.

Eberstadt, F. 'Twisted Sisters'. *The New York Times Magazine* 6 July 2008. http://www.nytimes.com/2008/07/06/magazine/06cocorosie-t.html.

Eburne, J. *Surrealism and the Art of Crime*. Ithaca and London. Cornell University Press, 2008.

Edkins, J. *Face Politics* (Interventions Series). London and New York: Routledge, 2015.

Edkins, J. 'Forensic Identification'. In *Missing: Persons and Politics*. Ithaca and London: Cornell University Press, 2011.

Edwards, C. *The Politics of Immorality in Ancient Rome*. Cambridge: Cambridge University Press, 1993.

Edwards, E. 'Entangled Documents: Visualized Histories'. In *Susan Meiselas: In History*, edited by K. Lubben, 330–41. New York: International Center of Photography, 2008.

Edwards, E. 'Photographs and History: Emotion and Materiality'. In *Museum Materialities: Objects, Engagements, Interpretations*, edited by S. H. Dudley, 21–38. London: Routledge, 2010.

Edwards, E. 'Photography and the Material Performance of the Past'. *History and Theory* 48.4 (2009): 130–50.

Edwards, E. and J. Hart, eds. *Photographs, Objects, Histories: On the Materiality of Images*. London: Routledge, 2004.

Ekman, P. and W. C. Friesen. *Facial Action Coding System: A Technique for the Measurement of Facial Movement*. Palo Alto: Consulting Psychologists Press, 1978.

Ekman, P., W. C. Friesen and J. C. Hager. *Facial Action Coding System: The Manual on CD ROM*. Salt Lake City: A Human Face, 2002.

Elkins, J. *The Object Stares Back: On the Nature of Seeing*. New York: Harcourt Inc, 1996.

Elkins, J. *Picturing the Body: Pain and Metamorphosis*. Stanford: Stanford University Press, 1999.

English, A. *The Saint Who Would be Santa Claus: The True Life and Trials of Saint Nicholas of Myra*. Waco, TX: Baylor University Press, 2012.

Epictetus. *Discourses*. Trans. W. C. Oldfather. Loeb Classical Library 218. Cambridge: Harvard University Press, 1928.

Etcoff, N. *Survival of the Prettiest: The Science of Beauty*. New York: Doubleday, 1999.

Evans, E. 'Roman Descriptions of Personal Appearance in History and Biography'. *Harvard Studies in Classical Philology* 46 (1935): 43–84.

Fanshawe, A. *The Memoirs of Ann, Lady Fanshawe, wife of the Right Honble. Sir Richard Fanshawe, Bart., 1600–72*. London: John Lane The Bodley Head, 1907.

Farrimond, J. and M. Morris. 'Knowing or not Knowing Before Birth: Parents' Experiences of Having a Baby with a Cleft Lip, with or without Cleft Palate'. BSc. dissertation, University of the West of England, Bristol, UK, 2004.

Fass, P. S. 'Childhood and Memory'. *Journal of the History of Childhood and Youth* 3.2 (2010): 155–64.

Federer, W. J. *There Really is a Santa Claus: The History of St. Nicholas and Christmas Holiday Traditions*. St. Louis: Amerisearch, Inc., 2002.

Fedosyutkin, B. A. and J. V. Nainys. 'The Relationship of Skull Morphology to Facial Features'. In *Forensic Analysis of the Skull*, edited by M. Y. Iscan and R. P. Helmer. New York: Wiley-Liss, 1993.

Fein, S. G. 'The Ghoulish and the Ghastly: A Moral Aesthetic in Middle English Alliterative Verse'. *Modern Language Quarterly* 48 (1987): 3–19.

Feragen, K. B., I. L. Kvalem, N. Rumsey and I. H. Anne Borge. 'Adolescents with and without a Facial Difference: The Role of Friendships and Social Acceptance in Perceptions of Appearance and Emotional Resilience'. *Body Image* 7 (2010): 271–9.

Fernandes, C. M. S., M. da Costa Serra, J. V. Lopes da Silva, P. Y. Noritomi, F. C. Alencar de Sena Pereira and R. F. Haltenhoff Melani. 'Tests of One Brazilian Facial Reconstruction Method Using Three Soft Tissue Depth Sets and Familiar Assessors'. *Forensic Science International*, 214 (2012): 211.e1–211.e7.

Fibiger, L., T. Ahlström, P. Bennike and R. J. Schulting. 'Patterns of Violence-Related Skull Trauma in Neolithic Southern Scandinavia'. *American Journal of Physical Anthropology* 150.2 (2013): 190–202.

Finucane, B. C., K. Manning and M. Touré. 'Prehistoric Dental Modification in West Africa: Early Evidence from Karkarichinkat Nord, Mali'. *International Journal of Osteoarchaeology* 18.6 (2008): 632–40.

Finucci, V. *The Prince's Body: Vincenzo Gonzaga and Renaissance Medicine*. Cambridge, MA: Harvard University Press, 2015.

Fiorato, V., A. Boylston and C. Knüsel. *Blood Red Roses: The Archaeology of a Mass Grave from the Battle of Towton, AD1461*, 2nd edn. Oxford: Oxbow Books, 2007.

Fitzgerald, E. R. 'Postmortem Transition in the Dynamic Mechanical Properties of Bone'. *Medical Physics* 4 (1977): 49–53.

Flusser, V. *Vilém Flusser: Writings*, edited by A. Ströhl, Trans. E. Eisel. Minneapolis: University of Minnesota Press, 2002.

Foltyn, J. L. 'The Corpse in Contemporary Culture: Identifying, Transacting, and Recoding the Dead Body in the 21st Century'. *Mortality* 13.2 (2008): 99–104.

Forster, E. M. *A Passage to India*. London: Penguin, 1924 Reprint 1961.

Forth, S. 'Antenatal Diagnosis of a Cleft Lip and Palate: Supporting the Parents'. *Journal of Neonatal Nursing* 9 (2003): 57–61.

Foucault, M. *The Birth of the Clinic: An Archaeology of Medical Perception*. New York: Vintage Books, 1973.

Fox, C. *Pattern and Purpose: A Survey of Early Celtic Art in Britain*. Cardiff: National Museum of Wales, 1958.

Frances, J. 'Damaged or Unusual Bodies: Staring, or Seeing and Feeling'. *Body, Movement and Dance in Psychotherapy: An International Journal for Theory, Research and Practice* 9.4 (2014): 198–210. doi:10.1080/17432979.2014.931887.

Freedberg, D. 'Feelings on Faces. From Physiognomics to Neuroscience' In *Rethinking Emotion: Interiority and Exteriority in Premodern, Modern, and Contemporary Thought*, edited by R. Campe and J. Weber. Berlin-Boston: Walter de Gruyter GmbH & Co KG, 2014.

Freedberg, D. *The Power of Images: Studies in the History and Theory of Response*. Chicago and London: University of Chicago Press, 1989.

Freedberg, D., and V. Gallese. 'Motion, Emotion and Empathy in Esthetic Experience'. *Trends in Cognitive Sciences* 11.5 (2007): 197–203.

Frey, L. 'A Survey of Restorative Prosthesis in Fractures of the Jaws and Mutilations of the Face'. *British Dental Journal* (War Supplement) 37.2 (1916), 7–11.

Friend, T. 'Could this be the Profile of a Queen?'. *Race and History Forum, USA Today* (15 August 2003). http://www.raceandhistory.com/cgi-bin/forum/webbbs_config.pl?md=read;id=1368 [accessed 31 March 2017].

Frowd, C. D., P. J. B. Hancock and D. Carson. 'EvoFIT: A Holistic, Evolutionary Facial Imaging Technique for Creating Composites'. *ACM Transactions on applied perception (TAP)*, 1.1 (2004): 19–39.

Frowd, C. D., V. Bruce, D. Ross, A. McIntyre and P. J. B. Hancock. 'An Application of Caricature: How to Improve the Recognition of Facial Composites'. *Visual Cognition*, 15 (2007): 954–84.

Frowd, C. D., P. J. B. Hancock, V. Bruce, F. C. Skelton, C. J. Atherton, L. Nelson, A. H. McIntyre, M. Pitchford, R. Atkins, D. C. I. Webster and J. Pollard. 'Catching More Offenders with EvoFIT Facial Composites: Lab Research and Police Field Trials'. *Global Journal of Human-Social Science* 11.3 (2011): 35–46.

Galen. *Claudii Galeni Opera Omnia*. Volumes 12 and 13. Trans. K. G. Kühn. Cambridge: Cambridge University Press, 2011.

Galen. *Mixtures*. Trans. P. N. Singer. Oxford: Oxford University Press, 1997.

Garland, R. 'The Mockery of the Deformed and the Disabled in Graeco-Roman Culture'. In *Laughter Down the Centuries*, edited by S. Jäkel and A. Timonen, 71–84. Turku: Turun Yliopisto, 1994.

Garland-Thomson, R. 'The Politics of Staring: Visual Rhetorics of Disability in Popular Photography'. In *Disability Studies: Enabling the Humanities*, edited by S. L. Snyder, B. J. Bruggemann and R. Garland-Thomson, 56–75. New York: Modern Language Association of America, 2002.

Garland-Thomson, R. *Staring: How We Look*. New York and Oxford: Oxford University Press, 2009.

Gehrhardt, M. *The Men with Broken Faces: Gueules Cassées of the First World War*. Oxford and New York: Peter Lang, 2015.

Gellius, A. *Attic Nights*. Trans. J. C. Rolfe. Loeb Classical Library 200. Cambridge, MA: Harvard University Press, 1927.

George, R. M. 'Anatomical and Artistic Guidelines for Forensic Facial Reconstruction'. In *Forensic Analysis of the Skull*, edited by M. Y. Iscan and R. P. Helmer. New York: Wiley-Liss, 1993.

Gerasimov, M. M. *The Face Finder*. London: Hutchinson, 1971.

Gilman, S. L. *Disease and Representation: Images of Illness from Madness to AIDS*. Ithaca: Cornell University Press, 1988.

Gilman, S. L. *Health and Illness: Images of Difference*. London: Reaktion, 1995.

Gilman, S. L. *Making the Body Beautiful: A Cultural History of Aesthetic Surgery*. Princeton: Princeton University Press, 1999.

Gladhill, B. 'The Emperor's No Clothes: Suetonius and the Dynamics of Corporeal Ecphrasis'. *Classical Antiquity* 31.2 (2012): 315–48.

Gleason, M. *Making Men: Sophists and Self-presentation in Ancient Rome*. Princeton: Princeton University Press, 1995.

Gleason, M. 'The Semiotics of Gender: Physiognomy and Self-Fashioning in the Second Century AD'. In *Before Sexuality*, edited by D. Halperin, J. Winkler and F. Zeitlin, 389–415. Princeton: Princeton University Press, 1991.

Goffman, E. *Stigma: Notes on the Management of Spoiled Identity*. Harmondsworth: Penguin, 1963.

Goldhill, S. 'Rhetoric and the Second Sophistic'. In *The Cambridge Companion to Ancient Rhetoric*, edited by E. Gunderson, 228–44. Cambridge: Cambridge University Press, 2009.

Goodley, D. *Disability Studies: An Interdisciplinary Introduction*. London: Sage, 2011.

Goodley, D. *Dis/ability Studies: Theorising Disableism and Ableism*. London and New York: Routledge, 2014.
Goodman, J. R. *Chivalry and Exploration 1298–1630*. Woodbridge: Boydell and Brewer, 1998.
Gorlin, R. J. and M. M. Cohen Jr. 'Frontometaphyseal Dysplasia: A New Syndrome'. *American Journal of Diseases of Children* 118 (1969): 487–94.
Gould, R. *A Poem Most Humbly Offered to the Memory of Her Late Sacred Majesty, Queen Mary*. London: Jacob Tonson and Francis Saunders, 1695.
Gowing, L. 'Gender and the Language of Insult in Early Modern London'. *History Workshop* 35 (1993): 1–21.
Gowing, L. 'Language, Power, and the Law: Women's Slander Litigation in Early Modern London'. In *Women, Crime and the Courts in Early Modern England*, edited by J. Kermode and G. Walker, 26–47. London: Routledge, 1994.
Gowing, L. 'Marked Bodies and Social Meanings'. In *A Cultural History of the Human Body in the Age of Enlightenment*, edited by C. Reeves, 133–53. Oxford: Bloomsbury Academic, 2010.
Gowing, L. 'Women, Status, and the Popular Culture of Dishonour'. *Transactions of the Royal Historical Society* 6th series 6 (1996): 225–34.
Grant, P. *Philanthropy and Voluntary Action in the First World War*. London: Routledge, 2014.
Grealy, L. *Autobiography of a Face*. New York: Harper Perennial, 1994.
Greek Anthology. Trans. W. R. Paton. Loeb Classical Library 85. Cambridge, MA: Harvard University Press, 1918.
Green, R. *A Disputation, Betweene a Hee Conny-catcher, and a Shee Conny-catcher Whether a Theefe or a Whoore, is Most Hurtfull in Cousonage, to the Common-wealth*. London: A. Jeffes for T. Gubbin, 1592.
Gregersen, M., J. Boldsen, H. Björn, L. W. Boel and P. Fromholt. 'Examination and Identification of a Danish 17th-Century Nobleman, Laurids Ebbesen: A Multi-Disciplinary Study'. *Forensic Science and Medical Pathology* 2 (2006): 51–8.
Greville, F. *The Life of the Renowned Sr Philip Sidney. With the true interest of England as it then stood in relation to all forrain princes*. London: Henry Seile, 1651.
Grmek, M. D. and D. Gourevitch. *Les maladies dans l'art antique*. Paris: Fayard, 1998.
Groebner, V. *Defaced: The Visual Culture of Violence in the Late Middle Ages*. Trans. P. Selwyn. New York: MIT Press, 2004.
Groebner, V. 'Losing Face, Saving Face: Noses and Honour in the Late Medieval Town'. *History Workshop Journal* 40 (1995): 1–15.
Grue, J. 'Interdependent Discourses of Disability: A Critical Analysis of the Social/Medical Model Dichotomy'. PhD thesis, University of Oslo, 2011.
Gunderson, E. *Staging Masculinity: The Rhetoric of Performance in the Roman World*. Ann Arbor: University of Michigan Press, 2000.
Hacking, P. 'A Pituitary Tumour in a Medieval Skull'. *International Journal of Osteoarchaeology* 5.4 (1995): 390–3.
Hadi, H. and C. M. Wilkinson. 'The Post-mortem Resilience of Facial Creases and the Possibility for use in Identification of the Dead'. *Forensic Science International* 237 (2014): 149–e1.
Haglund, W. 'Forensic "Art" in Human Identification'. In *Craniofacial Identification in Forensic Medicine*, edited by G. J. Clement and D. L. Ranson. Arnold: London, 1998.
Haglund, W. D. and D. T. Reay. 'Use of Facial Approximation Techniques in Identification of Green River Serial Murder Victims'. *The American Journal of Forensic Medicine and Pathology* 12.2 (1991): 132.

Hahn, T., ed. *Sir Gawain: Eleven Romances and Tales*. Kalamazoo, MI: Medieval Institute Publications, 1995.

Hall, S. 'The Spectacle of the "Other"'. In *Representation: Cultural Representations and Signifying Practices*, edited by S. Hall, 225–79. London: Sage, 1997.

Hammond, V., A. Fletcher, J. Hunt and S. Channon. 'Mothers' Experiences of Cleft Repair Surgery for their Babies born with a Cleft Lip and/or Palate'. Paper presented at the Cranio-Facial Society of Great Britain and Ireland, Annual Conference, Bristol, April 25th–27th, 2012.

Handwerk, B. 'King Tut's New Face'. *National Geographic News* (11 May 2005). http://news/nationalgeographic.com/news/2005/05/photogalleries/tut_mummy/photo4.html [accessed 31 March 2017].

Handwerk, B. 'St. Nicholas to Santa: The Surprising Origins of Mr. Claus'. *National Geographic* (20 December 2013) http://news.nationalgeographic.com/news/2013/12/131219-santa-claus-origin-history-christmas-facts-st-nicholas/ [accessed 31 March 2017].

Hanna, R. '*The Awntyrs off Arthure*: An Interpretation'. *Modern Language Quarterly* 31 (1970): 275–97.

Hansen, A. 'The Characterization of Monstrous Femininity in the *Testament of Cresseid* and *The Awntyrs off Arthure*'. MA diss., Queen's University, Toronto, 2009.

Haraway, D. 'A Cyborg Manifesto: Science, Technology and Socialist-Feminism in the Late Twentieth Century'. In *The Cybercultures Reader*, edited by D. Bell and B. M. Kennedy, 291–324. London and New York: Routledge, 1991.

Harms, S. and A. S. Schutz. Simons versus The State, (195/96) 1998, ZASCA (27 March 1998). http://www.saflii.org/za/cases/ZASCA/1998/31.html.

Haught, L. 'Ghostly Mothers and Fated Fathers: Gender and Genre in the *Awntyrs off Arthure*'. *Arthuriana* 20.1 (2010): 3–22.

Havstam, C., K. Laakso and K. C. Ringsberg. 'Making Sense of the Cleft. Young Adults' Accounts of Growing Up with a Cleft and Deviant Speech'. *Journal of Health Psychology* 16.1 (2011): 22–30.

Hayes, G. B. 'Discussion on War Injuries of the Jaw and Face'. *British Dental Journal* (War Supplement) 37.5–6 (1916), 80–2.

Hayes, S. and C. Tullberg. 'Police Witness Identification Images: A Geometric Morphometric Analysis* Police Witness Identification Images: A Geometric Morphometric Analysis'. *Journal of Forensic Sciences (Wiley-Blackwell)*, 57 (2012): 1487–94. https://doi.org/10.1111/j.1556-4029.2012.02168.x.

Hayes, S. and N. Milne. 'What's Wrong with This Picture? An Experiment in Quantifying Accuracy in 2D Portrait Drawing'. *Journal of Visual Communication* 10 (2011): 149–74. https://doi.org/10.1177/1470357211398442.

Health Experiences Research Group. *Researcher's Handbook Healthtalkonline and Youthtalk Modules, Version 27*. Oxford: University of Oxford, 2012.

Healthtalk.org. 'Overview'. http://www.healthtalk.org/about/overview.

Helmer, R. P., S. Rohricht, D. Petersen and F. Mohr. 'Assessment of the Reliability of Facial Reconstruction'. In *Forensic Analysis of the Skull*, edited by M. Y. Iscan and R. P. Helmer. New York: Wiley-Liss, 1993.

Henderson, K. U. and B. F. McManus. *Half Humankind: Contexts and Texts of the Controversy about Women in England 1540–1640*. Urbana: University of Illinois Press, 1985.

Henen, T. 'The Portrait Machine'. http://portraitmachine.blogspot.co.uk/.

Herodotus. *Histories*. Trans. A. D. Godley. Loeb Classical Library 119. Cambridge, MA: Harvard University Press, 1922.

Herzig, R. *Plucked: A History of Hair Removal*. New York and London: New York University Press, 2015.

Higgins, J. *Seeing Death: Portraiture in Contemporary Postmortem Photography*. Unpublished MAFA thesis. University of Cape Town, 2013.

Higgins, J. 'Subject and Object in Pieter Hugo's The Bereaved'. *Mortality* 21.4 (2014): 357–77. doi:10.1080/13576275.2016.1150977.

Higgins, J., K. Smith and P. Siopis. *Between Subject and Object: Human Remains at the Interface of Art and Science*. Exhibition catalogue, 2014. Available online at https://www.academia.edu/9826993/Between_Subject_and_Object_human_remains_at_the_interface_of_art_and_science.

Higonnet, A. *Pictures of Innocence: The History and Crisis of Ideal Childhood*. London: Thames and Hudson, 1998.

Hill, J. D. 'Wetwang Chariot Burial'. *Current Archaeology* 15.178 (2002): 410–12.

Hippocrates. *Aphorisms*. Trans. W. H. S. Jones. Loeb Classical Library 150. Cambridge, MA: Harvard University Press, 1931.

Hippocrates. *Nature of the Child*. Trans. P. Potter. Loeb Classical Library 520. Cambridge, MA: Harvard University Press, 2012.

His, W. *Johann Sebastian Bach: Forschungen uber dessen Brabstatte, Bebeine und Antlitz*. Bericht an den Rath der Stadt Leipzig. Leipzig: F. C. W. Vogel, 1895.

Hjorztsjö, C. H. *Man's Face and Mimic Language*. Trans. W. F. Salisbury. Lund: Studentlitteratur, 1969/1970. Available: http://diglib.uibk.ac.at/download/pdf/782346?name=Man's%20face%20and%20mimic%20language [accessed 30 March 2017].

Hole, The. 'Future Feminism'. The Hole NYC 2014. http://theholenyc.com/2014/08/15/future-feminism-2/.

Homer. *Iliad*. Trans. A. T. Murray. Loeb Classical Library 170. Cambridge, MA: Harvard University Press, 1924.

Hopper, E. 'The Social Unconscious in Clinical Work' [1996]. In *The Social Unconscious – Selected Papers*, edited by E. Hopper, 126–61. London: Jessica Kingsley, 2003.

Houston, V. and R. Bull. 'Do People Avoid Sitting Next to Someone who is Facially Disfigured?' *European Journal of Social Psychology* 24.2 (1994): 279–84.

Hughes, I. 'Is this the Real Face of Santa Claus? Presenting St Nicholas, the Inspiration for Father Christmas'. *The Mirror* (5 December 2015). http://www.mirror.co.uk/news/uk-news/real-face-santa-claus-presenting-4755042.

Hunt, O., D. Burden, P. Hepper and C. Johnston. 'The Psychosocial Effects of Cleft Lip and Palate: A Systematic Review'. *European Journal of Orthodontics* 27 (2005): 274–85.

Hunt, O., D. Burden, P. Hepper, M. Stevenson and C. Johnston. 'Parents' Reports of Psychosocial Functioning in Children with Cleft Lip and/or Palate'. *Cleft Palate-Craniofacial Journal* 44.3 (2007): 304–11.

Hunt, O., D. Burden, P. Hepper, M. Stevenson and C. Johnston. 'Self-reports of Psychosocial Functioning among Children and Young Adults with Cleft Lip and Palate'. *Cleft Palate-Craniofacial Journal* 43 (2006): 598–605.

Hyginus. *Fables*. Trans. M. Grant. Lawrence: University of Kansas Press, 1960.

Imrie, R. 'Disability and Discourses of Mobility and Movement'. *Environment and Planning* 32 (2000): 1641–56.

Ingham, P. *Sovereign Fantasies: Arthurian Romance and the Making of Britain.* Philadelphia: University of Pennsylvania Press, 2001.

Jacoby, A. 'Felt versus Enacted Stigma: A Concept Revisited: Evidence from a Study of People with Epilepsy in Remission'. *Social Science and Medicine* 38.2 (1994): 269–74.

Jeamson, T. *Artificiall Embellishments, or Arts best directions how to preserve beauty or procure it*. Oxford: William Hall, 1665.

Jesani, A., A. T. DiBase, M. T. Cobourne and T. Newton. 'Perceived Changes by Peer Group of Social Impact Associated with Combined Orthodontic-Surgical Correction of Class III Malocclusion'. *Journal of Dentistry* 42.9 (2014): 1135–42.

Johnston, R. A. and A. J. Edmonds. 'Familiar and Unfamiliar Face Recognition: A Review'. *Memory* 17.5 (2009): 577–96. doi:10.1080/0965821090297696.

Jones, C. W. *St Nicholas of Myra, Bari and Manhattan: Biography of a Legend*. Chicago: University of Chicago Press, 1988.

Jones, B. W. *The Emperor Domitian*. London: Routledge, 1992.

Jost, J. E. 'Margins in Middle English Romance: Culture and Characterization in *The Awntyrs off Arthure at the Tarn Wathelyne* and *The Wedding of Sir Gawain and Dame Ragnell*'. In *Meeting the Foreign in the Middle Ages*, edited by A. Classen, 133–52. New York and London: Routledge, 2002.

Joy, J. *Iron Age Mirrors: A Biographical Approach*. Oxford: Archaeopress, 2010.

Judd, M. A. 'One Accident Too Many'. *British Museum Studies in Ancient Egypt and Sudan* 3 (2002): 42–54.

Jurda, M., P. Urbanová and M. Králík. 'The Post-Mortem Pressure Distortion of Human Crania Uncovered in an Early Medieval Pohansko (Czech Republic) Graveyard'. *International Journal of Osteology* 25.4 (2015): 539–49.

Juvenal. *Satires*. Trans. S. M. Braund. Loeb Classical Library 91. Cambridge, MA: Harvard University Press, 2004.

Kaczorowski, J. 'The Good, the Average and the Ugly: Socio-Economic Dimensions of Physical Attractiveness'. MA Thesis, Dept of Sociology, McGill University, Montreal, 1989.

Kanz, F. and K. Grossschmidt. 'Head Injuries of Roman Gladiators'. *Forensic Science International* 160.2 (2006): 207–16.

Karim-Cooper, F. *Cosmetics in Shakespearean and Renaissance Drama*. Edinburgh: Edinburgh University Press, 2006.

Karim-Cooper, F. 'Fashioning the Face: Embodiment and Desire in Early Modern Poetry'. In *Shakespeare and the Power of the Face*, edited by J. A. Knapp, 29–42. Farnham: Routledge, 2015.

Kaufman, M. H., D. Whitaker and J. McTavish. 'Differential Diagnosis of Holes in the Calvarium: Application of Modern Clinical Data to Paleopathology'. *Journal of Archaeological Science* 24 (1997): 193–218.

Kazanjian, V. H. 'Immediate Treatment of Gunshot Fracture of the Jaws'. *British Dental Journal* (War Supplement) 37.2 (1916): 297–302.

Keary, A. M. 'Do We? Can We Look at the Disabled Body?' In *Contemporary Body Psychotherapy: The Chiron Approach*, edited by L. Hartley, 164–76. London: Routledge, 2009.

Keenan, T. 'Counter-forensics and Photography'. *Grey Room* 55 (2014): 58–77.

Keenan, T. 'Getting the Dead to Tell me what Happened: Justice, Prosopopoeia, and Forensic Afterlives'. Forensic Architecture (ed.), *Forensis: The Architecture of Public Truth*, 35–55. Berlin: Sternberg Press, 2014.

Keenan, T. and E. Weizman. *Mengele's Skull: The Advent of a Forensic Aesthetic*. Berlin and Frankfurt: Sternberg Press/Portikus, 2012.

Kellett, P. 'Advocacy in Anthropology: Active Engagement or Passive Scholarship?' *Durham Anthropology Journal* 16.1 (2009): 22–31, https://community.dur.ac.uk/anthropology.journal/vol16/iss1/kellett.pdf.

Kelso, R. *Doctrine for the Lady of the Renaissance*. Urbana: Illinois University Press, 1956.

Kemp, M. '"A Perfect and Faithful Record": Mind and Body in Medical Photography Before 1900'. In *Beauty of Another Order: Photography in Science*, edited by A. Thomas, 120–49. New Haven: Yale University Press, 1997.

Kerr, W. J. S. and J. M. O'Donnell. 'Panel Perception of Facial Attractiveness'. *British Journal of Orthodontics* 17.4 (1990): 299–304.

Kilgore, L., R. Jurmain and D. Van Gerven. 'Palaeoepidemiological Patterns of Trauma in a Medieval Nubian Skeletal Population'. *International Journal of Osteoarchaeology* 7.2 (1997): 103–14.

Kirkup, G. 'Introduction to Part One'. In *The Gendered Cyborg: A Reader*, edited by G. Kirkup, L. Janes, K. Woodward and F. Hovenden, 3–10. London and New York: Routledge, 2000.

Kjellström, A. 'A Sixteenth-Century Warrior Grave from Uppsala, Sweden: The Battle of Good Friday'. *International Journal of Osteoarchaeology* 15.1 (2005): 23–50.

Klausner, D. N. 'Exempla and *The Awntyrs off Arthure*'. *Mediaeval Studies* 34 (1972): 307–25.

Klaver, E., ed. *Images of the Corpse: From the Renaissance to Cyberspace*. Madison, WI: The University of Wisconsin Press, 2004.

Klaver, E. *Sites of Autopsy in Contemporary Culture* (SUNY Series in Postmodern Culture). New York: State University of New York, 2005.

Klein, G. *Reading into Racism: Bias in Children's Literature and Learning Materials*. London: Routledge and Kegan Paul, 1985.

Klum, S., H. Han, A. K. Jain and B. Klare. 'Sketch Based Face Recognition: Forensic vs Composite Sketches'. *International Conference on Biometrics* (2013): 1–8.

Korhonen, A. 'To See and to be Seen: Beauty in the Early Modern London Street'. In *Cultural History of Early Modern European Streets*, edited by R. Laitinen and T. V. Cohen, 141–66. Leiden: Brill, 2009.

Korhonen, A. 'Strange Things Out of Hair: Baldness and Masculinity in Early Modern England'. *The Sixteenth Century Journal* 41.2 (2010): 371–91.

Kramer, F.-J., C. Baethge, B. Sinikovic and H. Schliephake. 'An Analysis of Quality of Life in 130 Families having Small Children with Cleft Lip/palate using the Impact on Family Scale'. *International Journal of Oral and Maxillofacial Surgery* 36.12 (2007): 1146–52.

Kristeva, J. *Powers of Horror: An Essay on Abjection*. Trans. L. S. Roudiez. New York: Columbia University Press, 1982.

La Guardia, J. J. 'Developing Who I am: Self-Determination Theory Approach to the Establishment of Healthy Identities'. *Educational Psychologist* 44.2 (2009): 90–104.

Lacan, J. *Écrits: A Selection*. Trans. A. Sheridan. London: Routledge, 1996.

Laine Talley, H. *Saving Face: Disfigurement and the Politics of Appearance*. New York and London: New York University Press, 2014.

Laing, M. 'The *Lula girl* as "Sublime and Childlike": Nostalgic Investments in Contemporary Fashion Magazines'. *Critical Studies in Fashion and Beauty* 5.2 (2014): 271–93.

Larson, R., M. Richards, G. Moneta, G. Holmbeck and E. Duckett. 'Changes in Adolescents' Daily Interactions with their Families from Ages 10–18: Disengagement and Transformation'. *Developmental Psychology* 32 (1996): 744–5.

Lee, W. J., C. M. Wilkinson and H. S. Hwang. 'An Accuracy Assessment of Forensic Computerized Facial Reconstruction Employing Cone-Beam Computed tomography from Live Subjects'. *Journal of Forensic Sciences* 57.2 (2012): 318–27.

Lee, W., S. Mackenzie and C. Wilkinson. 'Facial Identification of the Dead' In *Forensic Anthropology 2000–10*, edited by S. Black and E. Ferguson. Boca Raton: Taylor & Francis, 2011.

Lehnert, G. 'Gender', In *Berg Encyclopedia of World Dress and Fashion*, Volume 8, West Europe. Oxford: Berg Fashion Library, 2010. doi:http://dx.doi.org/10.2752/BEWDF/EDch8075.

Lesnik-Oberstein, K. 'The Last Taboo: Women, Body Hair and Feminism'. In *The Last Taboo: Women and Body Hair*, edited by K. Lesnik-Oberstein. Manchester: Manchester University Press, 2006.

Levin, C. and C. Medici. 'Lady Mary Dudley Sidney (c.1531–86) and her Siblings'. In *The Ashgate Research Companion to the Sidneys, 1500–1700*, 2 vols, edited by M. P. Hannay, M. G. Brennan and M. E. Lamb, i, 31–40. Farnham: Ashgate, 2015.

Levinas, E. 'The Face'. In *Ethics and Infinity: Conversations with Philippe Nemo*. Pittsburgh: Duquesne University Press, 1985.

Levinas, E. 'Section III – Exteriority and The Face'. In *Totality and Infinity*. The Hague, Boston and London: Martinus Nijhoff Publishers, 1979.

Levine, M. M. 'The Gendered Grammar of Ancient Mediterranean Hair'. In *Off with Her Head! The Denial of Women's Identity in Myth, Religion, and Culture*, edited by H. Eilberg-Schwartz and W. Doniger, 76–130. Berkeley: University of California Press, 1995.

Lindauer, M. A. *Devouring Frida: The Art History and Popular Celebrity of Frida Kahlo*. Hanover and London: University Press of New England, 1999.

Lindsay, R., D. Ross, J. D. Read and M. P. Toglia, eds. *Handbook of Eyewitness Psychology (Vol. 2): Memory for People*. Mahwah, NJ: Lawrence Erlbaum & Associates, 2007.

Littleton, A. *A Sermon at the Funeral of the Right Honourable the Lady Jane Eldest Daughter to His Grace, William, Duke of Newcastle, and Wife to the Honourable Charles Cheyne, Esq*. London: John Macock, 1669.

Lockhart, E. 'Mental Health Needs of Children and Adolescents with Cleft lip and/or Palate'. *Clinical Child Psychology and Psychiatry* 8 (2003): 7–16.

Lomazzo, G. P. *A Tracte containing the Artes of Curious Paintinge Caruinge Buildinge*. Oxford: n.p., 1598.

Lowe, V. 'Folklore as a Unifying Factor in *The Awntyrs off Arthure*'. *Folklore Forum* 13 (1980): 199–223.

Lowry, J. 'Portraits, Still Video Portraits and the Account of the Soul'. In *Stillness and Time: Photography and the Moving Image*, edited by D. Green and J. Lowry, 65–78. Brighton: Photoworks, 2006.

Luthar, S. S., D. Cicchetti and B. Becker. 'The Construct of Resilience: A Critical Evaluation and Guidelines for Future Work'. *Child Development* 71.3 (2000): 543–62.

Lynnerup, N. 'Medical Imaging of Mummies and Bog Bodies, a Mini-Review'. *Gerontology* 56 (2010): 441–8.

MacCormack, Patricia. 'Faciality'. Paper presented at the 4th European Feminist Research Conference, Bologna, Italy. 2000. Online: http://orlando.women.it/cyberarchive/files/mac-cormack.htm [accessed 7 October 2015].

Macgregor, F. 'Facial Disfigurement: Problems and Management of Social Interaction and Implications for Mental Health'. *Aesthetic Plastic Surgery* 14.4 (1990): 249–57.

Macgregor, F. , T. M. Abe, A. Bry, E. Lave and S. Weissmann. *Facial Deformities and Plastic Surgery*. Springfield, IL: Thomas, 1953.

Macleod, I. B. and B. Hill. *Heads and Tales: Reconstructing Faces*. Edinburgh: National Museums of Scotland, 2001.

MacLin, O. H. and M. K. MacLin. 'The Effect of Criminality on Face Attractiveness, Typicality, Memorability and Recognition'. *North American Journal of Psychology* 6.1 (2004): 145–54.

Macrobius. *Saturnalia*. Trans. R. A. Kaster. Loeb Classical Library 510–2. Cambridge, MA: Harvard University Press, 2011.

Mahoney, G. and C. Wilkinson. 'Computer-Generated Facial Depiction'. In *Craniofacial Identification*, edited by C. Wilkinson and C. Rynn. Cambridge: Cambridge University Press, 2012 [Kindle edition].

Maples, W. R. 'Trauma Analysis by the Forensic Anthropologist'. In *Forensic Osteology: Advances in the Identification of Human Remains*, edited by K. J. Reichs, 218–28. Springfield, IL: Charles C. Thomas, 1986.

Marsin, M. *The Womens Advocate, or, Fifteen Real Comforts of Matrimony being in requital of the late fifteen sham-comforts: With satyrical reflections on whoring, and the debauchery of this age*. London: Benjamin Alsop and Thomas Malthus, 1683.

Martial. *Epigrams*. Trans. D. R. Shackleton Bailey. Loeb Classical Library 94, 95 and 480. Cambridge, MA: Harvard University Press, 1993.

Martin, C. G. '*The Awntyrs off Arthure*, an Economy of Pain'. *Modern Philology* 108.2 (2010): 177–98.

Martino, L. 'Ricognizione anatomica a studo antropometrico delle reliquie ossee di San Nicola di Bari'. *Bollettino di San Nicola, Special Issue* (April to December 1957). http://www.stnicholascenter.org/pages/anatomical-examination/

Masnari, O., C. Schiestl, J. Rossler, S. K. Gutlein, K. Neuhaus, L. Weibel, M. Meuli and M. A. Landolt. 'Stigmatization Predicts Psychological Adjustment and Quality of Life in Children and Adolescents With a Facial Difference'. *Journal of Pediatric Psychology* 38.2 (2013): 162–72.

Mathewson, J. T. 'Displacement of the Feminine in *Golagras and Gawane* and *The Awntyrs off Arthure*'. *Arthurian Interpretations* 1.2 (1987): 23–8.

Matsuda, T. '*The Awntyrs off Arthure* and the Arthurian History'. *Poetica* 19 (1984): 48–62.

McGough, L. J. 'Quarantining Beauty: The French Disease in Early Modern Venice'. In *Sins of the Flesh: Responding to Sexual Disease in Early Modern Europe*, edited by K. Siena, 211–37. Toronto: Centre for Reformation and Renaissance Studies, 2005.

McGowan, J. F. and T. L. Porter. *An Introduction to the Vocational Rehabilitation Process*. Washington DC: Rehabilitation Services Administration, 1967.

Meet the Ancestors, series 5, episode 1: The Chariot Queen (27 November 2007). http://www.bbc.co.uk/programmes/b008fj5s [accessed 31 March 2017].

Metzler, I. *A Social History of Disability in the Middle Ages: Cultural Considerations of Physical Impairment*. New York: Routledge, 2013.

Miller, J. *Fashion and Music*. Oxford and New York: Berg, 2011.

Miller, J. 'Heroes and Villains: When Men Wear Make-up'. In *Fashion Cultures Revisited*, edited by S. Bruzzi and P. Church Gibson, 341–51. Oxford and New York: Routledge, 2013.

Mitchell, W. J. T. *Image Science: Iconology, Visual Culture, and Media Aesthetics*. Chicago: University of Chicago Press, 2015.

Mitchell, D. T. 'Narrative Prosthesis and the Materiality of Metaphor'. In *Disability Studies: Enabling the Humanities*, edited by Sharon L. Snyder, Brenda Jo Bruggemann and R. Garland-Thomson, 15–30. New York: MLA, 2002.

Mnookin, J. L. 'The Image of Truth: Photographic Evidence and the Power of Analogy'. *Yale JL & Human* 10.1 (1998): 1–74.

Mojon-Azzil, S. M., A. Kunz and D. S. Mojon. 'Strabismus and Discrimination in Children: Are Children with Strabismus invited to Fewer Birthday Parties?' *British Journal of Ophthalmology* 95.4 (2010): 473–6. http://dx.doi.org/10.1136/bjo.2010.185793.

Moll, R. J. *Before Malory: Reading Arthur in Later Medieval England*. Toronto: University of Toronto Press, 2003.

Monaghan, L. F. and S. J. Williams. 'Stigma'. In *Key Concepts in Medical Sociology*, edited by J. Gabe and L. F. Monaghan, 58–62. London: Sage, 2013.

Morgan, L. 'Hair and Heroism According to Domitian'. *Classical Quarterly* 47.1 (1997): 209–14.

Mori, M. *The Uncanny Valley*, Trans. K. F. MacDorman and N. Kageki. IEEE Robotics and Automation, 19.2 (2012), 98–100. (Original work published in 1970). doi:10.1109/MRA.2012.2192811. Online: http://spectrum.ieee.org/automaton/robotics/humanoids/the-uncanny-valley [accessed 7 September 2016].

Morris, D. B. *The Culture of Pain*. Berkeley, Los Angeles and London: University of California Press, 1991.

Morton, C. 'The Ancestral Image in the Present Tense'. *Photographies* 8.3 (2015): 253–70.

Mossey, P. A., R. G. Julian Little, M. J. Munger Dixon and W. C. Shaw. 'Cleft Lip and Palate'. *The Lancet* 379.9703 (November 9 2009): 1173–8.

Mower, J. 'Deliberate Ante-Mortem Dental Modification and its Implications in Archaeology, Ethnography and Anthropology'. *Papers from the Institute of Archaeology* 10 (1999): 37–53.

Muir, W. *The Happy Hospital*. London: Simpkin, Marshall, Hamilton, Kent and Co, Ltd, 1918.

'Mummy Returns to Elgin'. *The Northern Scot* (25 September 2015). http://www.northern-scot.co.uk/Features/People-and-Places/Mummy-returns-to-Elgin-25092012.htm.

Munslow, A. *A History of History*. London: Routledge, 2012.

Musgrave, J. H., R. A. H. Neave and A. J. N. W. Prag. 'The Skull from Tomb II at Vergina: King Philip II of Macedon'. *Journal of Hellenic Studies* 104 (1984): 60–78.

N. H. *The Ladies Dictionary*. London: John Dunton, 1694.

Nabors, L. A., H. D. Lehmkuhl and J. S. Warm. 'Children's Acceptance Ratings of a Child with a Facial Scar: The Impact of Positive Scripts'. *Early Education and Development* 15.1 (2004): 79–92.

National Archives, London. PRO, SP 12/159, fol. 38^{r-v}.

Needham, C., C. Wilkinson and C. J. Knüsel. 'Reconstructing Visual Manifestations of Disease from Archaeological Human Remains'. *Journal of Audiovisual Media in Medicine* 26.3 (2003): 103–7.

Neilson, G. 'Cross-links Between *Pearl* and *The Awntyrs off Arthure*'. *Scottish Antiquary* 16 (1902): 67–78.

Nelson, P. A. 'Qualitative Approaches in Craniofacial Research'. *Cleft Palate and Craniofacial Journal* 46.3 (2008): 245–51.

Nelson, P. A., A. L. Caress, A. M. Glenny and S. A. Kirk. '"Doing the Right Thing": How Parents Experience and Manage Decision-making for Children's "Normalising" Surgeries'. *Social Science and Medicine* 74 (2012): 796–804.

Nelson, P. A., S. A. Kirk, A. L. Caress and A. M. Glenny. 'Parents' Emotional and Social Experiences of Caring for a Child through Cleft Treatment'. *Qualitative Health Research* 22.3 (2012): 346–59.

Nerlich, A., O. Peschel, A. Zink and F. W. Rösing. 'The Pathology of Trepanation: Differential Diagnosis, Healing and Dry Bone Appearance in Modern Cases'. *Trepanation History, Discovery, Theory* 17 (2003): 43–51.

Nevitt, M. 'The Insults of Defeat: Royalist Responses to Sir William Davenant's *Gondibert* (1651)'. *The Seventeenth Century* 24.2 (2009): 287–304.

Newland-Pedley, F. 'The Dental Needs of the Army'. *The Lancet* 2 (1915): 305.

Nichols, J. G., ed. *The Autobiography of Anne Lady Halkett*. London: Camden Society, 1875.

Niederland, W. G. 'Narcissistic Ego Impairment in Patients with Early Physical Malformations'. *Psychoanalytic Study of the Child* 20 (1965): 518–53.

Niederland, W. G. 'Scarred: A Contribution to the Study of Facial Disfigurement'. *Psychoanalytic Quarterly*, 44 (1975): 450–9.

Noar, J. 'Questionnaire Survey of Attitudes and Concerns of Patients with Cleft Lip and Palate and their Parents'. *Cleft Palate-Craniofacial Journal* 28 (1991): 279–84.

Northcroft, G. 'Section on Odontology. A Short Account of a Year's Work at One of the Jaw Injuries Centres of the London Command'. *Proceedings of the Royal Society of Medicine* 11.3 (1918): 6–26.

Nortje, A., C. G. Tredoux, K. Kempen and A. Vredeveldt. 'Applying Laboratory Techniques to a Real-life Case: What Insight can we Provide About the Station Strangler Case?' Poster presentation at the American Psychology-Law Society Annual Conference, Portland, United States, March 2013.

Nussbaum, V. 'Serial Künstler: Portrait of the Artist as a Malefactor' In *Elective Affinities: Testing Word and Image Relationships*, edited by C. MacLeod, V. Plesch and C. Schoell-Glass 6 (2009). Rodopi, 212.

O'Neill, M. 'Images of the Dead: Ethics and Contemporary Art Practice'. In *Cultural and Ethical Turns: Interdisciplinary Reflections on Culture, Politics and Ethics*, edited by B. Garner, S. Pavlenko, S. Shaheen and A. Wolanski. Oxford: Inter-Disciplinary Press, 2011.

O'Neill, M. 'Speaking to the Dead: Images of the Dead in Contemporary Art'. *Health* 15.3 (2011): 129–36.

Olah, N. 'Antony Hegerty: Future Feminism'. *DazedDigital*, 2013. http://www.dazeddigital.com/artsandculture/article/14121/1/antony-hegarty-future-feminism.

Olson, K. 'Masculinity, Appearance, and Sexuality: Dandies in Roman Antiquity'. *Journal of the History of Sexuality* 23.2 (2014): 182–205.

Omiya, T., M. Ito and Y. Yamzaki. 'The Process Leading to Affirmation of Life with Cleft Lip and Cleft Palate: The Importance of Acquiring Coherence'. *Japan Journal of Nursing Science* 9 (2012): 127–35.

Ordinary's Account, 6th September 1700. *The Proceedings of the Old Bailey*, https://www.oldbaileyonline.org/browse.jsp?id=OA17000906n14-2&div=OA17000906#highlight.

Orlemanski, J. 'Desire and Defacement in the Testament of Cresseid'. In *Reading Skin in Medieval Literature and Culture*, edited by K. L. Walter, 161–82. Basingstoke: Palgrave Macmillan, 2013.

Ortner, D. J. and M. F. Ericksen. 'Bone Changes in the Human Skull probably resulting from Scurvy in Infancy and Childhood'. *International Journal of Osteoarchaeology* 7.3 (1997): 212–20.

Ovid. *Amores*. Trans. G. Showerman. Loeb Classical Library 41. Cambridge, MA: Harvard University Press, 1914.

Ovid. *Art of Love*. Trans. J. H. Mozley. Loeb Classical Library 232. Cambridge, MA: Harvard University Press, 1929.

Ovid. *Metamorphoses*. Trans. F. J. Miller. Loeb Classical Library 42. Cambridge, MA: Harvard University Press, 1916.

Padfield, D. 'In Search of a Visual Language for Pain'. In *Perceptions of Pain*, edited by D. Padfield, 16–27. Stockport: Dewi Lewis Publishing, 2003.

Paglen, T. 'Invisible Images (Your Pictures are Looking at You)', *The New Inquiry*, 8 December 2016. Online https://thenewinquiry.com/essays/invisible-images-your-pictures-are-looking-at-you/ [accessed 8 December 2016].

Papagrigorakis, M. J., P. N. Synodinos, A. Antoniadis, E. Maravelakis, P. Toulas, O. Nillson and E. Baziotopoulou-Valavani. 'Facial Reconstruction of an 11-Year-Old Female Resident of 430BC Athens'. *The Angle Orthodontist* 81 (2011): 171–9.

Pappas, S. 'Did Richard III really have a Friendly Face?' *Live Science: History* (8 February 2013). http://www.livescience.com/26959-richard-iii-friendly-face.html.

Parales, J. 'Review: In "This is Acting," Sia Adhered to Her Victim-to-Victory Formula'. *The New York Times* (27 January 2016). https://www.nytimes.com/2016/01/28/arts/music/review-in-this-is-acting-sia-adheres-to-her-victim-to-victory-formula.html?smid=tw-nytimesarts&smtyp=cur.

Parkin, T. *Old Age in the Roman World*. Baltimore: Johns Hopkins University Press, 2003.

Parry, R. *Moderatus, the Most Delectable & Famous Historie of the Blacke Knight*. London: Richard Jhones, 1595.

Partridge, J. *Changing Faces: The Challenge of Facial Disfigurement*. London: Penguin, 1990.

Passmore, S. E. and S. Carter, eds. *The English 'Loathly Lady' Tales: Boundaries, Traditions, Motifs*. Kalamazoo, MI: Medieval Institute Publications, 2007.

Paterson, K. and B. Hughes. 'Disability Studies and Phenomenology: The Carnal Politics of Everyday Life'. *Disability and Society* 14.5 (1999): 597–610.

Patterson, D., A. Barnes and C. Duncan. *Fighting Shadows. Self-stigma and Mental Illness*. Auckland: Mental Health Foundation of New Zealand, 2009.

Paul of Aegina. *Medical Compendium in Seven Books*. Trans. F. Adams. London: The Sydenham Society, 1844.

Pausanias. *Description of Greece*. Trans. W. H. S. Jones. Loeb Classical Library 93. Cambridge, MA: Harvard University Press, 1918.

Payne, J. L. 'Discussion on War Injuries of the Jaws and Face'. *British Dental Journal* (War Supplement) 37.5–6 (1916): 81–7.

Pelling, M. 'Appearance and Reality: Barber-Surgeons, the Body and Disease'. In *London 1500–1700: The Making of the Metropolis*, edited by A. L. Beier and R. Finlay, 82–122. London: Longman, 1986.

Perez, S. I. 'Artificial Cranial Deformation in South America: A Geometric Morphometrics Approximation'. *Journal of Archaeological Science* 34.10 (2007): 1649–58.

Persson, M., G. Aniansson, B. Magnus and H. Svensson. 'Self-concept and Introversion in Adolescents with Cleft Lip and Palate'. *Scandinavian Journal of Plastic and Reconstructive Surgery and Hand Surgery* 36 (2002): 24–7.

Peter, J. and R. R. Chinsky. 'Sociological Aspects of Cleft Lip and Palate Adults. 1. Marriage'. *Cleft Palate Journal* 11 (1974): 295–309.

Peter, J. and R. R. Chinsky. 'Sociological Aspects of Cleft Lip and Palate Adults. 2. Social Integration'. *Cleft Palate Journal* 11 (1974): 304–10.

Peter, J. and R. R. Chinsky. 'Sociological Aspects of Cleft Lip and Palate Adults. 3. Education'. *Cleft Palate Journal* 11 (1974): 443–9.

Peter, J., R. R. Chinsky and M. J. Fishe. 'Sociological Aspects of Cleft Lip and Palate Adults. 4. Vocational and Economic Aspects'. *Cleft Palate Journal* 12 (1975): 193–9.

Petherbridge, D. *Witches and Wicked Bodies*. Edinburgh: National Galleries of Scotland, 2013.

Petridis, A. 'Tub Thumping'. *The Guardian* 12 November 2004. https://www.theguardian.com/music/2004/nov/12/popandrock1.

Phaedrus. *Fables*. Trans. B. E. Perry. Loeb Classical Library 436. Cambridge, MA: Harvard University Press, 1965.

Phillips, H. '*The Awntyrs off Arthure*: Structure and Meaning, a Reassessment'. *Arthurian Literature* 14 (1993): 71–88.

Phillips, H. 'The Ghost's Baptism in *The Awntyrs off Arthure*'. *Medium Ævum* 58.1 (1989): 49–58.

Philogelos. Trans. B. Baldwin. Amsterdam: Gieben, 1983.

Pierson, D. P. 'Evidential Bodies: The Forensic and Abject Gazes in *C.S.I.: Crime Scene Investigation*.' *Journal of Communication Inquiry* (2009): 1–20.

Pinney, C. 'Things Happen: Or, From Which Moment Does that Object Come?' In *Materiality*, edited by D. Miller, 256–72. Durham and London: Duke University Press, 2005.

Pliny the Elder. *Natural History*. Trans. H. Rackham. Loeb Classical Library 352 and 353. Cambridge, MA: Harvard University Press, 1940–2.

Pliny the Elder. *Natural History*. Trans. W. H. S. Jones. Loeb Classical Library 418. Cambridge, MA: Harvard University Press, 1963.

Plutarch. *How the Young Man Should Study Poetry*. Trans. F. C. Babbitt. Loeb Classical Library 197. Cambridge, MA: Harvard University Press, 1927.

Plutarch. *Lycurgus*. Trans. B. Perrin. Loeb Classical Library 46. Cambridge, MA: Harvard University Press, 1914.

Plutarch. *On Common Conceptions*. Trans. H. Cherniss. Loeb Classical Library 470. Cambridge, MA: Harvard University Press, 1976.

Plutarch. *Table Talk*. Trans. P. A. Clement and H. B. Hoffleit. Loeb Classical Library 424. Cambridge, MA: Harvard University Press, 1969.

Polemon. Physiognomy. Trans. R. Hoyland. In *Seeing the Face, Seeing the Soul: Polemon's Physiognomy from Classical Antiquity to Medieval Islam*, edited by S. Swain, 329–463. Oxford: Oxford University Press, 2007.

Ponce de León, M. S. and C. P. E. Zollikofer. 'New Evidence from Le Moustier 1: Computer-Assisted Reconstruction and Morphometry of the Skull'. *The Anatomical Record* 254.4 (1999): 474–89.

Posel, D. and P. Gupta, eds. 'The Life of the Corpse: Framing Reflections and Questions'. *African Studies* 68.3, December 2009 (Special edition): 299–309.

Powers, N. D. 'Cranial Trauma and Treatment: A Case Study from the Medieval Cemetery of St Mary Spital, London'. *International Journal of Osteoarchaeology* 15.1 (2005): 1–14.

Prag, J. and R. Neave. *Making Faces*. London: British Museum Press, 1999.

Price, J. L. 'The Radiology of Excavated Saxon and Medieval Human Remains from Winchester'. *Clinical Radiology* 26 (1975): 363–70.

Propertius. *Elegies*. Trans. G. P. Goold. Loeb Classical Library 18. Cambridge, MA: Harvard University Press, 1990.

Purkiss, D. 'Material Girls: The Seventeenth-Century Woman Debate'. In *Women, Texts and Histories 1575–1760*, edited by C. Brant and D. Purkiss, 69–101. London: Routledge, 1992.

Quatrehomme, G., T. Balaguer, P. Staccini and V. Alunni-Pettret. 'Assessment of the accuracy of a three-dimensional manual craniofacial reconstruction: A series of 25 controlled cases'. *International Journal of Legal Medicine* 121 (2007): 469–75.

Quintilian. *The Orator's Education*. Trans. D. A. Russell. Loeb Classical Library 494. Cambridge, MA: Harvard University Press, 2002.

Rackin, P. 'Foreign Country: The Place of Women and Sexuality in Shakespeare's Historical World'. In *Enclosure Acts: Sexuality, Property, and Culture in Early Modern England*, edited by R. Burt and J. M. Archer, 68–95. Ithaca: Cornell University Press, 1994.

Ramstad, T., E. Ottem and W. C. Shaw. 'Psychosocial Adjustment in Norwegian Adults who had undergone Standardised Treatment of Complete Cleft Lip and Palate. Part II. Self-reported Problems and Concerns with Appearance'. *Scandinavian Journal of Reconstructive Surgery and Hand Surgery* 29 (1995): 329–6.

Ray-Bellett, C. and J. Hohfield. 'Parental Diagnosis of Facial Clefts: Evaluation of Specialised Counselling'. *Swiss Medical Weekly* 134 (2004): 640–44.

Real Beauty Sketches (You're More Beautiful Than You Think). Dove US, 14 April 2013. https://www.youtube.com/watch?v=litXW91UauE.

Report by the Joint Committee and the Joint War Finance Committee of the British Red Cross Society and the Order of St John of Jerusalem in England in Voluntary Aid Rendered for the Sick and Wounded at Home and Abroad and to the British Pensioner of War, 1917–1919. London, 1921.

Rhine, J. S. 'Coming to Terms with Facial Reproduction'. *Journal of Forensic Sciences (JFSCA)* 35.4 (July 1990): 960–3.

Rhodes, G., L. W. Simmons and M. Peters. 'Attractiveness and Sexual Behavior: Does Attractiveness Enhance Mating Success?' *Evolution and Human Behavior* 26.2 (2005): 186–201.

Richman, L. C. 'The Effects of Facial Disfigurement on Teachers' Perception of Ability in Cleft Palate Children'. *Cleft Palate Journal* 15 (1978): 155–60.

Rickman, J. *Love, Lust, and License in Early Modern England*. Aldershot: Routledge, 2008.

Riehl, A. *The Face of Queenship: Early Modern Representations of Elizabeth I*. Basingstoke: Palgrave Macmillan, 2010.

Robertson, S. 'Otopalatodigital Spectrum Disorders'. *Gene Reviews* [Internet] 2005–13. https://www.ncbi.nlm.nih.gov/books/NBK1393/

Robertson, S. 'Otopalatodigital Syndrome Spectrum Disorders: Otopalatodigital Syndrome Types 1 and 2, Frontometaphyseal Dysplasia and Melnick-Needles Syndrome'. *European Journal of Human Genetics* 15 (2007): 3–9.

Robinson, M. '*The Awntyrs Off Arthure*: Barren Does and a Motherly Ghost as Authorial Agent of "Arthurial" Parable and Reform'. *Criterion* (2014): 32–40.

Robson, M. 'From beyond the Grave: Darkness at Noon in *The Awntyrs off Arthure*'. In *The Spirit of Medieval English Romance*, edited by A. Putter and J. Gilbert, 219–36. Harlow and New York: Pearson Educational Ltd., 2000.

Rogers, M. 'Beauty and the Concepts of the Ideal'. In *A Cultural History of the Human Body in the Renaissance*, edited by L. Kalof and W. Bynum, 125–48. Oxford: Bloomsbury Academic, 2010.

Rohrbacher, D. 'Physiognomics in Imperial Latin Biography'. *Classical Antiquity* 29.1 (2010): 92–116.

Roscoe, M. 'Differential Diagnosis of an Unknown 17th-Century Pathological Skull'. Honours Dissertation, University of Dundee, 2012.

Rousseau, J. *Émile*. London and Vermont: Everyman, 1993 [1762].

Rule, N. O. and N. Ambady. 'The Face of Success: Inferences from Chief Executive Officers' Appearance Predict Company Profits'. *Psychological Science* 19 (2008): 109–11.

Rumsey, N. 'Psychological Aspects of Face Transplantation: Read the Small Print Carefully'. *American Journal of Bioethics* 4.3 (2004): 22–5.

Rumsey, N. and D. Harcourt. 'Body Image and Disfigurement: Issues and Interventions'. *Body Image* 1.1 (2004): 83–97.

Rumsey, N., R. Bull and D. Gahagan. 'The Effect of Facial Disfigurement on the Proxemic Behavior of the General Public'. *Journal of Applied Psychology* 12.2 (1982): 137–50.

Rutter, M. 'Resilence, Competence and Coping'. *Child Abuse and Neglect* 31.3 (2007): 205–9.

Rynn, C., T. Balueva and E. Veselovskaya. 'Relationships Between the Skull and Face'. In *Craniofacial Identification*, edited by C. Wilkinson and C. Rynn. Cambridge: Cambridge University Press, 2012 [Kindle edition].

Sagne, S. and G. Olsson. 'Studies of the Periodontal Status of a Medieval Population'. *Dentomaxillofacial Radiology* 6.1 (1977): 46–52.

Salih, S. *Judith Butler*. Oxford: Routledge, 2002.

Samuel, H. 'Face of France's Good King Henri IV Reconstructed 400 Years After his Death'. *The Telegraph* (14 February 2013). http://www.telegraph.co.uk/news/worldnews/europe/9870454/Face-of-Frances-Good-King-Henri-IV-reconstructed-400-years-after-his-death.html.

Sassoon, J. 'Photographic Materiality in the Age of Digital Reproduction'. In *Photographs, Objects, Histories: On the Materiality of Images*, edited by E. Edwards and J. Hart, 186–202. London: Routledge, 2004.

Sauer, N. J. 'The Timing of Injuries and Manner of Death: Distinguishing among Antemortem, Perimortem and Postmortem Trauma'. In *Forensic Osteology: Advances in the Identification of Human Remains*, edited by K. J. Reichs, 321–32. Springfield, IL: Charles C. Thomas, 1986.

Saunders, C. *Magic and the Supernatural in Medieval English Romance*. Cambridge: D.S. Brewer, 2010.

Scarry, E. *The Body in Pain*. New York: Oxford University Press, 1985.

Schiff, R. P. 'Borderland Subversions: Anti-Imperial Energies in *The Awntyrs off Arthure* and *Gologras and Gawane*'. *Speculum* 84.3 (2009): 613–32.

Schulting, R. J. and M. Wysocki. 'In this Chambered Tumulus were found Cleft Skulls: An Assessment of the Evidence for Cranial Trauma in the British Neolithic'. *Proceedings of the Prehistoric Society* 71 (2005): 107–38.

Schuster, J. D., T. A. Rakusan, T. Chonmaitree and Q. T. Box. 'Tuberculous Osteitis of the Skull Mimicking Histiocytosis X'. *Journal of Pediatrics* 105.2 (1984): 269–71.

Schwartz, M. *Dead Matter: The Meaning of Iconic Corpses*. Minneapolis: Minnesota, 2015.

Schwartz, M. 'An Iconography of the Flesh: How Corpses Mean As Matter'. *communication+ 1* 2.1 (2013) http://scholarworks.umass.edu/cpo/vol2/iss1/1. doi:10.7275/R5TB14T7.

Searle, A., P. Neville and A. Waylen. 'Psychological Growth and Well-being in Individuals Born with Cleft: An Application of Self-determination'. *Psychology and Health* 32.4 (2017): 459–82.

Sekula, A. 'The Body and the Archive'. *October* 39 (Winter, 1986): 3–64.

Seneca. *Moral Epistles*. Trans. R. M. Gummere. Loeb Classical Library 77. Cambridge, MA: Harvard University Press, 1925.

Seneca. *Natural Questions*. Trans. T. H. Corcoran. Loeb Classical Library 450. Cambridge, MA: Harvard University Press, 1971.

Seneca. *On Firmness*. Trans. J. W. Basore. Loeb Classical Library 214. Cambridge, MA: Harvard University Press, 1928.

Sensi, L. 'Ornatus e status sociale delle donne romane'. *Annali della Faculta di Lettere e Filosofia Perugia-Sezione Studi Classici* n.s. 4 (1980–1): 55–102.

Shadwell, T. *Bury-fair a Comedy.* London: James Knapton, 1689.

Shakespeare, W. *Macbeth*, edited by K. Muir. London: Methuen, 1964 [1623].

Shepherd, S. '"Heathenic" Catechesis and the Source of Awntyrs B'. *Medium Ævum* 131.1 (2012): 1–17.

Shildrick, M. *Embodying the Monster: Encounters with the Vulnerable Self.* London: Sage Publications, 2002.

Shuttleton, D. 'A Culture of Disfigurement: Imagining Smallpox in the Long Eighteenth Century'. In *Framing and Imagining Disease in Cultural History*, edited by G. S. Rousseau, M. Gill, D. Haycock and M. Herwig, 68–91. Basingstoke: Palgrave Macmillan, 2003.

Shuttleton, D. *Smallpox and the Literary Imagination.* Cambridge: Cambridge University Press, 2007.

Siculus, D. *Library*. Trans. F. R. Walton. Loeb Classical Library 423. Cambridge, MA: Harvard University Press, 1967.

Siebers, T. *Disability Theory*. Ann Arbor: The University of Michigan Press, 2008. Reprint 2011.

Siegel, G. 'The Similtude of the Wound'. *Cabinet* 43 (2011): 95–100.

Siegel, J. T. 'George Simmel Reappears: "The Aesthetic Significance of the Face"'. *Diacritics* 29.2 (1999): 100–13.

Siena, K. P. '"The Foul Disease" and Privacy: The Effects of Venereal Disease and Patient Demand on the Medical Marketplace in Early Modern London'. *Bulletin of the History of Medicine* 75.2 (2001): 199–224.

Silverman, K. *The Miracle of Analogy: Or the History of Photography, Part 1*. California: Stanford University Press, 2015.

Sinason, V. *Mental Handicap and the Human Condition: New Approaches from the Tavistock.* London: Free Association, 1992.

Smelik, A. 'A Close Shave: The Taboo on Female Body Hair'. *Critical Studies in Fashion and Beauty* 6.2 (2015): 233–51.

Smith, E. 'The Local Popularity of *The Concealed Fansyes*'. *Notes and Queries* 53.2 (2006): 189–93.

Smith, H. 'Mary Astell, *A Serious Proposal to the Ladies* (1694), and the Anglican Reformation of Manners in Late-Seventeenth-Century England'. In *Mary Astell: Reason, Gender, Faith*, edited by W. Kolbrener and M. Michelson, 31–48. Aldershot: Routledge, 2007.

Smith, K. *Limits of Excess: Abjection in the Photographic Work of Joel-Peter Witkin*. Unpublished MAFA thesis, University of the Witwatersrand, 1997.

Smith, K. 2016. 'Under the Influence of… Paul Stopforth's Paul Stopforth's Biko painting called "Elegy"'. *The Conversation (Africa edition)*, 17 August 2016. Online: https://theconversation.com/under-the-influence-of-paul-stopforths-biko-painting-called-elegy-64031 [accessed 17 August 2016].

Smith, P. 'The 21 Worst Police Sketches Of All Time'. *BuzzFeed* 8 May 2014 https://www.buzzfeed.com/patricksmith/the-27-worst-police-sketches-of-all-time [accessed 21 May 2017].

Snape, D. and L. Spencer. 'The Foundations of Qualitative Research'. In *Qualitative Research Practice; A Guide for Social Scientists and Researchers*, edited by J. Ritchie and J. Lewis, 1–23. London: Sage Publications Ltd, 2003.

Snow, C. C., B. P. Gatliff and K. R. McWilliams. 'Reconstruction of facial features from the skull: An evaluation of its usefulness in forensic anthropology'. *American Journal of Physical Anthropology* 33.2 (1970): 221–7.
Sofaer, J. *The Body as Material Culture: A Theoretical Osteology*. Cambridge: Cambridge University Press, 2006.
Sofaer, J. 'Touching the Body: The Living and the Dead in Osteaoarchaeology and the Performance Art of Marina Abramović'. *Norwegian Archaeological Review*, 1–16. iFirst 2012.
Sontag, S. *Illness as Metaphor and AIDS and its Metaphors*. London: Penguin, 1991.
Sontag, S. *Regarding the Pain of Others*. London: Penguin, 2003.
Southern, P. *Domitian: Tragic Tyrant*. London: Routledge, 1997.
Spearing, A. C. 'Central and Displaced Sovereignty in Three Medieval Poems'. *Review of English Studies* 33 (1982): 247–61.
St Nicholas Centre. 'Anatomical Examination of the Bari Relics'. 2002–07. http://www.stnicholascenter.org/pages/anatomical-examination [accessed 31 March 2017].
Stallybrass, P. 'Patriarchal Territories: The Body Enclosed'. In *Rewriting the Renaissance: The Discourses of Sexual Difference in Early Modern Europe*, edited by M. W. Ferguson, M. Quilligan and N. J. Vickers, 123–42. Chicago: University of Chicago Press, 1986.
Standen, V. G. and B. T. Arriaza. 'Trauma in the Preceramic Coastal Populations of Northern Chile: Violence or Occupational Hazards?' *American Journal of Physical Anthropology* 112.2 (2000): 239–49.
Stephan, C. N. 'Facial Approximation – From Facial Reconstruction Synonym to Face Prediction Paradigm'. *Journal of Forensic Sciences* 60.3 (2015): 566–71.
Stephan, C. N. and M. Henneberg. 'Building Faces from Dry Skulls: Are They Recognized above Chance Rates?' *Journal of Forensic Sciences* 46.3 (2001): 432–40.
Stephan, C. N. and M. Henneberg. 'Recognition by Forensic Facial Approximation: Case Specific Examples and Empirical Tests'. *Forensic Science International* 156 (2006): 182–91.
Strabo. *Geography*. Trans. H. L. Jones. Loeb Classical Library 211. Cambridge, MA: Harvard University Press, 1928.
Strauss, R. P. '"Only Skin Deep": Health, Resilience and Craniofacial Care'. *Cleft palate-Craniofacial Journal* 38.30 (2001): 226–30.
Strauss, R. P., B. Ramsey, T. C. Edwards, T. D. Topolski, K. A. Kapp-Simon, C. D. Thomas, C. Fenson and D. I. Patrick. 'Stigma Experiences in Youth with Facial Differences: A Multi-site Study of Adolescents and their Mothers'. *Orthodontic Craniofacial Research* 10 (2007): 96–103.
Suetonius. *Julius Caesar, Gaius*. Trans. J. C. Rolfe. Loeb Classical Library 31. Cambridge, MA: Harvard University Press, 1914.
Suetonius. *Galba, Domitian, Otho, Vespasian, Titus*. Trans. J. C. Rolfe. Loeb Classical Library 38. Cambridge, MA: Harvard University Press, 1914.
Surgeons Hall Museum. 'The Enigma of the Neanderthaloid Skull (GC 9413)'. 8 October 2014. https://surgeonshallmuseum.wordpress.com/2014/10/08/the-enigma-of-the-neanderthaloid-skull-gc-9413/ [accessed 31 March 2017].
Svenson, A. *Unspeaking Likeness*. Sante Fe: Twin Palms, 2016.
Swain, S., ed. *Seeing the Face, Seeing the Soul: Polemon's Physiognomy from Classical Antiquity to Medieval Islam*. Oxford: Oxford University Press, 2007.
Swetnam, J. *The Araignment of Lewd, Idle, Froward, and Vnconstant Women or the Vanitie of them, Choose you Whether: With a Commendation of Wise, Vertuous and Honest Women*. London: George Purslowe for Thomas Archer, 1615.
Synesius. *Orations*. Trans. A. Fitzgerald. London: Oxford University Press, 1930.

Synnott, A. 'Truth and Goodness, Mirrors and Masks, Part 1: A Sociology of Beauty and the Face'. *The British Journal of Sociology* 40.4 (1989): 607–36.

Talley, H. L. *Saving Face: Disfigurement and the Politics of Appearance*. New York: New York University Press, 2014.

Taussig, M. *Defacement: Public Secrecy and the Labor of the Negative*. Stanford: Stanford University Press, 1999.

Taussig, M. *Mimesis and Alterity: A Particular History of the Senses*. London and New York: Routledge, 1993.

Taylor, J. *Body Horror: Photojournalism, Catastrophe and War*. New York: New York University Press, 1998.

Taylor, K. T. *Forensic Art and Illustration*. Boca Raton: CRC, 2001.

Taylor, L., R. Atkinson, R. Burrows and E. Spence. 'Thinking Allowed'. BBC Radio 4, 9 January 2017. http://www.bbc.co.uk/programmes/b086s8yv [accessed 7 February 2017].

Thompson, J. L. and B. Illerhaus. 'A New Reconstruction of the Le Moustier 1 Skull and Investigation of Internal Structures using 3-D-mCT Data'. *Journal of Human Evolution* 35.6 (1998): 647–65.

Thornton, A. *The Diary of Mrs Alice Thornton*. Durham: Andrews and Co., 1875.

Thornton, S. *Club Cultures: Music, Media and Subcultural Capital*. Cambridge: Polity, 1996.

Tillotson, A. C. Disaster Victim Identification using Craniofacial Analysis. PhD dissertation, University of Dundee, Dundee, 2011.

Tinwell, A. *The Uncanny Valley in Games and Animation*. Boca Raton: CRC Press, 2015.

Todorov, A., A. N. Mandisodza, A. Goren and C. C. Hall. 'Inferences of Competence from Faces Predict Election Outcomes'. *Science* 308 (5728): 1623–6.

Torres-Rouff, C. 'Shaping Identity: Cranial Vault Modification in the Pre-Columbian Andes'. PhD thesis, University of California, 2003.

Toulalan, S. and K. Fisher. 'Introduction'. In *The Routledge History of Sex and the Body 1500 to the Present*, edited by S. Toulalan and K. Fisher, 1–20. London: Routledge, 2013.

Tredoux, C., D. Nunez, O. Oxtoby and B. Prag. 'An Evaluation of ID: An Eigenface Based Construction System: Reviewed Article'. *South African Computer Journal* 37 (2006): 90–7.

Tronick, E. Z. 'Things Still To Be Done on the Still-Face Effect'. *Infancy* 4.4 (2003): 475–82.

Troyer, J. 'Embalmed vision'. *Mortality* 12.1 (2007): 22–47.

Tuke, T. *A Discourse against Painting and Tincturing of Women Wherein the Abominable Sinnes of Murther and Poysoning, Pride and Ambition, Adultery and Witchcraft are set foorth & discouered*. London: Thomas Creede and Bernard Alsop for Edward Marchant, 1616.

Turner, D. M. 'The Body Beautiful'. In *A Cultural History of the Human Body in the Age of Enlightenment*, edited by C. Reeves, 113–31. Oxford: Bloomsbury Academic, 2010.

Turner, D. M. 'Introduction: Approaching Anomalous Bodies'. In *Social Histories of Disability and Deformity*, edited by D. M. Turner and K. Stagg, 1–16. London: Routledge, 2006.

Turner, J. G. *Libertines and Radicals in Early Modern London: Sexuality, Politics,* and *Literary Culture, 1630–85*. Cambridge: Cambridge University Press, 2002.

Turville-Petre, T. '*Summer Sonday, De Tribus Mortus*, and *The Awntyrs off Arthure*'. *Review of English Studies* 25 (1974): 1–14.

Two, Krista Sue-Lo. '*The Awntyrs off Arthure at the Terne Wathelyne*: Reliquary for Romance'. *Arthurian Literature* 20 (2003): 103–22.

Van Deventer, R. The State versus Norman Sarfaraaz Simons, High Court of South Africa, Good Hope Provincial Division SS.104/94 (1995): 1404–5.
Varro. *On Agriculture*. Trans. W. D. Hooper and H. B. Ash. Loeb Classical Library 283. Cambridge, MA: Harvard University Press, 1934.
Verano, J. W. 'Trepanation in Prehistoric South America: Geographic and Temporal Trends over 2000 Years'. *Trepanation History, Discovery, Theory* 17 (2003): 223–36.
Vergil. *Georgics*. Trans. H. R. Fairclough. Loeb Classical Library 63. Cambridge, MA: Harvard University Press, 1999.
Verloes, A., S. Lesenfants, M. Barr, D. K. Grange, H. Journel, J. Lombet, G. Mortier, and E. Roeder. 'Fronto-otopalatodigital Osteodysplasia: Clinical Evidence for a Single Entity Encompassing Melnick-Needles Syndrome, Otopalatodigital Syndrome Types 1 and 2, and Frontometaphyseal Dysplasia'. *American Journal of Medical Genetics* 90 (2000): 407–22.
Victor, A. *On the Caesars*. Trans. H. W. Bird. Liverpool: Liverpool University Press, 1994.
Vlahogiannis, N. 'Curing Disability'. In *Health in Antiquity*, edited by H. King, 180–91. London: Routledge, 2005.
Walker, G. *Crime, Gender and Social Order in Early Modern England*. Cambridge: Cambridge University Press, 2009.
Wallace-Hadrill, A. *Suetonius: The Scholar and his Caesars*. London: Duckworth, 1983.
Wardman, A. E. 'Description of Personal Appearance in Plutarch and Suetonius: The Use of Statues as Evidence'. *Classical Quarterly* 17 (1967): 414–20.
Warner, M. *Monuments and Maidens*. London: Vintage, 1985.
Watkins, C. V. 'On Pandering'.
2015. http://www.tinhouse.com/blog/41314/on-pandering.html.
Weber, G. W. and F. L. Bookstein. *Virtual Anthropology: A Guide to a New Interdisciplinary Field*. Vienna: Springer, 2011.
Weiss, D. L and V. Møller-Christensen. 'Leprosy, Echinococcosis and Amulets: A Study of a Medieval Danish Inhumation'. *Medical History* 15.3 (1971): 260–7.
Wheatley, H. B., ed. *The Diary of Samuel Pepys*. Vol. 8 London: George Bell and Sons, 1893, Reprint 1946.
Whyman, S. E. *Sociability and Power in Late Stuart England: The Cultural Worlds of the Verneys 1660–1720*. Oxford: Oxford University Press, 1999.
Wilfing, H., H. Seidler, D. Nedden, G. Weber, H. Preuschoft, W. Platzer, R. Knapp, G. Haouser and W. A. Murphy. 'Cranial Deformation of the Neolithic Man from the Hauslabjoch'. *Collegium Anthropologicum* 18.2 (1994): 269–82.
Wilkinson, C. M. 'A Review of Forensic Art'. In *Research and Reports in Forensic Medical Science* 5 (2015): 17–24.
Wilkinson, C. 'Computerized Forensic Facial Reconstruction: A Review of Current Systems'. *Forensic Science, Medicine, and Pathology*, V1–3 (2005): 173–7. doi:10.1385/Forensic Sci. Med. Pathol.:1:3:173.
Wilkinson, C. 'Facial Reconstruction: Anatomical Art or Artistic Anatomy?' *Journal of Anatomy* 216.2 (2010): 235–50.
Wilkinson, C. 'The Facial Reconstruction of Ancient Egyptians'. In *Egyptian Mummies and Modern Science*, edited by R. David, 162–80. Cambridge: Cambridge University Press, 2008.
Wilkinson, C. *Forensic Facial Reconstruction*. Cambridge: Cambridge University Press, 2004.
Wilkinson, C. 'The Man Himself: The Face of Richard III'. *The Ricardian Bulletin* (September 2013): 50–55.

Wilkinson, C. 'A Review of the Changing Culture and Social Context Relating to Forensic Facial Depiction of the Dead'. *Forensic Science International*, 245 (2014): 95–100.
Wilkinson, C. and R. Neave. 'The Reconstruction of a Face Showing a Healed Wound'. *Journal of Archaeological Science* 30.10 (2003): 1343–8.
Wilkinson, C. and C. Rynn, eds. *Craniofacial Identification*. Cambridge: Cambridge University Press, 2012.
Wilkinson, C., C. Rynn, H. Peters, M. Taister, C. H. Kau and S. Richmond. 'A Blind Accuracy Assessment of Computer-Modelled Forensic Facial Reconstruction using Computer Tomography Data from Live Subjects'. *Forensic Science, Medicine and Pathology* 2.3 (2006): 179–87.
Williams, C. *Roman Homosexuality*. Oxford: Oxford University Press, 1999.
Williamson, G. C., ed. *Lady Anne Clifford, Countess of Dorset, Pembroke and Montgomery, 1590–1676. Her Life, Letters and Work*. Kendal: Titus Wilson and Son, 1922.
Wiseman, R. *Severall Chirurgicall Treatises*. London: E. Flesher and J. Macock, for R. Royston, 1676.
Wittenburg, J. *Disorderly Women and Female Power in the Street Literature of Early Modern England and Germany*. Charlottesville: University of Virginia Press, 1992.
Wittkowski, J., K. J. Doka, R. A. Neimeyer and M. Vallerga. 'Publication Trends in Thanatology: An Analysis of Leading Journals'. *Death Studies* 39.8 (2015): 453–62.
Wood, S. 'Public Images of the Flavian Dynasty: Sculpture and Coinage'. In *A Companion to the Flavian Age of Imperial Rome*, edited by A. Zissos, 129–47. Chichester and Malden: John Wiley and Sons Ltd, 2016.
Wuensch, K. L., W. A. Castellow and C. H. Moore. 'Effects of Defendant Attractiveness and Type of Crime on Juridic Judgement'. *Journal of Social Behaviour and Personality* 6.4 (1991): 399–402.
Yurchak, A. 'Bodies of Lenin: The Hidden Science of Communist Sovereignty'. *Representations* 129.1 (2015): 116–57.
Zamora, M. *Frida Kahlo: The Brush of Anguish*. London: Art Data, 1990.
Zanker, P. *The Mask of Socrates: The Image of the Intellectual in Antiquity*. Berkeley: University of California Press, 1995.
Žižek, S. *Looking Awry: An Introduction to Jacques Lacan through Popular Culture* Cambridge, MA: MIT Press, 1991).
Zolkowski, S. M. and L. M. Bullock. 'Resilience in Children and Youth: A Review'. *Children and Youth Services Review* 34.12 (2012): 2295–2303.

Index

CPSIA information can be obtained
at www.ICGtesting.com
Printed in the USA
LVOW13*1925050618
579665LV00011B/201/P

9 781350 028296